Exercise Physiology for Health Care Professionals

Frank J. Cerny, PhD
University at Buffalo, State University of New York

Harold W. Burton, PhD
University at Buffalo, State University of New York

Human Kinetics

Library of Congress Cataloging-in-Publication Data

Cerny, Frank J., 1946-
 Exercise physiology for health care professionals / Frank J. Cerny, Harold W. Burton.--
1st ed.
  p. cm.
 Includes bibliographical references and index.
 ISBN 0-88011-752-4
  1. Exercise--Physiological aspects. 2. Exercise therapy. I. Burton, Harold, 1951-II.
Title.

QP301 .C46 2001
612'.044--dc21   2001023046

ISBN: 0-88011-752-4

Permission notices for material reprinted in this book from other sources may be found on pages 351-352.

Acquisitions Editor: Michael S Bahrke, PhD; **Developmental Editor:** Elaine Mustain; **Assistant Editor:** Maggie Schwarzentraub and Sandra Merz Bott; **Copyeditor:** Joyce Sexton; **Proofreader:** Sarah Wiseman; **Indexer:** Marie Rizzo; **Permission Manager:** Dalene Reeder; **Graphic Designer:** Nancy Rasmus; **Graphic Artist:** Yvonne Griffith; **Photo Manager:** Clark Brooks; **Cover Designer:** Nancy Rasmus; **Photographer (cover):** Dept. of PTES, University at Buffalo, SUNY; **Photographer (interior):** Tom Roberts unless otherwise noted; **Art Manager:** Craig Newsom; **Illustrator:** Mic Greenberg; **Printer:** Edwards Brothers

Printed in the United States of America   10 9 8 7 6 5 4 3 2 1

Human Kinetics
Web site: www.humankinetics.com

United States: Human Kinetics
P.O. Box 5076
Champaign, IL 61825-5076
800-747-4457
e-mail: humank@hkusa.com

Canada: Human Kinetics
475 Devonshire Road Unit 100
Windsor, ON N8Y 2L5
800-465-7301 (in Canada only)
e-mail: orders@hkcanada.com

Europe: Human Kinetics
Units C2/C3 Wira Business Park
West Park Ring Road
Leeds LS16 6EB, United Kingdom
+44 (0) 113 278 1708
e-mail: humank@hkeurope.com

Australia: Human Kinetics
57A Price Avenue
Lower Mitcham, South Australia 5062
08 8277 1555
e-mail: liahka@senet.com.au

New Zealand: Human Kinetics
P.O. Box 105-231, Auckland Central
09-523-3462
e-mail: hkp@ihug.co.nz

To a supportive family, including my mother and father, brothers and sisters, and especially to my wife, Nancy, and our three boys, Hans, Peter, and Scott. Also to those who taught me about the excitement of exercise physiology, including Jack Bachman, Bruno Balke, Jerry Dempsey, Bill Reddan, and Fran Nagle. Finally, to those graduates from the University of Wisconsin who have shared in this lifelong excitement.

Frank Cerny

To my patient and understanding wife, Barbara, and my children, Davis, Evan, Brianne, and Marshal, who make everything worthwhile. Also, to all my mentors and colleagues who encouraged and inspired me through the years, especially Brian Hamilton, Jack Barclay, and John Faulkner.

Harold Burton

Contents

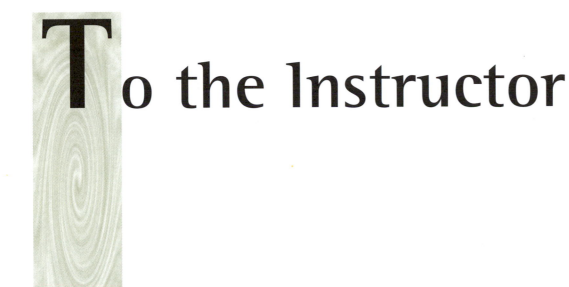

To the Instructor

This text has evolved from our teaching of exercise physiology to physical therapists, exercise scientists, medical and dental students, and other health professionals over the past 15 years. The longer we have taught, the more clear it has become that several requirements are unique to these students.

Exercise is a cornerstone of many of the treatment modalities available to health professionals. But problem solving in the wide variety of patient groups that are treated requires more than the memorization of facts about the fundamental concepts of exercise physiology: it also requires the ability to apply knowledge. Thus in this text we present basic physiological concepts as they apply to exercise, minimizing factual overload and maximizing application of the basic principles to exercise responses and adaptations under many different conditions.

The second requirement for a text on exercise physiology for health professionals is that it present concepts of exercise physiology as they apply to both healthy and unhealthy populations. Exercise physiology must be understood across the spectrum of individuals—from those with disease to those with exceptional physical attributes. Most texts of exercise physiology emphasize the application of exercise responses and adaptations to the latter end of the spectrum, although some texts present a brief discussion of such topics as cardiac rehabilitation, diabetes, obesity, and so on. Our combined 43 years' experience teaching health professionals has convinced us that how a topic is taught is as important as what is taught. Thus we have presented each topic with an eye toward understanding the clinical applications of the material.

With these requirements in mind, we emphasize a conceptual, integrative approach to understanding exercise physiology. Exposure to the material is within a clinical context, and where there are significant disease-exercise interactions (e.g., diabetes or heart disease), a separate chapter is devoted to relevant issues.

Although many exercise physiology courses require a prerequisite course in physiology, we have not made the assumption that this is so or that, if it is, the students have retained critical knowledge. However, each major section begins with a brief summary of basic physiology. If more review is required, this text can be used in conjunction with one of the many excellent textbooks of physiology.

The case studies in each chapter can provide the instructor the opportunity to problem-solve with students. This interactive learning can make the study of exercise physiology exciting to more students as they begin to peer through the window of clinical application. The book includes other learning aids that are described in the "To the Student" section.

To the Student

Physiology is the study of how the body functions. Exercise physiology, then, is the study of how exercise affects these body functions. Physiologists define themselves as "pulmonary physiologists," "cardiovascular physiologists," and so on. These physiologists study, with increasing specialization, various aspects of their chosen field. "Exercise physiologists," on the other hand, frequently find it difficult to concentrate on only a small part of the physiological responses to increased energy expenditure because of the multiple effects that exercise has on virtually all of the body's systems. For instance, while some may define their specialization as "pulmonary exercise physiology," it is difficult to understand the pulmonary responses to exercise without also understanding the cardiovascular or metabolic responses to exercise, since these systems have a profound effect on the pulmonary system. Thus students of exercise physiology must understand the integrative nature of the reactions of the body's multiple systems to exercise.

What You Will Learn in This Text

As a student of exercise physiology your understanding must begin with knowledge of the basic physiology of each system. The study of exercise physiology reveals how the equilibrium of physiological systems, established at rest, is altered during acute (running for the bus, walking to class, performing tasks of daily living) or chronic (athletic conditioning, rehabilitation, work hardening) exercise. The next time you exercise, try to imagine the changes in body function that are occurring. First you might simply notice the sensation of movement. Your muscle and nervous systems are interacting to initiate movement and provide feedback as to direction, intensity, and so on. Next you may observe that you are breathing harder. Again, the movements themselves and the carbon dioxide (CO_2) being produced in the muscles provide information to the brain centers to control ventilation so that adequate oxygen (O_2) is delivered to the muscles. In this situation, you experience the integration of the muscular, metabolic, nervous, and pulmonary systems. Next, you may become aware that your heart is beating faster. Of course, this faster heartbeat and harder contraction of the heart deliver the oxygenated blood to the muscles and carry away the CO_2. What you may not sense are the vascular changes taking place to ensure that the O_2 is delivered to the muscles, which are producing the movement. Here we see the integration of the muscular, metabolic, nervous, and pulmonary systems with the cardiovascular system. You may next notice that you are sweating and that your skin is reddening to dissipate the heat produced by the working muscles. Now you are experiencing the integration of the muscular, metabolic, nervous, cardiovascular, and hormonal systems. Finally, you may experience fatigue, or even exhaustion—the partial or complete failure of one or more of the body's systems.

It is clear, then, that understanding your responses to exercise requires an understanding of each of the body's physiological systems and, most importantly, an understanding of how these systems work together to support increased levels of exercise intensity.

In patient populations, your understanding of the integrative nature of physiological responses during exercise must be extended. Consider the following case study.

Case 1

On July 5 a patient with pulmonary disease is referred to the Miami Regional Rehabilitation and Work Hardening Center. You are asked to increase the exercise capacity of this patient so that she can return to her job as a postal worker.

History: Jane is a 1.67 m (5 ft 6 in.), 61 kg (134 lb) 48-year-old. She has smoked cigarettes since she was 18. As a mail carrier, she is required to walk, carrying a bag of 5 to 10 kg (11 to 22 lb) for 3 to 5 hr per day. Her physical exam shows that she is not short of breath at rest but becomes dyspneic (breathless) at mild to moderate levels of exertion. Heart sounds are normal, but lung sounds suggest congestion. Her x-ray shows some patchiness of the lung, indicating early emphysema, and some enlargement of the heart. Pulmonary function tests (PFTs) are also indicative of mild chronic obstructive pulmonary disease (COPD).

Test	Actual	%Predicted
Forced vital capacity (FVC)	2.70 L	79
Forced expired volume in 1 s (FEV_1)	1.69 L	65
Forced expiratory flow between 25% and 75% of FVC ($FEF_{25-75\%}$)	1.71 L/s	58

A treadmill exercise test has been performed. Jane stopped exercise at a load that was slightly less than 75% of her predicted load. In addition, monitoring of arterial O_2 levels using the ear oximeter showed a drop in % hemoglobin saturation from 95% at rest to 89% during peak exercise.

As you develop Jane's rehabilitation program, you need to consider the effects of the lung disease on her ability to exchange gases and on the ability of the cardiovascular system to deliver O_2 to the working muscles. Since gas exchange is abnormal, as indicated by the exercise-induced decrease in arterial O_2, there is a likelihood that her peripheral O_2 delivery is somewhat compromised. Finally, working in the heat will place increased demands on the circulatory system, which in turn could affect gas exchange and peripheral O_2 delivery.

Your ability to prescribe a rehabilitation program depends on your ability to discern how disease affects the interactions among the physiologic systems and how these interactions will affect the exercise response of patients such as Jane.

Case 2

Since you have developed a local reputation for working with athletes, a group of cyclists have come to you for advice. They have heard that altitude training can have a positive effect on cycling performance. They understand that the effects of altitude training can be ascribed to an increase in red blood cell (RBC) concentration and an increase in blood volume, which increases the O_2-carrying capacity of the blood.

They are aware of the apparent success of certain cyclists who use "blood doping" (removing blood, then reinfusing the blood several weeks later) to increase the RBC and blood volume. Since it is not possible for them to go to altitude for an extended period, they are asking you for your advice regarding the possibility of blood doping.

Aside from the moral dilemma this situation presents, to properly advise these athletes you need to understand the physiologic interactions between changes in RBC volume, hematocrit, and O_2 delivery and the responses of the cardiopulmonary system—and particularly how these interactions may affect performance.

These cases reinforce the need to understand the basic concepts of neuromuscular, metabolic, and cardiopulmonary exercise physiology, as well as the need to know how disease affects these systems, in order to prescribe exercise properly. By the end of this course, you should be able to adequately answer questions related to these two cases.

How to Use This Text

This textbook emphasizes the understanding of concepts and their clinical applications as opposed to the rote learning of facts. There are several aids in each chapter to draw your attention to these concepts and to test your understanding.

• We have prepared **case studies** throughout the text as a means of highlighting the clinical applications of the physiological concepts being presented. Many of these studies involve solving problems. Try to solve the problem on your own before reading the solution. Finding out whether or not you can solve these problems will enable you to evaluate your understanding of the critical concepts presented. These features can be quickly found for review by the vertical lines of color that appear along their left margins.

• Throughout each chapter you will find **key concepts** highlighted like this:

Key statements look like this and are distillations of key concepts.

These key statements will focus your attention on the most important concepts to be understood. We prefer that you understand these concepts and their clinical application rather than simply memorize facts.

• As a further aid to direct your attention to critical concepts, we boldface **key terms** in color throughout the text. These words are defined in the **glossary** at the end of the book. You should be sure that you understand the meaning of these key terms before reading on.

• The glossary also contains definitions of words that many of you will already know but that are included in case you need to refresh your memory. Such words are used without explanation in the text itself, but you will find definitions of these terms in text that looks like this:

➤ Definitions of words that you may know from previous courses look like this. They are easy to spot, and can thus easily be skipped over or read, depending on your needs.

• We list **what you need to know** at the end of each chapter. Use these statements as a checklist to determine if you understand and can apply the concepts listed.

• Finally, there are **review questions and problems** at the end of each chapter to help you review and problem-solve using key concepts. Some chapters have more questions than others because the intent of the questions is not to stress rote memorization but to elicit understanding of concepts. Answers are provided in an appendix at the end of the book.

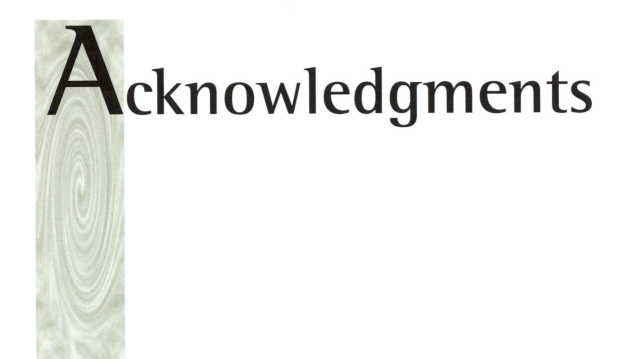

Acknowledgments

We would like to thank Rainer Martens for his early encouragement in our pursuit of this book. Thanks also to Mike Bahrke, Elaine Mustain, Maggie Schwarzentraub, and others at Human Kinetics for their hard work and politeness in reminding us of our deadlines. We would especially like to thank Dr. Susan Bennett for her contribution to the chapter on neural diseases. Susan's expertise in this area is something neither of us could ever hope to match. Finally, thanks are owed to colleagues who provided helpful feedback, but mostly put up with grumbling and other missed deadlines because we were working on this project.

Frank Cerny
Harold Burton

Bioenergetics

This chapter will discuss

➤ sources of energy;
➤ oxygen consumption;
➤ work, power, and energy; and
➤ metabolic calculations.

Understanding energy transformations is critical to understanding the body's responses to the demands of exercise. This is so because the level of energy demand ultimately determines the physiologic responses to exercise. It is readily apparent that the harder you exercise (increasing energy demand), the more you breathe and the faster your heart beats. It is also apparent that the body's responses to increases in energy demand are quite different among groups of people. Moreover, appropriate prescription requires some knowledge of the energy requirements of the variety of tasks that you are preparing your client to perform, as well as of the exercises and activities you will use in rehabilitation. For example, determining the ability to return to work necessitates some knowledge of the energy required for that work. Understanding energy demands, then, is essential for understanding individual responses to exercise. This chapter focuses on the basic sources of energy and on ways that the health professional can measure and estimate the amount of energy a particular task requires.

Sources of Energy

Performing exercise requires energy. The ultimate source of energy for muscle contraction is **adenosine triphosphate (ATP)**, which is an adenosine molecule with three phosphates attached. Adenosine triphosphate supplies energy not just for muscle contraction, but also for the maintenance of all cellular functions. For instance, the cell membrane requires continuous maintenance and repair, which rely on energy available through ATP.

All metabolic energy processes result in the production of heat such that heat production reflects the amount of energy involved in a metabolic process. The standard measure of energy is the **calorie (cal)**, usually expressed as the **kilocalorie** (**kcal** = 1000 cal); **Cal**, with uppercase "C," is sometimes used to indicate kcal, defined as the heat required to raise one gram of water one degree centigrade. Heat production can be measured directly with use of a calorimeter (figure 1.1).

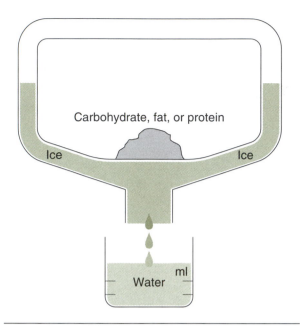

Figure 1.1 A calorimeter. This instrument can be used to measure the heat produced by burning carbohydrate, fat, or protein. The heat produced melts the ice, allowing us to determine the heat production, which can be related to Calories.

The ultimate source of energy for muscle contraction is adenosine triphosphate (ATP).

In its simplest form the calorimeter is an ice-encased chamber large enough to accommodate the energy process being observed. Energy, in kcal, is estimated by measuring the amount of ice melted. If we want to measure the amount of energy produced by burning 1 g of carbohydrate, we could place 1 g of carbohydrate in the chamber, burn it, and measure the amount of ice melted. We would find that burning 1 g of carbohydrate produces 4.2 kcal of energy; burning 1 g of fat would produce about 9.4 kcal, and burning 1 g of protein would produce about 5.7 kcal (table 1.1).

Table 1.1	Energy Equivalents for Carbohydrates, Fats, and Proteins	
Fuel	Energy/g	Energy/L O_2
Carbohydrate	4.2	5.0
Fat	9.4	4.7
Protein	5.7	4.5

We could also use the calorimeter to measure the metabolic energy, in kcal, produced by living organisms. In this case we would put the organism into the box, and the amount of ice melted would be a direct indication of heat produced and energy expended. In this form, the calorimeter theoretically could be used to measure the energy expended by humans at rest and during exercise. However, there is considerable potential for error in this measurement, particularly during exercise. These errors include heat absorbed by the large calorimeter, heat produced by the ergometric device (i.e., treadmill or cycle ergometer), and heat lost to the air being circulated to the subjects. Early scientists noticed that living organisms require oxygen (O_2) and that the amount of O_2 consumed increased as energy expenditure increased. The measurement of **oxygen consumption ($\dot{V}O_2$)** allows indirect determination of energy production.

Measuring oxygen consumption ($\dot{V}O_2$) allows one to determine energy production indirectly.

Oxygen Consumption

Measurements of the air circulating in a calorimeter led to the observation that O_2 was continuously consumed and carbon dioxide (CO_2) was produced during the measurement of energy production. It soon became obvious that the amount of O_2 consumed was proportional to the energy expended and that measuring the amount of O_2 consumed was a better way to determine the energy expenditure of living organisms, particularly during exercise, than using the calorimeter. With this measurement, we determine the volume, in liters, of O_2 consumed per minute. Briefly, we measure O_2 consumption (volume of O_2 consumed/min = $\dot{V}O_2$) by subtracting the volume of O_2 exhaled each minute from the volume of O_2 inhaled each minute, expressing this value as L/min.

$$\dot{V}O_2 = \dot{V}O_2 \text{ inhaled} - \dot{V}O_2 \text{ exhaled}$$

In its simplest form, the volume of O_2 inhaled is calculated by multiplying the percentage O_2 in inspired air times the volume of air inhaled in 1 min (minute ventilation). Likewise, the volume of O_2 exhaled is calculated by multiplying the percentage of O_2 in exhaled air times the exhaled minute ventilation. (See appendix A for specific

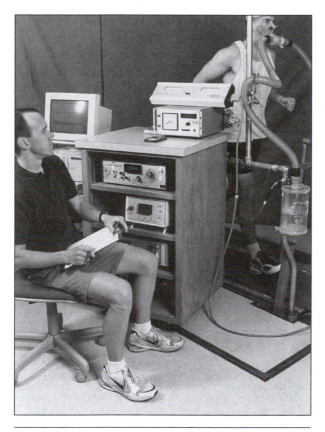

Equipment used for measuring maximum oxygen consumption. The expired air from the subject is collected in a mixing chamber from which samples are sent to oxygen and carbon dioxide analyzers. The measurements from these instruments are used to calculate O_2 uptake and CO_2 production.

information on the measurements and calculations necessary for determining O_2 consumption.) Measuring the kilocalories produced in relation to the amount of O_2 consumed when carbohydrates, fats, or proteins were incinerated revealed the relationships shown in the third column of table 1.1. For living organisms on a diet of a mixture of carbohydrates, fats, and proteins, approximately 4.8 kcal of energy is produced for every liter of O_2 consumed. For purposes of most calculations, using the conversion 5 kcal/L O_2 allows the estimate of energy, in kcal, from measurements of O_2 consumption. The expression 5 kcal/L O_2 is important to remember because this conversion allows us to use measurements of $\dot{V}O_2$ to estimate the amount of energy consumed, in kcal.

Remember the expression 5 kcal/L O_2 because this conversion allows us to use measurements of $\dot{V}O_2$ to estimate the amount of energy consumed, in kcal.

Calculations:

- What is the energy expended, in kcal, if we measure an O_2 consumption of 1 L O_2/min?

$$1 \text{ L } O_2/\text{min} \times 5 \text{ kcal/L } O_2 = 5 \text{ kcal/min}$$

- Our resting $\dot{V}O_2$ is about 3.5 ml O_2/kg body weight/min. For a 60 kg person, how much energy, in kcal, is expended every minute?

$$3.5 \text{ ml } O_2/\text{kg/min} \times 60 =$$
$$210 \text{ ml } O_2/\text{min total resting } \dot{V}O_2/\text{min}$$

$$210 \text{ ml} = 0.210 \text{ L/min so } .210 \times 5 \text{ kcal} =$$
$$1.05 \text{ kcal/min}$$

Work, Power, and Energy

In physical terms, **work** is measured as the amount of force produced times the distance ($F \times D$) over which that force is exerted. If we generate force sufficient to lift one kilogram one meter, we have produced 1 kilogram-meter (**kgm**) of work. The rate at which work is done, or the **power**, is calculated as $F \times D/\text{time}$, expressed as **Watts (W)**. You will recognize this as the same term used to describe electrical power use in your home. The conversion of power and work is performed by use of the factor 1 W = 6.12 kgm/min.

Calculations:

- If we produce 5 W of power, how much work is being done?

$$5 \text{ W} \times 6.12 \text{ kgm/min} = 30.6 \text{ kgm}$$

- Conversely, if we do 10 kgm of work in a minute, what is the power output?

$$10 \text{ kgm/min} \times 1/6.12 = 1.63 \text{ W}$$

These relationships allow us to describe, in the same terms,

- heat, or energy, produced (kcal);
- work performed (kgm); and
- power (W).

Linking measures of mechanical work to $\dot{V}O_2$ provides a powerful tool, allowing the exercise scientist to measure the physiologic cost of human activity and to relate that cost to settings on the treadmill or cycle ergometer. The relationships between physiologic cost, as reflected in

$\dot{V}O_2$, and mechanical terms have some practical implications for prescribing exercise in rehabilitation using mechanical devices such as treadmills or cycle ergometers. In many clinical situations, we may measure a person's peak exercise capacity on a treadmill. We can then calculate the relationship between actual mechanical work performed and the energy required to perform that work (see next section on metabolic calculations) so that we can express exercise capacity, measured on the treadmill, as a percentage of peak $\dot{V}O_2$. For example, we may know that a patient has a peak, or maximum, capacity to produce energy of 3 L O_2/min and may wish to have the person exercise at 50% of that capacity on a cycle ergometer. The well-defined relationship between $\dot{V}O_2$ and work allows us to calculate the workload setting on the ergometer that would elicit a $\dot{V}O_2$ of 1.5 L/min. The next section presents formulas that allow the estimation of $\dot{V}O_2$ at any speed and elevation on the treadmill or any resistance setting on a cycle ergometer. Using these formulas we can establish settings on these devices that will elicit the desired O_2 consumption or energy expenditure.

Different areas within the health care industry use different terms to describe relationships among energy, work, power, and O_2 consumption. Exercise physiologists, particularly those involved in athletics, tend to express activity in terms of O_2 consumption. The O_2 consumption during a particular activity may be measured as 3.4 L/min. If the person performing that activity has a measured maximum O_2 consumption of 4 L/min, she is exercising at 85% of her maximum. In industry, it is much more common to express energy in terms of kilocalories. For instance, the National Institute of Occupational Safety and Health has developed worker guidelines for the maximum amount of heat-load exposure based on the energy expenditure, in kcal, of the activity.

Case Study

The personnel manager for a large auto manufacturing company has contacted you regarding the ability of an employee to return to work as a punch-press operator. This task is classified as a "very heavy effort task," requiring between 6.0 and 10.0 kcal/min. How can you determine whether this worker has the physical capacity to return to work?

To ascertain the worker's ability to perform the required task, you first need some estimate of his peak work capacity. Let's assume that you have a treadmill and perform a standard incremental, progressive work capacity test (see chapter 18). The worker reaches his peak capacity at 10% grade walking at 3.0 mph. On the basis of metabolic equations (see the next section), you calculate that his peak O_2 consumption is 34.0 ml O_2/kg/min. He weighs 70 kg, so his total peak O_2 consumption is 2.38 L/min (34 ml O_2/kg/min $\times$ 70 kg). Since 5 kcal/L O_2 $\times$ 2.38 L/min = 11.9 kcal/min, his peak work capacity is higher than the demands of the task. To do the task, however, he would have to work at between 50% (6 kcal/min $\div$ 11.9 kcal/min) and 84% (10 kcal/min $\div$ 11.9 kcal/min) of his peak capacity. You would be able to recommend that he return to work, but that he reduce his work intensity toward the low end of the job requirement so that he is working closer to 50% of his peak capacity.

Another useful convention is to describe energy expenditure in terms of multiples of resting $\dot{V}O_2$. The average adult expends about 3.5 ml O_2/kg/min at rest, which is defined as **1 MET**, or one resting metabolic equivalent. A $\dot{V}O_2$ of 14 ml O_2/kg/min during exercise could be expressed as 4 METs.

---

1 MET, or one resting metabolic equivalent, is 3.5 ml O_2/kg/min, or what the average adult expends at rest.

Metabolic Calculations

The energy costs of walking are associated with forces moving both horizontally and vertically. For instance, the cost to move 1 kg of body weight in a horizontal direction at a speed of 1 meter/min (m/min) is 0.1 ml O_2/kg/min; thus, we can calculate the $\dot{V}O_2$ of walking at different speeds if we know how fast someone is walking. Of course, the complete energy expenditure for walking also must consider that some vertical movement of the body takes place, that movement may be up a grade, and that some energy is expended to support resting metabolism. The general equations for calculating $\dot{V}O_2$ must sum the energy involved in horizontal and vertical movements as well as resting energy expenditure. Specific examples of the equations are included in appendix B. Here are the equations:

- **Horizontal.** Use these conversion factors to calculate the energy cost of horizontal movement during walking or running:

0.1 ml O_2/kg/min for every m/min walking (50-100 m/min)

0.2 ml O_2/kg/min for every m/min running (>134 m/min)

- **Vertical.** Use these conversion factors to calculate the energy cost of vertical movement during walking or running:

1.8 ml O_2/kg/min for every m/min speed walking (50-100 m/min)

0.9 ml O_2/kg/min for every m/min speed running (>134 m/min)

And multiply the product by % grade, as a fraction.

- **Rest.** Now we need to add the resting energy expenditure. Resting expenditure normalized for body mass = 3.5 ml O_2/kg/min.

Thus the complete equation for walking becomes:

$$\dot{V}O_2 \text{ ml / kg / min} =$$
$$\left(0.1 \frac{\text{ml / kg / min}}{\text{m / min}} \times \text{speed}\right) +$$
$$\left(\text{grade} \times \text{speed m / min} \times 1.8 \frac{\text{ml / kg / min}}{\text{m / min}}\right) +$$
$$3.5 \text{ ml / kg / min}$$

Similar equations have been developed for the cycle ergometer, with two important differences from the equations for walking or running. In the case of a cycle ergometer there is no horizontal component, and the vertical component is based on the power output (kgm/min). In addition, you will have noticed that expressions of energy expenditure during running and walking are **normalized** for body mass (ml O_2/kg/min). This is necessary because the energy expenditure during weight-bearing activities is dependent on the weight of the person; that is, the more mass that must be moved, the greater the total energy expended. On the other hand, energy expenditure for an activity that is not weight bearing, such as

cycling, is not dependent on body mass; so energy expenditure during these activities is expressed in **absolute** terms (L O_2/min). This generalization is true for a reasonably normal population. For a person who is obese and who must move ponderous limbs on the cycle ergometer, energy expenditure is dependent, in part, on body mass. When we wish to compare O_2 among people of differing body mass, we express $\dot{V}O_2$ in normalized terms (per kilogram).

Use the following factors for calculating energy expenditure during cycle ergometer exercise:

- **Horizontal.** No conversion factors.
- **Vertical.** Use these conversion factors to calculate the energy cost of vertical movement during cycling:

2.0 ml O_2/min for each kgm/min

where kgm/min = kg resistance × meters × revolutions per minute

(Note: the distance per revolution is 6 m for Monark ergometers and 3 m for Tunturi and Bodyguard ergometers.)

- **Rest.** Resting expenditure = 3.5 ml O_2/kg/min × body mass in kg

The complete equation for cycling becomes:

$$\dot{V}O_2 \text{ ml / min} =$$
$$\left(2\frac{\text{ml}}{\text{kgm}} \times \text{load kgm / min}\right) +$$
$$(3.5 \text{ ml / kg / min} \times \text{weight kg})$$

To compare O_2 among people of differing body mass, you still must express $\dot{V}O_2$ in normalized terms (per kilogram).

Case Study
Susan is a 67-year-old, 145 lb female recovering from a heart attack. She has some damage to the left ventricle, but her recent stress test showed no ECG abnormalities. Her physician has referred her to your rehabilitation facility. Her discharge exercise test measured her peak work capacity as 5 METs. Her initial exercise prescription is for exercise at 50% of her maximum capacity on

either the cycle ergometer or the treadmill. You need to determine the ergometer settings for this prescription.

Treadmill

$$5 \text{ METs} \times 3.5 \text{ ml O}_2/\text{kg/min} = 17.5 \text{ ml O}_2/\text{kg/min peak } \dot{V}O_2$$

$$17.5 \text{ ml O}_2/\text{kg/min} \times 50\% = 8.75 \text{ ml O}_2/\text{kg/min targeted exercise } \dot{V}O_2$$

Walking on treadmill at 0% grade:

$$8.75 \text{ ml O}_2/\text{kg/min} =$$
$$\frac{0.1 \text{ ml O}_2/\text{kg/min}}{\text{m/min}} \times x \text{ m/min} +$$
$$(\text{grade factor}) + 3.5 \text{ ml O}_2/\text{kg/min}$$

Subtract 3.5 from both sides and set grade factor to 0:

$$5.25 \text{ ml O}_2/\text{kg/min} =$$
$$\frac{0.1 \text{ ml O}_2/\text{kg/min}}{\text{m/min}} \times x \text{ m/min}$$

Divide both sides by 0.1:

$$52.5 = \text{m/min or } 52.5 \text{ m/min} \div 26.8 = 2 \text{ mph}$$

Cycle Ergometer

$$145 \text{ lb} \div 2.2 = 66 \text{ kg}$$

$$8.75 \text{ ml O}_2/\text{kg/min} \times 66 \text{ kg} = 574 \text{ ml O}_2/\text{min targeted exercise } \dot{V}O_2$$

Assume 6 m/revolution and 60 revolutions per minute (rpm):

$$574 \text{ ml O}_2/\text{min} = 2 \times x \text{ kg} \times 6 \text{ m} \times 60 \text{ rpm}$$

$$574 \text{ ml O}_2/\text{min} = 2 \times x \text{ kg} \times 360 + (3.5 \text{ ml O}_2/\text{kg/min} \times \text{kg})$$

Rearrange:

$$574 \text{ ml O}_2/\text{min} = x \text{ kg} \times 2 \times 360 + 231 \text{ ml O}_2/\text{min}$$

$$343 \text{ ml O}_2/\text{min} = x \text{ kg} \times 720$$

Divide both sides by 720:

$$0.5 \text{ kg resistance on the ergometer}$$

Workers in an auto plant. Physical therapists are frequently called on to determine a worker's ability to return to his or her job.

Case Study

You have used the treadmill to test a 72-year-old male patient with chronic obstructive pulmonary disease. You are unable to measure $\dot{V}O_2$ and must estimate his exercise capacity based on the treadmill speed and grade attained during the test. At 3.0 mph he reached a grade of 3%. You observe that he held onto the guardrails during the test. This is not unusual for pulmonary patients, who may use this strategy to stabilize the thoracic cage for better breathing mechanics. What is this patient's peak $\dot{V}O_2$?

$$\dot{V}O_2 = 0.1 \text{ ml } O_2/\text{kg}/\text{min} \times 80 \text{ m}/\text{min} + (.03 \times 80 \times 1.8) + 3.5 \text{ ml } O_2/\text{kg}/\text{min}$$

$$\dot{V}O_2 = 8 + 4.3 + 3.5 = 15.8 \text{ ml } O_2/\text{kg}/\text{min}$$

In this case, however, the fact that the patient held onto the rails would likely reduce the actual energy cost. Your prescription must take this into account by reducing the predicted $\dot{V}O_2$ by about 5%.

The equations presented here are appropriate for most clinical situations but must be used with caution. They were developed on the basis of data collected from a healthy population of normal weight. As already noted, obesity is a significant factor in altering the accuracy of the equations. These equations also should be used with caution in field situations where terrain and surface may vary.

Individual differences in movement economy ("efficiency") also affect the accuracy of these equations. At the upper range of speeds for the walking equation (100 m/min) and the lower range for the running equation (134 m/min), movement economy can vary considerably. In fact, the energy cost of walking above 100 m/min may be greater than the energy cost of running at that speed; that is, running may be more economical. Chapter 3 includes a more complete discussion of the effect of movement economy on energy expenditure (see discussion and figure 3.1, pages 23-27). The gap in the walking-running equations between 100 m/min and 134 m/min suggests that neither the walking nor the running equation can be used accurately in this range.

Movement economy also becomes an important issue in work with the pediatric population (see chapter 15). Children move less economically than adults do, such that the equations presented here underestimate the true energy cost in children by about 0.5 ml/kg for each year under 18 years.

What You Need to Know From Chapter 1

Key Terms

1 MET

absolute and normalized energy expenditure

adenosine triphosphate (ATP)

calorie (cal) and kilocalorie (kcal, Cal)

oxygen consumption ($\dot{V}O_2$)

power (Watts)

work (kilogram-meters, kgm)

Key Concepts

1. You should understand the relationships among energy (kcal), O_2 consumption ($\dot{V}O_2$), work (kgm), and power (Watts).

2. You should understand the concept of absolute and relative, or normalized, $\dot{V}O_2$.

3. You should have a basic understanding of how to use the equations for calculating $\dot{V}O_2$ for work on the treadmill or cycle ergometer.

Review Questions

1. Peter is a 67-year-old patient in cardiac rehabilitation. He is 173 cm (5 ft 8 in.) tall and weighs 61 kg (134 lb). He has been exercising on the cycle ergometer so that his electrocardiogram and blood pressure can be easily monitored. The load has been set at 200 kgm. He is now ready to exercise on the treadmill.

a. Calculate the treadmill speed required to attain the same exercise intensity as on the cycle ergometer.

b. You find that the speed you calculated is faster than Peter can walk, so you need to

(continued)

calculate the elevation required to give you the desired intensity at a speed of 1.5 mph or less.

c. Peter's physician has called to check on his progress. He wishes to know the MET level Peter is currently working at. What is it?

d. Finally, Peter is trying to watch his diet and asks you how many kilocalories he is consuming in his 20 min exercise session.

2. You have been assigned the task of selecting a squad of soccer players to represent your area at the state level. You have two athletes of equal ability and must choose one. Since this competition requires playing one or two games a day for five days, you know that you need athletes with the highest exercise capacity. Sue, who weighs 64 kg (141 lb), has a $\dot{V}O_2$ capacity ($\dot{V}O_2$ maximum) of 3.58 L O_2/min; Sara, who weighs only 61 kg (134 lb), has a $\dot{V}O_2$ capacity of 3.54 L O_2/min. To which one would you give the edge?

3. Your patient's chart shows that he has a peak exercise capacity of 10 METs, and you are supposed to have him exercise at 50% of that. He weighs 75 kg (165 lb). What setting on the cycle ergometer will be correct? Assume a 6 m/revolution distance and 60 rpm.

Bibliography

American College of Sports Medicine. *ACSM's guidelines for exercise testing and prescription (GETP)*, 5th ed., ed. Kenney WL, Humphrey RH, and Bryant CX. Baltimore, MD: Williams & Wilkins, 1995.

Daniels J and Daniels N. Running economy of elite male and elite female runners. *Med Sci Sports Exerc* 1992, 24:483–489.

Eastman Kodak Company, Human Factors Section. *Ergonomic design for people at work.* Volume I. *A sourcebook for human factors practitioners in industry including safety, design, and industrial relations personnel and management.* Rochester, NY, 1983.

Greiwe JS, Kaminsky LA, Whaley MH, and Dwyer GB. Evaluation of the ACSM submaximal ergometer test for estimating VO₂max. *Med Sci Sports Exerc* 1995, 27:1315–1320.

Krahenbuhl GS and Williams TJ. Running economy: Changes with age during childhood and adolescence. *Med Sci Sports Exerc* 1992, 24:462–466.

Lockwood PA, Yoder JE, and Duester PA. Comparison and cross-validation of cycle ergometry estimates of VO₂max. *Med Sci Sports Exerc* 1997, 29:1513–1520.

Morgan DW and Craib M. Physiological aspects of running economy. *Med Sci Sports Exerc* 1992, 24:456–461.

National Institute for Occupational Safety and Health. *NIOSH occupational exposure to hot environments, criteria for a recommended standard.* Cincinnati: U.S. Department of Health, Education, and Welfare, National Institute for Occupational Safety and Health, 1972.

Nutrition

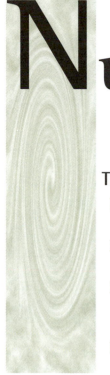

This chapter will discuss

➤ carbohydrate, fat, and protein structure and function;

➤ categories of carbohydrate (mono-, di-, and polysaccharides, including glycogen);

➤ categories of fats (fatty acids, triglycerides, and complex fats, including lipoproteins and sterols); and

➤ nutrient requirements for carbohydrates, fats, proteins, vitamins, and minerals.

The nutrients that we consume are critical in determining the availability of energy for movement. Nutrients that provide the fuel for energy include carbohydrates, fats, and proteins. Other nutrients, such as vitamins and minerals, are critical in regulating chemical reactions in the body, conducting neural impulses, and controlling many cellular functions. With the exception of water and minerals, all nutrients contain carbon. Nutrition affects our growth, our ability to defend against and recover from disease, and our ability to perform activities of daily living. Although there is no evidence that athletic performance can be improved by modifying a basically sound diet, considerable evidence supports the serious consequences of poor nutrition on health and performance. It is therefore critical that health professionals understand basic concepts of nutrition. This chapter provides an overview of basic nutritional information to help you understand some of the relationships between nutrition and energy production in health and disease.

Carbohydrates

Carbohydrates are grouped into three categories based on the number of sugar units:

- Monosaccharides
- Disaccharides
- Polysaccharides

Table 2.1 gives examples of foods containing each of these forms of carbohydrate.

Simple Carbohydrates

Mono- and disaccharides are classified as simple carbohydrates. **Monosaccharides**, such as glucose, fructose, and galactose, are the simplest forms and contain only one $C_6H_{12}O_6$ unit, (figure 2.1). Fructose and galactose can be converted to form glucose. Among the carbohydrates, glucose is the most important for energy production. When two simple sugars are combined, the result is a second form of carbohydrate, a **disaccharide**.

Table 2.1	Categories of Carbohydrates	
Form of carbohydrate	Examples	Found in
Monosaccharide	Glucose, fructose, galactose	Grains, fruits, vegetables
Disaccharide	Sucrose, lactose, maltose	Table sugar, milk, grains
Polysaccharide	Starch, cellulose, glycogen	Plants and animals

We derive approximately 25% of our total calories from the disaccharide sucrose, or table sugar.

Glucose is the most important carbohydrate for energy production.

Complex Carbohydrates

Polysaccharides, or complex carbohydrates, are formed when three or more monosaccharides are combined. The polysaccharide glycogen, a long-chained carbohydrate stored in the liver and within muscle, is a critical source of glucose during activity. The glycogen in muscles is a ready source of glucose during periods of highly intense exercise, avoiding the need for the muscle to import glucose from the blood. The glycogen stored in the liver becomes an important source of blood glucose during other types of exercise. Liver glycogen is an important part of the mechanisms required to maintain appropriate levels of blood glucose, critical to the nervous system, which relies on carbohydrates for energy. The storage of glucose, as glycogen, in the liver and muscle plays a critical role in energy production both as a direct source of fuel for metabolism and as part of a system that ensures the availability of other forms of glucose and lipids for energy production. (See chapter 4 for details on metabolism.)

Glucose stored as glycogen in the liver and muscle serves both as a direct source of fuel for metabolism and as part of a system that ensures the availability of other forms of glucose and lipids for energy production.

Another common polysaccharide, starch, is an important source of energy in most diets and forms the base of the food pyramid (figure 2.2). Starches were at one time avoided in the diet because they were thought to contribute to obesity. While this might be true if one consumed an excess of calories, carbohydrates contribute fewer calories per gram than fat does and should therefore receive emphasis in a healthy diet.

A final category of polysaccharides that plays an important role in our diets is the **fiber** group. Fibers are complex carbohydrates whose structure, in most cases, makes them extremely difficult for human digestive enzymes to break apart into glucose for use as energy. Fibers are readily identified as the rather tough strands running through fruits, vegetables, and grains. Compared to diets with little fiber, diets with recommended amounts of fiber are associated with a lower risk of cardiovascular disease and certain forms of cancer. Because of their relative indigestibility, fibers tend to carry fats such as cholesterol, as well as ingested potential cancer-causing agents,

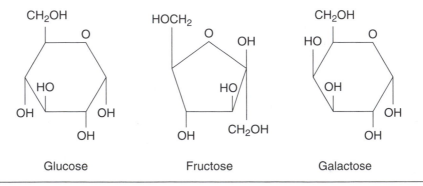

Glucose   Fructose   Galactose

Figure 2.1   Glucose, fructose, and galactose. These monosaccharides have the same number of elements, but different molecular configurations.

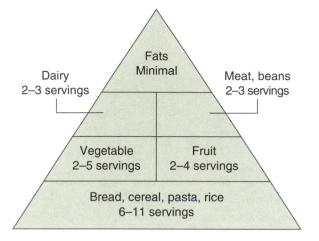

Figure 2.2   The food pyramid. This well-known chart presents the important sources of nutrients. Foods at the base should be eaten in larger quantities than those at the peak.

through the digestive system, preventing their absorption. Diets with adequate amounts of fiber also may play a role in weight control. Fibers tend to contribute to a feeling of fullness, decreasing total caloric intake and displacing high-calorie fats and sweet foods from the diet.

It is tempting to consider all carbohydrates equal in terms of dietary consumption. Eating too much food that is high in refined sugar content, however, can result in nutritional deficits because sweet foods tend to displace more nutrient-dense items. Sweet foods are often called "empty calories" because they have very few nutrients needed by the body other than calories. Candy and donuts are classic empty-calorie foods. In addition, food using refined flour—unless it has been reinfused with nutri-

ents (enriched)—also tends to have low content of nutrients such as phosphorus and potassium. For this reason, whole-grain breads and pastas tend to be healthier than similar foods made with white flour.

Fats

Like carbohydrates, fats (lipids) are composed of carbon, hydrogen, and oxygen; but the ratio of hydrogen to oxygen is much higher. This elevated H:O ratio explains why we get more energy per gram of fat than per gram of carbohydrate (table 1.1). The greater portion of our adenosine triphosphate (ATP) is produced through the oxidation of compounds carrying hydrogen ions (see chapter 4). Thus fats, with their greater number of hydrogens, have a higher potential for producing ATP.

> Because most of our ATP is produced through the oxidation of compounds carrying hydrogen ions, fats, with their greater number of hydrogens, have a higher potential for producing ATP than other fuels.

Fats include diverse chemical compounds such as triglyceride fats, fatty acids, steroids, and others. As with carbohydrates, fats are grouped into three primary categories:

- Simple
- Complex
- Sterols

Table 2.2 shows categories of food that each type of fat is found in.

Table 2.2   Categories of Fats

Form of fat	Examples		Found in
Simple	Saturated	Palmitic acid	Animal fat
	Unsaturated	Palmitoleic acid	Macadamia nuts
	Polyunsaturated	Linoleic acid	Vegetable oils
Complex	Lipoproteins, glycoproteins, phospholipids		Blood, tissue
Sterols	Steroids, cholesterol		

Simple Fats

The basic components of simple fats are the **fatty acids.** Fatty acids are formed with a chain of 2 to >25 carbons with hydrogen and oxygen attached (figure 2.3). If all of the carbons are associated with two hydrogens, the fat is **saturated**—whereas if one or more carbons are without a hydrogen, the fat is called a mono- or polyunsaturated fat, respectively. People who consume a high-fat diet, particularly if it is high in saturated fats, are at higher risk than others for cardiovascular diseases. Animal fats, which are typically saturated fatty acids, are solid at room temperature. In general, vegetable fats are **unsaturated** fatty acids and are liquid at room temperatures. Two exceptions to this general rule are coconut and palm oil, which contain a high proportion of saturated fatty acids. For general cardiovascular health, the American Heart Association recommends that diets contain less than 30% fat and also that the amount of saturated fat be kept to a minimum. To determine the fat content of food, simply find the total calories and the calories from fat as specified on the food label and divide the calories from fat by the total.

The American Heart Association recommends that diets contain less than 30% fat and minimum amounts of saturated fat.

The body can synthesize most fatty acids from carbon, hydrogen, and oxygen. Those fatty acids that cannot be synthesized are called **essential fatty acids** and must be included in a healthy diet. The essential fatty acids linoleic acid and linolenic acid are members of groups of fatty acids called omega-6 and omega-3 fatty acids, respectively. Ingestion of these appears to offer some protection from cardiovascular disease. A deficiency of omega-3 fatty acids may be associated with diminished vision related to aging. Seafood is an excellent source of omega-3 fatty acids, while leafy vegetables, nuts, and grains can be good sources of both omega-3 and omega-6 fatty acids. Although omega-3 fatty acids in the form of fish oil have been promoted for their potential cardiovascular health benefits, ingesting too much poses a potential risk. Because they are highly unsaturated, these fatty acids are susceptible to oxidation and free radical formation. Free radicals react with lipids in cell membranes, causing membrane destruction. Attempts to attenuate free radical damage by ingesting antioxidants such as vitamin E have not been successful.

Most fat in the human body is stored in the form of triglycerides. Glycerol forms the backbone of these simple fats (figure 2.4). As fatty acids, represented by the RCOO group, are added, mono-, di-, or triglycerides are formed. Multiple fatty acids on a single glycerol are not always the same fatty acid. The fatty acids associated with triglycerides are an important source of energy and represent a considerable energy reserve for most humans. In addition to being an important

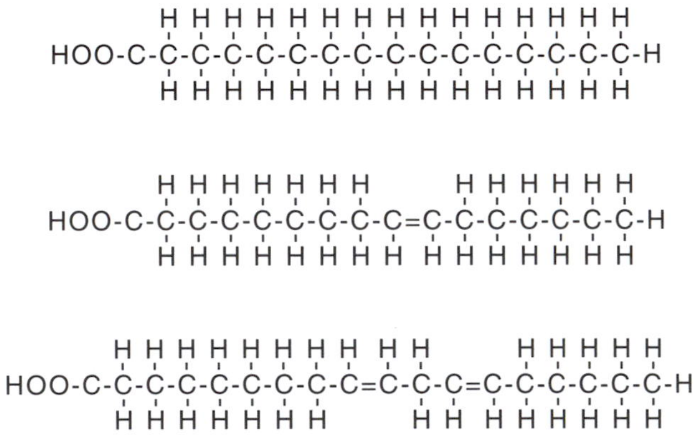

Figure 2.3 The structure of a saturated (palmitate), an unsaturated (palmitoleic), and a polyunsaturated (linoleic) fatty acid. Notice that palmitic acid, or palmitate, has 16 carbons, all of which are associated with at least two hydrogens. Palmitate is a saturated fatty acid. Palmitoleic acid, however, has two carbons joined by a double bond such that this fatty acid is monounsaturated; linoleic acid has two sets of carbons joined by a double bond, making it a polyunsaturated fatty acid.

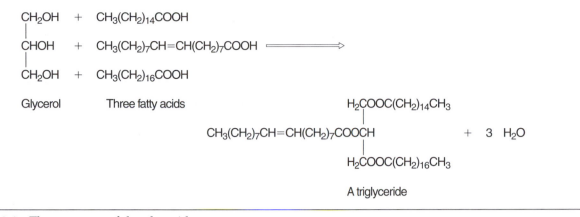

Figure 2.4   The structure of the glycerides.
Reprinted from Houston 1995.

energy reserve, fats are ideal for energy storage because of their structure. Fats are much more malleable than glycogen and can therefore be stored much more compactly. The water insolubility of fats also provides a storage advantage because, unlike glycogen, they are not stored with water; nor is there a tendency for them to become dissolved and transported away from the storage site. Finally, since fat is such a poor conductor of heat, it can act as an excellent insulator, particularly when stored subcutaneously.

The fatty acids associated with triglycerides are an important source of energy and constitute a considerable energy reserve for most humans.

Complex Fats

Complex, or compound, fats are important components of fatty acid metabolism in the body and of body structure. These fats include glycolipids (i.e., carbohydrates and lipids combined), phospholipids (i.e., phosphorus and lipids combined), and **lipoproteins** (i.e., protein and lipids combined). Glycolipids and phospholipids serve both physiologic and structural functions. High concentrations of glycolipids and phospholipids are found in nervous tissue. Glycolipids are involved in neural cell adhesion. Phospholipids are part of the structure of most mammalian cell membranes, forming a fat-protein bilayer (figure 2.5). The protein component of this bilayer is intimately involved in regulating the passage of elements into and out of the cell.

The lipoproteins transport fatty acids in the blood and are found in several forms, including

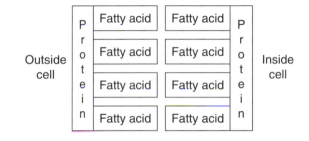

Figure 2.5   A typical cell wall. In most cells, fat, in combination with proteins, forms a bilayer membrane wall.

- **chylomicrons,**
- **very low density lipoproteins (VLDL),**
- **low-density lipoproteins (LDL),** and
- **high-density lipoproteins (HDL).**

These lipoproteins are actually a mixture of triglycerides, phospholipids, and cholesterol surrounded by protein. A high fat content reduces the density of the lipoprotein. Since fats are not soluble in water, lipoproteins are among the more important mechanisms by which fats are transported from the digestive system to storage sites, including the liver, and from storage sites to areas of the body where they are used as a fuel, such as muscles.

➤ A chylomicron is a complex composed of lipid and protein that is responsible for transporting the water-insoluble lipids in the blood from the intestines to storage sites. Other lipoproteins include the high–density and low–density lipoproteins.

High blood levels of VLDL or LDL are associated with a higher risk of cardiovascular disease whereas high blood levels of HDL, which is involved in cholesterol breakdown, are associated with a lower risk of cardiovascular disease. Likewise, low ratios of LDL to HDL are related to a reduced risk of cardiovascular disease. An LDL:HDL ratio of 2.65 is considered normal. Increased risks of coronary heart disease are associated with ratios >3.55 for males and >3.22 for females. Reducing dietary intake of fat results in lower VLDL and LDL levels, while regular aerobic exercise increases levels of HDL. In light of these facts, in addition to the American Heart Association recommendation that <30% of the total caloric intake come from fat, the American College of Sports Medicine recommends the accumulation of 20 to 30 min of activity every day.

> The American College of Sports Medicine recommends the accumulation of 20 to 30 min of activity every day for strong cardiovascular health.

Sterols

The final class of fats is the sterols. Included in this class are cholesterol, sex hormones, and adrenal cortex hormones. Thus the sterols have a wide range of physiologic actions.

Cholesterol, a component of brain and nervous tissue, is found in the skin and adrenal gland and is implicated in the formation of atherosclerosis and the development of coronary heart disease. It appears that regular exercise and reduction of dietary fat can attenuate the risk associated with high levels of blood cholesterol. For this reason, dietary levels of cholesterol intake should be kept below 300 mg/day, and blood cholesterol levels should be kept below 200 mg/dl (dl = deciliter or 100 ml blood). As just noted, it appears that the risk associated with high cholesterol levels is increased in the presence of elevated LDLs, while HDLs help break down cholesterol. Thus, a cholesterol:HDL ratio of <4.5 for males and <4.0 for females is desirable to reduce the risk of coronary heart disease.

> Cholesterol intake should be kept below 300 mg/day, and blood cholesterol levels should be kept below 200 mg/dl (dl = deciliter or 100 ml blood).

Proteins

Proteins form

- hormones,
- enzymes involved in control of cellular metabolism,
- contractile elements of muscle,
- elements of cell structure, and
- part of the energy system for cell contraction.

Proteins perform vital roles in acid-base and cellular fluid regulation and in the circulatory and immune systems (table 2.3). Proteins are made of combinations of 20 amino acids. Of the 20 amino acids required for protein formation, the human

Table 2.3   Proteins Support Many Functions in the Body

Enzymes	Metabolic functions: facilitate all energy-producing reactions in the body
Hormones	Regulatory functions: insulin, glucagon, thyroid
Structure	Cell membrane formation: skin, muscle, airway; tissue formation: ligaments, tendons
Disease protection	Immunity: antibodies
Transport	Oxygen transport: hemoglobin; lipid transport: lipoprotein
Fluid and acid-base regulation	Fluid and electrolyte balance and cell fluid content

body can form 11. The remaining 9 amino acids are referred to as "essential" amino acids, meaning that we need to ingest them in our diet. Animal protein from meat, eggs, and dairy products is a good source of protein because these products contain all of the amino acids. For any one amino acid to be useful, all others must be available in an adequate quantity. The availability of all amino acids in meat gives meat an advantage over plant proteins. Plant protein may be lacking one or more amino acids, and overreliance on plant protein can lead to an amino acid and protein deficiency. It is important, then, to eat a combination of plant products to ensure that amino acid deficiencies in one plant can be balanced with a good supply of those amino acids in another. For instance, beans and wheat are good complements because the amino acids that are low in beans are higher in wheat, and vice versa. Good sources of plant protein are cereals, legumes (beans, peas), and nuts.

According to dietary recommendations, we should consume 0.8 g of protein/kg body weight to obtain the amino acids we require. For young children, the recommendation is about 1.0 g/kg. In either case, these recommendations would be met if between 8% and 10% of our diet were protein, whereas in actual practice our diets contain between 10% and 15% protein. It is important to recognize that if we are consuming sufficient calories to meet our metabolic needs with a healthy diet, we are likely exceeding the dietary recommendations for protein. Dietary supplements of amino acids are rarely needed.

A healthy diet with sufficient calories to meet our metabolic needs usually exceeds the dietary recommendations for protein, so amino acid dietary supplements are rarely needed.

Case Study

A wrestler who wishes to increase his muscle mass has asked you for advice regarding the use of amino acid supplements. He read in a popular fitness magazine that amino acid supplements are absolutely necessary if a muscle-building program is to be successful. He thought the idea of supplementation sounded very reasonable since muscles are composed of proteins, which are amino acids, and since his intent was to increase the amount of muscle protein in the body. What is your advice?

You first must determine whether the athlete is consuming sufficient calories. This is of particular concern in a wrestler, since these athletes are frequently trying to increase muscle mass while keeping total body weight as low as possible. You also need to estimate his daily caloric need. You ask him to keep a dietary record for the next week. He should record each item ingested along with an approximate amount. In addition, he should keep a record of his daily activities. From these records you can estimate his caloric intake and his caloric expenditures. If he is consuming a healthy diet and meeting his caloric needs, he should be ingesting sufficient proteins/amino acids to provide the building blocks for increasing muscle mass. You should, however, determine clearly whether his diet is balanced enough to ensure adequate protein intake.

Your analysis of dietary caloric intake indicates that this athlete consumes 3450 Cal a day and that he expends roughly the same. On this basis you could assume that he does not require amino acid supplements. Further analysis of his diet, however, indicates that he consumes a largely vegetarian diet to minimize fat intake. This athlete requires further counseling to improve his dietary intake of amino acids. He should consider consuming some quantity of meat and/or some quantity of the vegetables that contain the essential amino acids—for example, cereals, legumes (beans, peas), and nuts.

Because of the importance of proteins in the body, dietary deficiencies in any of the amino acids become readily apparent. Although protein losses from body tissues are able to make up dietary deficiencies for short periods, chronic amino acid deficiencies will show up as failure to grow in children and as edema in adults. In adults, protein deficiencies result in decreased blood albumin levels—which, because of changes in colloid osmotic pressures, lead to water retention. Finally, liver and pancreas pathologies are frequently observed in both adults and children with chronic protein deficiencies because of the high amino acid turnover in these organs.

➤ Colloid osmotic pressure is the pressure exerted on a semipermeable membrane by unequal concentrations of particles on the two sides. When the concentration is greater on one side, fluid is drawn from the other side.

Case Study

Dietary counseling is an important part of cardiac rehabilitation. Since lowering fat intake is a general goal of dietary advice, you may be frequently confronted with questions regarding this issue. A common question is "What about avoiding meats altogether?"

You might consider dealing with this question on two levels. The first level might include the consideration that avoiding meat altogether carries with it a certain risk of not ingesting sufficient amino acids. At this level, your advice should be to read food labels carefully and avoid foods with more than 20% fat. This, in combination with moderate ingestion of lower-fat meat products—for example, fish and poultry—should keep fat intake at appropriate levels while supplying adequate protein. For instance, a meal with a small portion of poultry along with portions of grains, nuts, and beans can supply as much usable protein as a single serving of steak.

On another level, you might advise that a diet with no meat but with sufficient dairy product and egg intake, in combination with careful attention to vegetable protein intake, should ensure adequate amino acid intake. A 56 g (2 oz) piece of low-fat cheese can supply approximately 30% of an average adult's protein allowance. Dairy products tend to have high concentrations of several amino acids that are low in cereal grains, making this an excellent complementary combination. As an alternative, since eggs have a nearly ideal balance of amino acids, they can be used to complement the grains.

A vegetarian diet must be carefully planned to ensure good nutrition. One should become a vegetarian only after careful study or in consultation with a professional.

Vitamins and Minerals

Vitamins are essential to proper functioning of many bodily processes such as digestion and metabolism. Vitamins must be ingested or must be formed in the body by ingestion of precursors. Vitamins are grouped into two broad categories based on their solubility characteristics: fat solubility and water solubility. Table 2.4 lists vitamins by category, along with information on their functions, effects of vitamin deficiencies and excesses, and common sources. As might be expected, the fat-soluble vitamins are found in the fats and oils of food and are easily stored in body fat deposits. A diet that contains the recommended quantities of vegetables, particularly of the leafy green variety, helps to ensure adequate intake of vitamins.

A healthy diet, following the general guidelines depicted in the food pyramid, requires little or no vitamin supplementation. Exceptions are for

- pregnant women, or those in the postnatal period if breast-feeding;
- people who are dieting;
- persons who are elderly;
- people with known deficiencies; and
- people who may have poor dietary habits.

Vitamin supplementation has been stressed for pregnant women since the discovery that folate deficiency during pregnancy may result in neural tube defects. Since these defects arise before a woman may realize that she is pregnant, anyone considering pregnancy should begin vitamin supplementation.

A diet that follows the guidelines in the food pyramid requires little or no vitamin supplementation.

Minerals, when dissolved in body fluids, are important components of cell and body fluid and electrolyte and acid-base balance. In addition, each mineral serves important particular functions. Although table 2.5 lists the minerals that play a major role in body function, it is not a complete listing of minerals; others, such as iodine, sulfur, and zinc, also play important roles in body regulation.

Table 2.4   Names, Functions, Effects of Deficiencies and Excesses, and Important Sources of Vitamins

	Function	Effects of deficiency (name)	Effects of excess	Source
Fat soluble				
Vitamin A (retinol)	Vision; skin; bone; hormone production; immune function	Corneal degeneration/blindness; rashes; poor bone growth; tooth decay; anemia; infections	Night blindness; dry skin; hair loss; diminished growth; weight loss; nosebleeds; amennorhea	Fortified milk, dairy products, dark leafy green vegetables, orange fruits and vegetables
Vitamin D (calciferol)	Bone growth	Soft, deformed bones (rickets, osteomalacia)	Increased blood calcium; kidney stones	Fortified dairy products, exposure to sunlight
Vitamin E (tocopherol)	Antioxidant	Neural degeneration, anemia		Plant oils, green leafy vegetables, wheat germ
Vitamin K (phylloquinone)	Promotion of blood-clotting proteins and blood calcium regulation	Hemorrhaging	Hindrance of anti-clotting medications	Liver, green leafy vegetables, milk
Water soluble				
Vitamin B *Thiamin* (vitamin B$_1$)	Metabolism	Edema, heart failure, muscle weakness, confusion (beriberi)		Meats, whole grains
Riboflavin (vitamin B$_2$)	Metabolism; neural function	Light sensitivity; rash		Dairy products, leafy green vegetables, whole grains
Niacin (vitamin B$_3$, nicotinic acid)	Metabolism	Diarrhea; dry, flaky skin on exposure to sun (pellagra)	Diarrhea; painful skin and rash	Dairy products, poultry, fish, whole grains
B$_6$ (pyroxidine)	Amino acid and fat metabolism	Anemia; muscle twiches; convulsions; kidney stones	Fatigue; loss of coordination	Meats, whole grains, leafy vegetables, legumes, fruits

(continued)

Table 2.4 (continued)

	Function	Effects of deficiency (name)	Effects of excess	Source
Water soluble (cont.)				
Folate (folic acid)	Synthesis of cells	Anemia; immune dysfunction		Leafy green vegetables, legumes
B_{12}	Synthesis of cells; maintenance of neural cells	Anemia; fatigue; paralysis		Dairy products, meats
Pantothenic acid	Metabolism	Fatigue		Prevalent in most foods
Biotin	Metabolism	Myocardial dysfunction; weakness; fatigue		Prevalent in most foods
Vitamin C (asorbic acid)	Synthesis of collagen; antioxidant; metabolism; enhancement of immune function	Anemia; muscle degeneration; immune dysfunction; bone disorders; bleeding gums (scurvy)	Fatigue; headache; insomnia; upset stomach	Citrus fruits, most vegetables

Adapted from Whitney and Sizer, 1994.

Sharing a meal that follows the guidelines of the food pyramid. Such a meal requires little or no vitamin supplementation.

Table 2.5 Major Minerals and Their Function, Effects of Deficiencies, and Important Sources

	Function	Effects of deficiency/excess	Source
Calcium	Bone and tooth structure; ion transport; blood pressure; skeletal, smooth, and heart muscle contraction; secretion of neurotransmitters and hormones	Deficiency: bone loss, osteoporosis; inadequate bone formation in children	Dairy products
Potasssium	Fluid and electrolyte balance; neural transmission; heartbeat; blood pressure	Deficiency: depressed thirst, interference with neural function. Excess: vomiting, cardiovascular deregulation	Most *fresh* foods (lost from cells when cooked), vegetables and fruits
Sodium	Fluid and acid-base balance; nerve transmission; muscle contraction; blood pressure control	Deficiency: rare. Excess: elevated blood pressure in predisposed individuals, increased thirst	Salt
Magnesium	Energy release; muscle relaxation; enzymatic reactions	Deficiency: acute—convulsions; chronic—increased susceptibility to cardiac dysfunction, increased cholesterol deposition	Whole grain, seeds and nuts, peas, leafy green vegetables
Phosphorus	Fluid and acid-base balance; genetic component of cell formation; cellular energy transduction	Deficiency: rare	Meat, dairy products
Chloride	Fluid and acid-base balance; digestion	Deficiency: muscle cramps	Salt
Iron	Oxygen transport (hemoglobin and myoglobin)	Deficiency: anemia. Excess: increased risk of infections and heart attacks	Meat, legumes

What You Need to Know From Chapter 2

Key Terms

amino acids

cholesterol

chylomicrons

complements

essential amino acids

essential fatty acids

fatty acids—saturated and unsaturated

fiber

lipoproteins—high (HDL), low (LDL), and very low density (VLDL)

mono-, di-, and polysaccharides

Key Concepts

1. You should know the various forms of carbohydrates and fats and know how these substrates are used as a fuel for energy production.

2. You should understand the general role of vitamins and minerals in body functions and show some understanding of the implications of deficiencies in these nutrients.

3. You should know the elements of the food pyramid, the reason each element is put where it is, and the potential role of a food deficiency in energy production.

Review Questions

1. Can you name the form of carbohydrate used for metabolic energy production?

2. What is the form of glucose stored in the muscle and liver, and what important roles do these glucose stores play in regulation of metabolism?

3. In what form are lipids transported in the blood?

4. How are fatty acids stored, and what important role do they play in metabolic regulation?

5. What would you recommend to clients as healthy limits for fat intake?

6. What would you recommend to clients as healthy blood levels of cholesterol for reducing the risk of cardiovascular disease?

7. What is the role of diet and exercise, and of their interaction, in risks for cardiovascular disease?

8. Why are dietary supplements for amino acids generally not recommended?

9. What are the important considerations for people who wish to reduce fat intake by eliminating meat from their diet?

Bibliography

American Diabetes Association and American Dietetic Association. *Exchange lists for meal planning*. Alexandria, VA: American Diabetes Association; Chicago, IL: American Dietetic Association, 1986.

American Dietetic Association. Nutrition for physical fitness and athletic performance for adults. *J Am Dietetic Assn* 1987, 87:933–939.

Austin MA and Hokanson JE. Epidemiology of triglycerides, small dense low-density lipoproteins, and lipoprotein (a) as risk factors for coronary heart disease. *Med Clin North Am* 1994, 78:99–115.

Barron R and Vanscoy G. Natural products and the athlete: Facts and folklore. *Ann Pharmacother* 1993, 27:607–615.

Despres JP and Lamarche B. Low-intensity endurance exercise training, plasma lipoproteins and risk of coronary heart disease. *J Intern Med* 1994, 236:7–22.

Grundy SM and Denke MA. Dietary influences on serum lipids and lipoproteins. *J Lipid Res* 1990, 31:1149–1172.

Horswill CA. Weight loss and weight cycling in amateur wrestlers: Implications for performance and resting metabolic rate. *Int J Sports Nutr* 1993, 3:245–260.

Jenkins DJA, Thomas DM, Wolever MS, Taylor RH, Barker H, Fielden H, Baldwin JN, Bowling AC, Newman HC, Jenkins AL, and Goff DV. Glycemic index of foods: A physiological basis for carbohydrate exchange. *Am J Clin Nutr* 1981, 43:362–366.

Lemon P. Effect of exercise on protein requirements. *J Sport Sci* 1991, 9:53–70.

Lowry R, Galuska DA, Fulton JE, Wechsler H, Kann L, and Collins JL. Physical activity, food choice, and weight management goals and practices among U.S. college students. *Am J Prev Med* 2000, 18:18–27.

Ludwig DS, Pereira MA, Kroenke CH, Hilner JE, Van Horn L, Slattery ML, and Jacobs DR. Dietary fiber, weight gain, and cardiovascular disease risk factors in young adults. *JAMA* 1999, 282:1539–1546.

Plata-Salaman CR. Cytokines and Feeding. *News Physiol Sci* 1998, 13:298–304.

Sizer F and Whitney E, eds. *Nutrition: Concepts and controversies.* 6th ed. Minneapolis/St. Paul: West, 1994.

Tarnopolsky M, Atkinson S, MacDougall J, Chesley A, Phillips S, and Schwarcz N. Evaluation of protein requirements for training strength athletes. *J Appl Physiol* 1992, 73:1986–1995.

United States Department of Agriculture, Human Nutrition Information Service. *USDA's food guide pyramid.* Home and Garden Bulletin no. 249. Washington, D.C.: U.S. Government Printing Office, April, 1992.

Wee S, Williams C, Brown S, and Horbain J. Influence of high and low glycemic index meals on endurance running capacity. *Med Sci Sports Exerc* 1998, 31:393–399.

Energy Balance

This chapter will discuss

➤ energy balance toward an understanding of the energy requirements for physical activity (including walking, running, and water activities);

➤ the ways in which such elements as movement economy, impairment or disability, and assistive devices affect energy balance;

➤ special energy requirements for disease states;

➤ weight gain and weight loss as they affect being overweight and overfat, and the ways in which these conditions relate to health; and

➤ the measurement of body composition.

We have already discussed some aspects of energy expenditure and nutrition. This chapter begins to bring these two topics together. A balance between energy expenditure and consumption is necessary to ensure optimal performance for work and play. To enable you to counsel people seeking to attain this balance, the first section of this chapter deals with the energy requirements of specific tasks and movements. Because disease can affect energy requirements, we next consider how disease impacts energy requirements for movement. Next we discuss the role of energy balance in weight-loss or weight-control problems; and finally, we provide information on assessing current levels of body fat and physical activity.

Physical Activity

The physiological responses to activity—whether that activity relates to tasks of daily living, job requirements, or athletics—are dependent on the energy requirements of the activity. Therefore understanding the energy requirements of activity is critical to decisions regarding patient exercise rehabilitation programs, work-hardening programs, or the physiologic ability to perform tasks.

Energy Requirements for Physical Activity

Energy expenditure during exercise can vary from as little as 5 ml O_2/kg/min for typing to more than 60 ml O_2/kg/min for cross-country ski racing. These energy expenditures translate into energy expenditures slightly higher than resting (5 ml O_2/kg/min = 1.5 METs) for the former and over 17 times resting (60 ml O_2/kg/min = 17 METs) for the latter. The energy requirements for a broad array of activities are listed in appendix C. While such a compendium of energy costs for physical activity is extremely useful for estimating energy expenditures for a variety of tasks, you should be aware of several potential problems with such a listing. In general, values in the compendium are averages and must be used with some caution when one is applying energy expenditure levels to any given individual. People perform tasks in sufficiently different ways as to alter the energy requirement for the task. Factors such as movement economy (see page 27) and choice of movements affect the energy cost of an activity. These individual differences become extremely important in working with people who have temporary or permanent disability. Obviously, using crutches or a knee brace increases the energy required for walking. Spasticity can double the energy required for simple motor tasks (see page 28).

> Movement economy, choice of movements, and similar factors affect the energy cost of an activity and are extremely important when working with those with a temporary or permanent disability.

Case Study

The patient is a 68-year-old female with a history of heart disease. Her angina had been suppressed by medication (a combination of nitrates and calcium channel blockers) for several years, but she had a heart attack six months ago. Her rehabilitation has been progressing satisfactorily. Her most recent graded treadmill exercise test showed that angina is present at heart rates above 115 beats/min. This heart rate was observed at a speed of 3.0 mph and a grade of 2.5%. She wishes to return to her job as a waitress in a local diner. You are asked for advice.

You could give the best advice if you knew her heart rate on the job and compared that to her safe limit (<115 beats/min). Logistically this is difficult, so you need to estimate the job requirements and compare those to estimated heart rates for that type of job.

To obtain an estimate of the energy requirements for the patient's job, you look in the compendium in appendix C under "Occupation." The description closest to waitressing falls between "Walking, 3.0 mph, moderately and carrying light objects less than 25 pounds" and "Walking, 3.5 mph, briskly and carrying light objects less than 25 pounds." The estimated energy requirement is between 4.0 and 4.5 METs. Of course, this translates to an energy requirement of between 14 and 18 ml O_2/kg/min (3.5 ml O_2/kg/min × 4.0 or × 4.5 METs). Now you can calculate the energy expenditure of walking at 3 mph, 2.5% grade using the formulas in chapter 1 to determine whether the job requirement is below that level.

Horizontal

$$\dot{V}O_2 = 0.1\ ml\ \frac{O_2/kg/min}{m/min} \times 80\ m/min = 8\ ml\ O_2/kg/min$$

Vertical

$$1.8\ ml\ O_2/kg/min \times 80\ m/min \times .025 = 3.6\ ml\ O_2/kg/min$$

Rest

$$3.5\ ml\ O_2/kg/min$$

Total

$$15.1\ ml\ O_2/kg/min$$

The $\dot{V}O_2$ would be 15.1 ml O_2/kg/min, which falls in the middle of the estimated requirement for the patient's job. Your advice at this time would have to be that the patient's present job requirements may sometimes push her beyond her physiologically safe level and that she should continue rehabilitation. In addition, you might examine alternative jobs for her, such as cashiering.

An examination of the compendium in appendix C shows that within a task grouping the energy requirement can vary considerably. Depending on the exact task, for instance, tailoring can require between 2.5 and 4.0 METs, recreational walking between 2.0 and 4.5 METs, competitive or very fast walking between 6.5 and 7 METs, running between 7.0 and 18 METs, and recreational swimming between 6 and 10 METs.

An activity recommendation should recognize this variability.

In Walking and Running

Figure 3.1 shows the relationship between energy requirement and walking or running speed. For speeds less than 130 m/min (5 mph), the energy requirements for walking generally are less than those for running. As walking speed increases beyond 130 m/min, however, it is less costly to jog/run than to walk. The energy required to walk increases in an almost linear fashion until approximately 130 m/min (figure 3.1), beyond which it increases more than would be expected given continued increases in speed. This nonlinear increase in energy requirement can be attributed to less economical walking movements at the higher speeds. Anyone who has tried to mimic the speeds attained during race walking can attest to the increases in energy associated with walking at these speeds. At least a part of competitive-running success also is attributed to the economy of movement. Two runners with the same physiologic capacities, as reflected in $\dot{V}O_2$max, may perform differently during competition, with the more economical runner likely winning because he or she is spending less energy on movement.

> Two runners with the same physiologic capacities may perform differently during competition. The more economical runner will likely win because she is spending less energy on movement.

► $\dot{V}O_2$max is the measured maximum amount of O_2 that can be consumed during exercise. This also is called the maximum aerobic capacity.

Does walking a mile require the same energy as running a mile? The answer, of course, depends on the speed of walking or running. In the range where walking and running overlap (figure 3.1, between 115 and 160 m/min), running the mile would require less energy than walking the mile. Since the relationship between walking speed and energy requirement is linear under 115 m/min, and the relationship between running speed and energy requirement is linear above 120 m/min, the energy cost to move 1 mile in these ranges is nearly equal. The total cost to run 1 mile at 150 m/min is approximately the same as that for running 1 mile at 200 m/min, since the times are different.

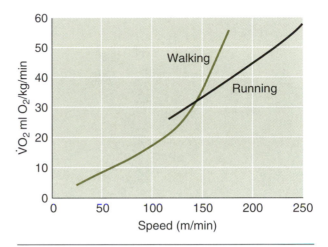

Figure 3.1   Energy cost ($\dot{V}O_2$) plotted against speed of walking and running.
Reprinted from Falls and Humphrey 1976.

With Loads

The addition of an upper-body load that is approximately 20% of body mass increases the energy requirement for walking at a moderate speed by 10% to 20%. This increase in requirement depends somewhat on the placement of the load, as the additional load may alter the position of the body in relation to the center of gravity. A load placed on the shoulders increases the energy requirement about the same as the increase in body mass associated with obesity. Loading the lower extremities, such as with a leg cast, increases the energy cost of moving more than does applying the same load on the upper extremity. This exaggerated energy requirement likely reflects the increased energy expended to accelerate and decelerate the casted limb. Finally, unloading, by use of a cane or other supportive device, may somewhat decrease the high energy requirement for movement in patients with a casted lower limb because it restores movement patterns in the direction of normal.

With Assistive Devices

Crutches and other assistive devices, such as walkers, are associated with significant increase in energy cost for moving in most patient groups (figure 3.2). The energy costs of crutch walking depend on the problem necessitating the crutches; people who have paraplegia show a higher cost (>150% above expected for the speed) than those with a fracture (>60% above expected). The use of a walker enables ambulation in many patients who otherwise would need to stay in a chair or

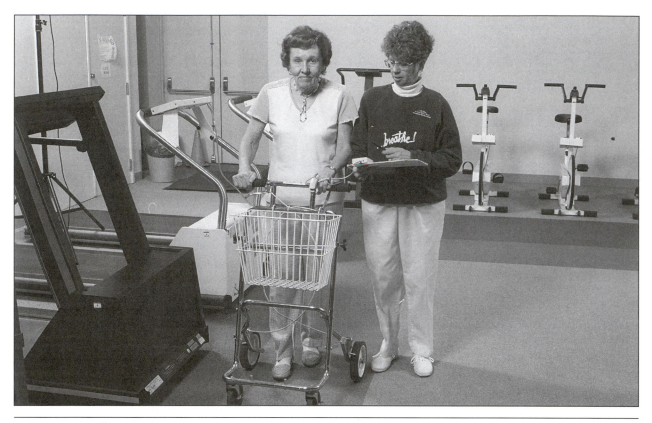

Figure 3.2   Walking with assistive devices. Because walking with assistive devices increases energy needs, the health care professional must take that additional energy requirement into account when prescribing exercises.

bed. The use of a walker, however, also increases the cost of moving at any given speed relative to normal gait. People who use assistive devices to ambulate are likely working very close to their peak work capacity, particularly if they have any underlying cardiopulmonary disease that may limit their work capacity. The increased energy demands in these cases mean that caloric intake must be increased above normal to avoid caloric deficits.

> Crutches and other assistive devices, such as walkers, are associated with significant increases in energy cost for moving in most patient groups.

In Water

The energy requirements for exercising in the water are greatly affected by the resistance of the water. The energy expenditure during swimming is also affected by the swimmer's skill, buoyancy, and the water temperature. You must take these factors into consideration when contemplating water activity or swimming as part of an exercise or rehabilitation program. Energy expenditure for cycling exercise is 30% to 40% higher and for walking is 15% to 30% higher in the water than for the same exercise done on land. This increased energy expenditure is reflected in a higher heart rate during exercise in the water. A large portion of the increased energy demand is attributable to the significantly lower efficiency of exercise in water (2-10% efficiency in water vs. 15-16% on land). Mechanical efficiency ranges from 8% to 15% in well-skilled swimmers and is likely considerably lower for less skilled swimmers. Swimming exercise therefore should be prescribed with some knowledge of the person's swimming ability. Because the energy expenditure for swimming in a less skilled swimmer may be very high, this person may not be able to swim long enough at lower intensities to attain the desired cardiovascular benefit. Being a less skilled swimmer does not, however, preclude the use of water exercise as part of a prescribed activity program. The use of flotation devices and snorkels to avoid awkward breathing mechanics can enable many less skilled swimmers to take advantage of water exercise.

As Affected by Other Factors

The energy requirement for movement also depends on factors such as the surface or terrain, footwear, body mass, and many others. In general, moving over smooth surfaces requires less energy, at the same speed, than moving over uneven surfaces. A surface like carpeting, as compared to linoleum, can increase the energy requirement for someone walking with a shuffle. Likewise, firm, supportive, but lightweight footwear can encourage more economical, and therefore less energy-requiring, movement in many people, and can be particularly important for persons who are elderly and others with altered gait.

Economy of Movement

The preceding description of energy requirements for physical activity implies that the energy requirement varies considerably depending on **movement economy** (efficiency). Efficiency is classically defined as the ratio of the energy output to the energy input. For our purposes, several other definitions are useful, including the work done divided by energy expended, and the mechanical work divided by metabolic energy expended. Because of the problems associated with measuring mechanical work, or work done, many scientists now avoid the term "efficiency," preferring the term "economy."

Mechanically, many factors may affect the economy of movement, including

- stride length and frequency,
- body mass and its distribution, and
- flexibility.

Stride length is an important determinant of energy cost for walking and running; changes in frequency of movement also alter the cost of these activities, as well as arm or leg cranking. Stride length and frequency are, in turn, affected to differing degrees by variables such as age and disease, which you must consider in prescribing activity for a person. Healthy people generally move at the stride length, velocity, or pedal frequency that is the most economical (figure 3.3). Notice in figure 3.3 that the energy cost increases as each variable moves away from the chosen level for a given task. There may be considerable differences in the energy requirements to perform the same task among different people, but most people move in the most economical way given their particular mechanical characteristics. This means that energy expenditure for movement for most people falls in the lowest part of the curves presented in figure 3.3. You must consider this self-selected economical movement pattern when attempting to train or retrain someone's movement pattern. For instance, significantly altering a person's gait may temporarily increase the energy cost. Continued training may or may not result in an eventual progression toward a more economical movement pattern, so it is important to continually evaluate the effectiveness of a retraining program. These issues are discussed in the next section.

Special Requirements for Disease States

Resting energy expenditure, and therefore nutritional requirements, are increased by certain

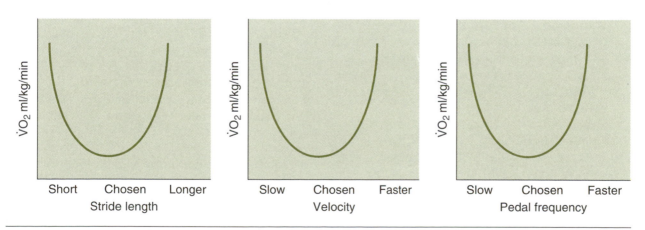

Figure 3.3 Energy cost ($\dot{V}O_2$) plotted against (a) walking stride length, (b) velocity, and (c) cycling pedal frequency.

diseases. Basal metabolism is more than 50% greater in patients with chronic obstructive lung disease (COPD) than in reference populations with no lung disease. Most of this increase in resting energy expenditure can be attributed to the increased work of breathing in these patients; but other factors, such as increased **thermogenesis** (heat production through energy expenditure, such as when shivering) and medication also may contribute. Patients with low subcutaneous body fat, including those with COPD and others who may be malnourished, must increase thermogenesis through increases in energy expenditure to balance the increased tendency to lose body heat.

Patients with disease-related alterations in movement patterns include those with cerebral palsy, arthritis, anterior cruciate ligament deficiency, leg-length discrepancy, stroke, and lower-limb skeletal defects. Increases in energy expenditure range from 30% to >200% and vary with the severity and type of pathology. The reasons for increased energy expenditure vary somewhat with the disease state. The spasticity associated with cerebral palsy, among other states, contributes to increases in energy expenditure both by altering gait mechanics and by increasing the energy requirements for muscles in a constant state of contraction. The increased energy cost of ambulation for these patients frequently contributes to complaints of fatigue. Altered gait is the primary contributor to increased energy requirements for movement in patients with orthopedic dysfunction. Finally, patients with neurological deficits, including stroke, also may alter gait sufficiently to increase movement energy costs.

Case Study

Your patient is a 59-year-old man recovering from a stroke that has left him with residual left-sided paralysis and asymmetrical gait. He has progressed from walking with a walker to using a cane, but appears to be ready to walk without assistance.

In a test trial, the patient is able to walk the length of the hallway (25 m [27 yd]) in 1.5 min. Your notes indicate that he struggled and was short of breath at this slow walking rate. The energy expenditure for this pace (25 m ÷ 1.5 min = 16.7 meters/min; 16.7 × 0.03728 = 0.62 miles/hr is estimated from the compendium

in appendix C at less than 2 METs or <7 ml/kg/min. You are puzzled by his response to this low estimated energy expenditure and refer him to the gait laboratory. An actual measurement of his energy expenditure at this pace shows a $\dot{V}O_2$ of 12 ml/kg/min or approximately 3.5 METs. In addition, his respiratory exchange ratio (RER) is 0.92, indicating that the exercise intensity is quite high for this person (see chapter 4). The laboratory also reports a $\dot{V}O_2$ of 8 ml/kg/min for walking with a cane at this, or a slightly faster, pace.

This patient's energy expenditure to move, while low for walking in general, is extremely high for this pace. This high energy cost is reinforced by the high RER. Until he can move more economically unassisted, the patient should be encouraged to continue to use the cane.

➤ Respiratory exchange ratio is the ratio of CO_2 produced to O_2 consumed ($\dot{V}CO_2{:}\dot{V}O_2$) and reflects the substrates being used to produce energy. A ratio of 1 indicates only carbohydrate use, and a ratio of 0.7 indicates only fat. Ratios that lie between these values indicate the proportion of each substrate that is being consumed.

The effect of altered gait on both total energy cost and economy is an important consideration when you are making decisions about rehabilitation and return to normal activity. You must also weigh the issue of movement economy when considering training someone to move with a more normal gait pattern. In some cases, the pathology-induced gait pattern, while abnormal, may be the most energy-economical pattern. Training a patient with altered gait to walk with a more normal pattern may be aesthetically desirable, but if the energy cost of movement increases significantly, goals need to be reexamined. Alternative approaches such as the use of aids, which may promote a more normal gait pattern without increasing the energy cost of moving, should be considered.

The effect of altered gait on both total energy cost and economy is an important consideration when you are making decisions about rehabilitation and return to normal activity.

Case Study

This patient with hip replacement still walks with an asymmetrical gait. You wish to retrain her so that she can achieve a more normal gait pattern. $\dot{V}O_2$ at 4.8 km/hr (3 mph) with her self-selected pattern is 14 ml O_2/kg/min, which is higher than predicted. The $\dot{V}O_2$ increases to 14.5 ml O_2/kg/min when you coach the patient to walk with a more symmetrical pattern, and you note that she is distinctly uncomfortable with this pattern. For this patient, the self-selected pattern, although abnormally high in energy cost, is more economical than a forced, more normal, pattern.

You have a concern that the patient's self-selected gait may place abnormal stress on her unoperated hip and thereby may predispose her to future problems. An orthotic placed in her shoe promotes a more normal gait pattern and reduces the cost of walking at 3 mph to <13 ml/kg/min. In this case you have found a solution that both allows a more normal gait pattern and reduces the cost of movement.

Weight Management

It is a great paradox that in the United States we seem to be obsessed with weight loss, but as a nation have also shown dramatic increases in body weight and fat over the past decade. It is estimated that over 40% of the population, including children, are overweight.

Weight management depends on maintaining a balance between energy intake (diet) and energy expenditure. If these two factors are equal, weight will be constant; but if caloric intake increases above expenditure, or if expenditure decreases below caloric intake, a weight gain will ensue. Those factors that affect intake and expenditure vary among people and in many cases are poorly understood. Clearly, however, weight loss is effective in the long term only under a program of reduced intake and increased expenditure.

An increase in body mass due to an increase in body fat is associated with an increased risk for several diseases. The risks of cardiovascular diseases—including atherosclerosis, hypertension, stroke, and myocardial infarction—rise dramatically with increases in body fatness. The increased work performed by the heart to pump blood through fat mass, particularly if associated with hypertension, contributes to cardiac failure in many persons who are obese. Finally, the risk of developing non-insulin-dependent diabetes is high in people with increased body fat. As many as 65% of adults and children may be classified as being overfat.

Overweight Versus Overfat

It is important to distinguish clearly between overweight or overfat. Standard height, age, and weight tables as established by population surveys define **overweight**. There are two problems with using this definition in discussion of weight management:

• In most height, age, and weight tables used in North America, the population measured to develop the norms was selected from a healthy population whose weight may be greater than that acceptable for public health. In other words, these tables may define "normal," but may not define healthy.

• It is possible for a person to be overweight as defined by the tables but have a low percent body fat. A professional rugby player, for example, may have a higher-than-expected body mass for his height but lower-than-expected body fat. Since the health risk is associated with the amount of body fat, this "overweight" person with low body fat would have a decreased risk compared to a person with the same mass but higher body fat.

Overfat is defined as an amount of body fat, expressed as a percent of total body mass (% body fat), above that associated with low risk of disease (see table 3.1). As table 3.1 shows, the acceptable % body fat is higher in females than in males. The greater female % body fat is due to greater accumulations of subcutaneous fat thought to be an evolutionary necessity for reproduction. The table also shows that fat tends to accumulate with age. This age-related increase in % body fat is due to changes in muscle mass with age and, to a great extent, to decreases in energy expenditure with little or no decrease in energy intake with age (see figure 3.4). In most cases it is more important to know whether someone is overfat than whether he or she is overweight, since you can be overweight but not overfat, or overfat but not overweight. Being overfat is associated with a high risk of cardiovascular disease, diabetes, and other diseases.

Table 3.1 Percent Body Fat Classified by Age and Health–Associated Categories

Males					
Age	Ideal	Good	Moderate	Fat	Obese
<19	12	12.5-17.0	17.5-22.0	22.5-27.0	27.5+
20-29	13	13.5-18.0	18.5-23.0	23.5-28.0	28.5+
30-39	14	14.5-19.0	19.5-24.0	24.5-29.0	29.5+
40-49	15	15.5-20.0	20.5-25.0	25.5-30.0	30.5+
50+	16	16.5-21.5	21.5-26.0	26.5-31.0	31.5+
Females					
Age	Ideal	Good	Moderate	Fat	Obese
<19	17	17.5-22.0	22.5-27.0	27.5-32.0	32.5+
20-29	18	18.5-23.0	23.5-28.0	28.5-33.0	33.5+
30-39	19	19.5-24.0	24.5-29.0	29.5-34.0	34.5+
40-49	20	20.5-25.0	25.5-30.0	30.5-35.0	30.5+
50+	21	21.5-26.5	26.5-31.0	31.5-36.0	36.5+

Adapted from Hoeger 1998.

Being overfat is associated with a high risk of cardiovascular disease, diabetes, and other diseases.

Estimating Body Fat

Several methods can be used to estimate a person's body fat (appendix D). These include

- hydrostatic weighing,
- skinfold-thickness measurement,
- bioelectrical impedance measurement,
- infrared interactance measurement,
- body mass index, and
- waist-to-hip ratio.

The technique used as the "gold standard" is underwater, or hydrostatic, weighing. (For specific description of the techniques and calcula-tions, see appendix D.) This technique assumes that the density of the various tissues of the body is reflected in the displacement of water, and therefore the weight measured when the body is immersed. When immersed, a person with greater fat will displace more water—and therefore show a greater difference between weight on land and weight while under water—than a person with less fat and greater muscle mass. Of course, air left in the lungs also affects the weight under water. Subjects minimize the air in the lungs by blowing out as much air as possible, but air left in the lungs at that point (residual volume) must be estimated. While still considered the criterion measurement used to determine whether other methods of ascertaining % body fat are valid and accurate, hydrostatic weighing also must be in-terpreted with an awareness of its limitations. The calculation of body fat using the hydrostatic

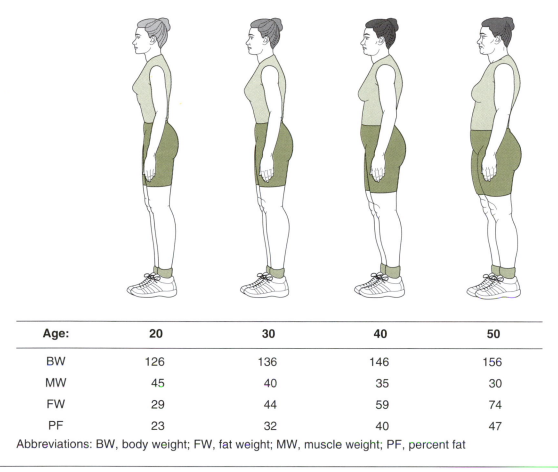

Age:	20	30	40	50
BW	126	136	146	156
MW	45	40	35	30
FW	29	44	59	74
PF	23	32	40	47

Abbreviations: BW, body weight; FW, fat weight; MW, muscle weight; PF, percent fat

Figure 3.4 Body weight and body composition changes during adult life.
Adapted from Westcott 1996.

technique assumes a consistent density within each tissue type (i.e., bone, muscle, fat). The density of various body tissues has been estimated using only a few cadaver samples, and continued research in the area indicates that there is greater individual variation in the densities of body tissue than previously realized. One should take this variation into account when using % body fat calculated by this technique.

A second source of error is in the estimation of residual volume. Residual volume can be measured directly using inert gas techniques or estimated based on height and age from a reference population. In either case, additional error in the underwater weight arises. From a practical point of view, the hydrostatic weighing technique requires complete cooperation by subjects in submersing themselves and exhaling completely. This cooperation is difficult to achieve without some practice on the subject's part, and it's also difficult to verify.

➤ Residual volume is the air left in the lungs after a complete exhalation.

The method that is easiest to administer in the field, and the one most commonly used to estimate body fat, is the skinfold-thickness method (figure 3.5). A specially designed spring-loaded caliper is used to measure subcutaneous fat at two or more sites on the body. These measurements are then used in equations to predict % body fat. The equations are validated against hydrostatic weighing. Accurate skinfold measurement requires skill that comes only from being taught appropriate techniques and practice. When done properly, skinfold measurements give values for % body fat that closely reflect measurements from hydrostatic weighing. But you must interpret skinfold measurements, too, taking the limitations of the technique into account. The most common source of error is the use of inappropriate technique. Persons attempting to measure

% body fat using calipers should learn the technique from an experienced mentor and should practice before attempting measurements on their own. Using a prediction equation that is inappropriate for the population being measured also contributes to error. Because of the individual variations in body composition, prediction equations have been developed for a variety of populations such as children and wrestlers. The standard equations (appendix D) were developed on the basis of individuals of average height and weight. People who are extremely thin or overfat

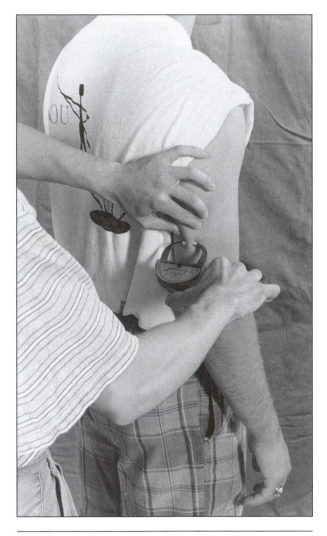

Figure 3.5 Skinfold thickness measurement of percentage of body fat. This type of body composition measurement must be administered by a highly trained person. The development of equations that apply to specific populations for interpreting the measurements obtained have resulted in a high degree of accuracy for this method of body fat estimation.

fall out of the range of these equations, and this decreases the accuracy of their predictions.

The skinfold-thickness method is the easiest test to administer in the field and is most commonly used to estimate body fat.

High-technology body fat-measuring techniques have been developed to eliminate the need for hydrostatic weighing facilities and to minimize human error. These technology-based techniques include impedance measurements and infrared interactance. The bioelectrical impedance measurement of fat is based on the principle that an electrical current passes more readily through hydrated tissue than dehydrated tissue. Since fat tends to be more hydrated than lean tissue, a person with a high fat content has a lower resistance, or impedance, to electrical current. Bioelectrical impedance machines pass a small amount of current through the body and monitor the impedance to that current. The impedance measurement is used to predict % body fat according to reference values based on hydrostatic weighing. Although this is another easy-to-use technique to estimate body fat, the technology can be abused. Since the impedance measurement is dependent on water content of body tissues, any change in hydration greatly affects the measurement. Ingestion of diuretics such as coffee or other caffeine products, as well as the menstrual cycle, alters body water and therefore impedance measurements. Factors affecting body water should be strictly controlled to minimize this potential source of error.

Infrared interactance was developed to estimate the fat content of animal carcasses in the meat-packing industry. This measurement uses a probe, emitting infrared light, that is placed over a selected single muscle area or over several sites. The reflected light is measured such that the amount absorbed indicates the amount of fat in the tissue below. This technique has some appeal because it purportedly measures actual fat in the underlying tissue. But current information suggests that it can lead to gross errors in body fat estimates, particularly in children and in persons who are obese. As with the other techniques, these measurements should be used with caution.

➤ Impedance is the resistance to electrical current flow in an alternating-current system.

The **body mass index (BMI)** is an index of body fatness. The BMI is calculated as the ratio of body mass, in kg, to the square of height, in m.

$$BMI = mass\ in\ kg \div height\ in\ m^2$$

A ratio between 22 and 24 is desirable for both males and females. Values above 25 are associated with increased risk of disease, including cardiovascular disease, and ratios >27 are generally considered to represent obesity. Ratios >45 are associated with a very high risk of disease and represent morbid obesity. Because the BMI uses the same information as the height and weight tables already discussed, this ratio has many of the same problems. Extremely muscular people, such as many athletes, will present with BMI values that are high but that do not reflect the same risk factors as those for a sedentary person with the same BMI.

The **waist-to-hip circumference ratio** also provides a useful field measurement of body fatness. Careful measurement of the circumference of the waist and hips is required to obtain accurate numbers to determine this ratio. For males, a ratio <0.8 is associated with a low health risk; for females, a ratio <0.7 is associated with a low risk. Values >0.9 represent a high health risk in either group.

Body Fat Location

Most body fat is stored in fat cells called **adipocytes**. The number of fat cells is established largely during infancy, with the greatest gain in number occurring during the first year of life. Beyond the age of 10 years, few fat cells are formed. There is little evidence to suggest that, short of surgical removal of these cells, the number of cells can be influenced past puberty; nutrition and physical activity alter the adipocyte number to a degree during the early growth period but appear to have little effect on it in adulthood. Restricted caloric intake or increased physical activity during infancy is associated with a lower-than-expected number of fat cells. Since fat cell number is relatively constant after adolescence, increases in adult weight are related to changes in the amount of fat stored within each cell **(hypertrophy)** rather than an increase in cell number **(hyperplasia)**. These data further suggest that people who accumulate a large number of fat cells in infancy are predisposed to filling those cells with fat deposits and therefore have a tendency to become obese as adults.

Since fat cell number is relatively constant after adolescence, increases in adult weight are related to changes in the amount of fat stored within each cell rather than an increase in cell number.

Body fat is categorized as either essential or storage fat. **Essential fat** is higher in females (10-12% of body weight) than in males (3-5% body weight). The greater essential fat in females is fat associated with reproduction, such as that in breast tissue and reproductive organs. **Storage fat** is found primarily subcutaneously and is a body insulator and a repository for lipids that may be required for metabolism. Distribution of storage fat is variable. In general, people tend to store fat either in the abdominal area or around the hips and thighs. Fat stored in the abdominal area produces a somewhat apple-shaped body form, which is associated with a high risk for cardiovascular disease compared to the more pear-shaped body form characteristic of those whose fat is stored at the hips and thighs. As might be expected, females, who are more pear-shaped, have a lower risk for cardiovascular disease than males, who are more apple-shaped. The waist-to-hip ratio provides a good index for categorizing pears (a ratio toward the low end) versus apples (a ratio toward 1.00).

The reason for the different risks associated with the distribution of body fat may relate to the differences in the type of fat that tends to be stored at each site. Body fat stored at the waist is more active (has a higher turnover rate) and is allied with the health risks associated with obesity; in contrast, the fat stored at the hips is less active and is allied with factors associated with some protection for cardiovascular disease, for example increased high-density lipoprotein (HDL).

While overfatness is associated with increased health risks, regular exercise—even in persons who are obese—appears to reduce many of these risks. Increasing evidence shows that a moderately obese person who exercises regularly can significantly reduce health risks associated with overfatness. This is particularly true if the exercise is associated with a loss of abdominal fat, but it is not possible to target a specific area such as the abdomen or the thighs for weight loss or to "spot reduce." In general, fat is mobilized for energy equally from all storage sites regardless of the activity performed. Weight-loss programs should include exercise as part of their regimen and should focus less on total weight loss than on improving exercise habits.

A moderately obese person who exercises regularly can significantly reduce health risks associated with overfatness.

Weight Loss

From what we have said, it should be clear that weight loss is a matter of unbalancing the calorie intake versus calorie expenditure equation. Since 1 lb (0.45 kg) of fat is equivalent to 3500 kcal, a person can lose 1 lb by accumulating a caloric deficit of this amount through reduced intake, or increased expenditure, or both. The unbalancing of this equation toward weight gain can be seen in overweight people of both sexes at all ages.

Overeating Versus Inactivity

Weight gain is generally thought to be the result of overeating. Although in some cases this is true, in most cases weight gain is more a matter of a decrease in activity levels. Ample evidence indicates that decreased childhood activity is associated with weight gains in this young population. Likewise, the weight gain associated with middle age and beyond relates more to a decrease in activity and a loss of muscle mass with little or no change in diet than to a simple increase in caloric intake with age.

Exercise, when combined with proper diet, is an effective means to lose weight. Dietary manipulation should occur under the supervision of a trained nutritionist to ensure that the reduction in calories is not associated with a reduced intake of nutrients required for health. In general, however, the dietary manipulation should ensure that the intake of fat calories decreases to between 20% and 25% of the total caloric intake. As discussed earlier in this chapter, any exercise, regardless of intensity or duration, will lead to improvements in health and can contribute to fat loss. Activity of higher intensity and longer duration (>20 min) will result in dramatic improvements in the health profile as well as decreases in body fat. A selection of activities chosen by the person is more effective in promoting long-term

compliance than are activities chosen by the counselor. Adding a resistance-exercise component to the weight-reduction program can ensure maintenance of or increases in muscle mass, but the emphasis of a weight-loss program must remain on aerobic activities.

Exercise combined with the proper diet is an effective means to lose weight.

As illustrated by the next case study, increasing exercise is clearly the most desirable means to unbalance the equation. Unfortunately, most weight-loss programs emphasize reducing caloric intake with little or no focus on exercise. The advice to simply reduce intake may be counterproductive in light of current information.

Case Study Part A

Sharon has a weight problem. Over the past 20 years she has attempted to lose weight through self-initiated or organized commercial weight-loss programs. The commercial programs typically weigh each client, determine a weight-loss goal, and restrict caloric intake. Sharon's normal caloric intake is ~ 2,600 kcal/day, which allows her to maintain a stable weight. Her weight-loss programs typically reduce this to <600 kcal. Within one to two months she is able to attain her goal and then returns to her normal caloric intake. Within several weeks, her weight begins to increase, and over a period of months she has exceeded her weight prior to the diet. This cycle makes subsequent weight loss even more difficult.

The problem here is that the severe caloric restriction has resulted in a lowered metabolic rate so that Sharon's usual isocaloric diet now exceeds her requirements, causing weight gain.

Case Study Part B

Sharon is a 40-year-old woman who has been referred to you for a weight-loss program. She describes decades of weight-loss programs that have led to a roller-coaster weight pattern. She is 165 cm (5 ft 5 in.) tall and weighs 80 kg (176 lb). Her cholesterol is 258 mg/dl; HDL is 31 mg/dl; LDL is 144 mg/dl. Resting blood pressure is 134/91. Although Sharon doesn't like to exercise, she realizes that dieting alone hasn't worked and is willing to try to develop the discipline for regular exercise. Skinfold calculation of body fat measurement is 36%, giving her a body fat weight of 29 kg (80 × .36) (63 lb) or a **lean body mass (LBM)** of 51 kg (113 lb). Your initial goal is to reduce her % body fat to 26% (.26). Her ideal body weight (IBW), then, is calculated using the formula:

$$IBW = LBM \div (100\% - 26\%)$$

$$or\ IBW = 51 \div .74 = 69\ kg$$

Sharon's needed weight loss is approximately 11 kg (24 lb), assuming that her LBM will not change. Your goal for the first week is for her to lose .9 kg (2 lb), requiring a caloric deficit of 7000 Cal. Sharon can attain this if she decreases intake by 1000 Cal a day, something she has tried with limited success, or by increasing expenditure by the same amount. Thirty minutes of walking will result in an energy expenditure of ~950 Cal. She thinks that this is a reasonable goal and contracts with you to walk for 30 min a day every day for the next seven days.

Over the next several months you reduce Sharon's caloric intake, primarily by decreasing dietary fat, and increase the intensity and duration of her walking program. Her evaluation at the end of this time shows a body weight of 72 kg (158 lb) and a body fat of 27%. Her fat weight is now 20 kg (43 lb), a reduction of 9 kg (20 lb) fat, while her LBM has increased by .5 kg (1 lb). As an added benefit, her cholesterol is now at 210 mg/dl; HDL is 39 mg/dl; LDL is 128 mg/dl. Resting blood pressure is 128/89. Sharon is extremely pleased with these results and finds that walking is likely to be the beginning of a long-term commitment.

The Set Point

Body weight appears to be regulated through the hypothalamus in the brain. The hypothalamus is involved in the regulation of food intake and the storage of fat. Part of this regulation of intake and storage suggests that each person may have a **set point** for body weight. The set point is analogous to a setting on a thermostat: any attempt to consciously move away from this set-point weight is resisted such that if food intake is reduced, basal metabolism is reduced or appetite is increased to bring body weight back to the set point. Conversely, if a person should consciously increase food intake, metabolism is increased or appetite is reduced to maintain body weight closer to the set point. The reduction of metabolism in response to caloric restriction explains, in part, why people who diet will often gain weight when they return to their normal caloric intake. The return to a normal intake, combined with the diet-induced reduction in metabolism, results in weight gain. The individual set point also may explain why it is more difficult for some people to lose weight than for others.

Assessing Physical Activity

Prescribing exercise for health and weight loss requires some knowledge of the person's normal level of daily physical activity. In addition, developing injury-reduction and -prevention or work-hardening programs requires an assessment of a person's normal daily activity, since your program is going to be an addition to your client's normal pattern of activity.

We discussed earlier the various techniques available for estimating task demands, including direct measurement of $\dot{V}O_2$ and the use of the "Compendium of Physical Activities" (appendix C). These techniques provide a reasonable estimate of task demands. Your conditioning program, however, also should include some assessment of the patient's level of daily activity. Commonly employed techniques for determining current levels of physical activity include having the patient keep a detailed diary of activity and the use of questionnaires and interviews. The selection of an appropriate questionnaire depends on the reason for gathering the information. The better questionnaires are reproduced in *Measuring Physical Activity and Energy Expenditure*, which the interested reader should consult for selecting a questionnaire. Two useful questionnaires are reproduced in appendix E.

What You Need to Know From Chapter 3

Key Terms

adipocytes

body mass index (BMI)

essential fat

hyperplasia

hypertrophy

lean body mass (LBM)

movement economy, movement efficiency

overfat

overweight

set point

storage fat

thermogenesis

waist-to-hip circumference ratio

Key Concepts

1. You should understand why various activities require different energy expenditures.

2. You should know some of the factors that will change energy expenditure for various movements, in health and disease.

3. You should understand the factors responsible for weight gain or weight loss and the ways in which interventions can influence these factors.

Review Questions

1. What do the following situations have in common? What might your advice be to each person?

    a. A track coach has been working with an athlete to help her change her stride length in an attempt to improve performance in the 5000 m (3.1 mile) run; but the athlete's performance has deteriorated, and she complains that she fatigues sooner than she used to.

    b. A stroke patient has developed a severe weakness in the left leg that has altered his gait, and he has come to you with complaints of how short of breath he has become. He is worried about possible lung disease.

    c. A client with her right leg in a walking cast is depressed about how difficult it is to get around and complains about severe breathlessness.

2. Mr. "Unwilling to Give Up Visions of Grandeur" played American football in secondary school and college. He is now a 35-year-old, 188 cm (6 ft 2 in.), 111 kg (245 lb) regular exerciser in your facility. As his personal trainer you suggest that he would be healthier if he were to lose weight. His response is that he weighs the same as he did during his college playing days. What is your response?

3. How would you monitor Mr. "Unwilling"?

4. How is the set point related to the "yo-yo" effect of weight loss by severe dieting?

5. A patient with cerebral palsy is contemplating surgery to make her gait more effective. The surgeon has advised her to receive exercise therapy prior to surgery in the hope that this will speed recovery. This patient, 12 years old and 35 kg (77 lb), is reasonably active under the circumstances. You have developed an exercise program for her that is calculated to increase her daily energy requirement by 20 kcal a day, and you give her dietary recommendations to accommodate this increase. On two-week follow-up, her mother is concerned about her daughter's weight loss. What did you miss?

6. You are asked to determine whether a large number of students in a particular school are overweight/overfat. You must get the information for a report to the county health office within 24 hr. What do you do?

Bibliography

Ainsworth BE, Haskell WL, Leon AS, Jacobs DS Jr., Montoye HJ, Sallis JF, and Paffenbarger AS Jr. Compendium of physical activities: Classification of energy costs of human physical activities. *Med Sci Sports Exerc* 1993, 25:71–80.

Brownell KD and Foreyt JP, eds. *Handbook of eating disorders.* New York: Basic Books, 1986.

Campaigne BN. Body fat distribution in females: Metabolic consequences and implications for weight loss. *Med Sci Sports Exerc* 1990, 22:291–297.

Cavanagh PR and Kram R. Mechanical and muscular factors affecting the efficiency of human movement. *Med Sci Sports Exerc* 1985, 17:326–331.

Clark RR, Kuta JM, Sullivan JC, Bedford WM, Penner JD, and Studesville EA. A comparison of methods to predict minimal weight in high school wrestlers. *Med Sci Sports Exerc* 1993, 25:151–158.

Conley DL and Krahenbuhl GS. Running economy and distance running performance of highly trained athletes. *Med Sci Sports Exerc* 1980, 12:357–360.

Gordon PM, Heath GW, Holmes A, and Christy D. The quantity and quality of physical activity among those trying to lose weight. *Am J Prev Med* 2000, 18:83–86.

Howley ET and Glover ME. The caloric cost of running and walking one mile for men and women. *Med Sci Sports Exerc* 1974, 6:235–237.

Hubert HB, Feinleib M, McNamara PM, and Castelli WP. Obesity as an independent risk factor for cardiovascular disease: A 26-year follow-up of participants in the Framingham heart study. *Circulation* 1983, 76:968–977.

Kraemer WJ, Volek JS, Clark KL, Gordon SE, Incledon T, Puhl SM, Triplett-McBride NT, McBride JM, Putukian M, and Sebastianelli WJ. Physiological adaptations to a weight-loss dietary regimen and exercise programs in women. *J Appl Physiol* 1997, 83:270–279.

Lohman TG. *Advances in body composition assessment.* Champaign, IL: Human Kinetics, 1992.

Manson JE, Colditz GA, Stampfer MJ, Willett WC, Rosner B, Monson RR, Speizer FE, and Hennekens GH. A prospective study of obesity and risk of coronary heart disease in women. *N Engl J Med* 1990, 322:822–889.

Mokdad AH, Serdula MK, Dietz WH, Bowman BA, Marks JS, and Koplan JP. The spread of the obesity epidemic in the United States, 1991-1998. *JAMA* 1999, 282:1519–1522.

Montoye HJ, Kemper HCG, Saris WHM, and Washburn RA, eds. *Measuring physical activity and energy expenditure.* Champaign, IL: Human Kinetics, 1996.

Must A, Spadano J, Coakley EH, Field AE, Colditz G, and Dietz WH. The disease burden associated with overweight and obesity. *JAMA* 1999, 282:1523–1529.

Oppliger RA, Nielsen DH, and Vance CG. Wrestlers' minimal weight: Anthropometry, bioimpedance, and hydrostatic weighing compared. *Med Sci Sports Exerc* 1991, 23:247–253.

Robinson TN. Reducing children's television viewing to prevent obesity. *JAMA* 1999, 282:1561–1567.

Wei M, Kampert JB, Barlow CE, Nichaman MZ, Gibbons LW, Paffenbarger RS, and Blair SN. Relationship between low cardiorespiratory fitness and mortality in normal-weight, overweight, and obese men. *JAMA* 1999, 282:1547–1553.

Metabolism

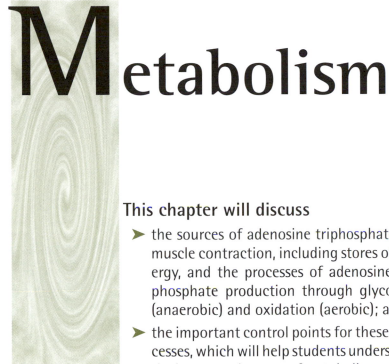

This chapter will discuss

➤ the sources of adenosine triphosphate for muscle contraction, including stores of energy, and the processes of adenosine triphosphate production through glycolysis (anaerobic) and oxidation (aerobic); and

➤ the important control points for these processes, which will help students understand the integrated nature of metabolic production of adenosine triphosphate.

Metabolism is the sum of processes resulting in the production (anabolism) and breakdown (catabolism) of **adenosine triphosphate (ATP)**. When ATP is broken down, the energy associated with one of the high-energy phosphate bonds is released for use by a variety of processes, including muscle contraction. This chapter concentrates on ATP production for muscle contraction, but you should remember that ATP also is required for many other purposes in the body.

A thorough understanding of the metabolic processes is important for the health professional. Whether you are preparing an athlete for optimal performance or rehabilitating a patient, one of the primary objectives must be to ensure that you are enhancing each person's metabolic capacity. Without this enhancement it is impossible to optimize muscle performance, whether the person is an athlete or not. We have discussed the importance of understanding the energy requirements of activity (chapter 3). Now it is important to understand how these energy requirements are met within the contracting muscles.

Sources of Adenosine Triphosphate

The ATP required for activity is derived from several metabolic processes:

- high-energy stores within the muscle,
- non-oxidative production of ATP (**anaerobic**), and
- oxidative (**aerobic**) production of ATP (figure 4.1).

As discussed in chapters 2 and 3, the "fuels" or **substrates** that are broken down for energy production are carbohydrates, fats, and, to a certain extent, proteins.

The small amounts of ATP stored in muscle could sustain only a few seconds of activity if the other metabolic pathways were not available to replenish these supplies. Critical to exercise metabolism is the fact that the metabolic pathways for ATP production ensure against this kind of depletion of ATP levels in the muscle. As soon as

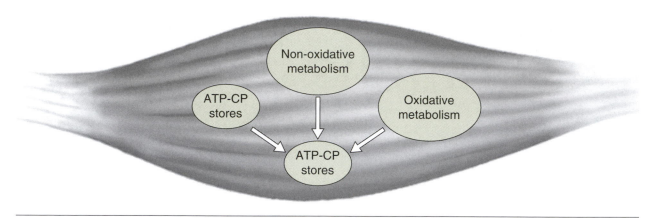

Figure 4.1   The three sources of adenosine triphosphate for muscle contraction.

ATP is broken down to release its energy, several processes are immediately initiated to re-form ATP.

The first mechanism available for re-forming ATP is to access energy from **creatine phosphate (CP)**, also stored in the muscle. First, ATP is broken down, with the release of energy, into **adenosine diphosphate (ADP)** and phosphate. Then the high-energy phosphate from CP is used to resynthesize ATP from ADP. Again, however, the limited supplies of CP in the muscle cannot maintain ATP at high enough levels for sustained activity. The continued use of ATP during exercise requires increased energy production from the non-oxidative and oxidative energy pathways.

Figure 4.2 is a schematic representation of the sources of ATP within the muscle. Each of these sources can be thought of as a reservoir that can be tapped to ensure a steady supply of ATP to the muscle. As soon as some of the ATP stored in the muscle is used for its energy, the second reservoir, filled with CP, is tapped. Note that there is no control valve at this reservoir, so the resynthesis of ATP takes place immediately. At the same time that CP stores are being used to resynthesize ATP, the valves on the other two pathways (non-oxidative and oxidative) are opened so that new ATP will be formed through these pathways. The mechanisms by which these pathways are controlled and integrated are discussed in detail later in this chapter. As depicted in figure 4.2, the integration of these pathways is controlled by several crucial enzymes, shown as valves. It is important to

Rising from a chair. A short-term movement such as this relies heavily on muscular ATP and CP stores. Health care professionals can work more intelligently with their patients if they understand which chemical pathways are involved in what kinds of movements.

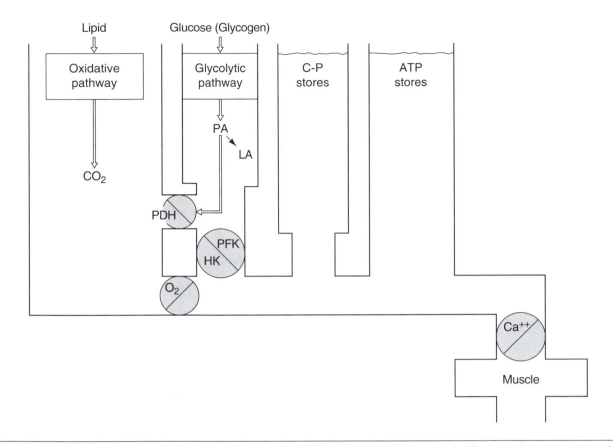

Figure 4.2   Schematic depiction of the primary pathways of adenosine triphosphate for muscle contraction, the substrates for each pathway, and the interactions of each pathway. Valves indicate important control points. PA = pyruvic acid; LA = lactic acid.
Reprinted from Stegemann 1977.

remember these enzymes, since they are critical control points in the integration of the several systems responsible for maintaining cellular ATP levels.

Glycolysis

Glycolysis, the non-oxidative energy pathway, also is called the anaerobic or glycolytic pathway (figure 4.3). The reactions of this pathway take place in the cytoplasm, the fluid substance of the cell. The fuel for this pathway is glucose from the blood or glucose stored in the muscle in the form of glycogen. The end product of the pathway is pyruvic acid (also referred to as pyruvate), which may be further processed to produce energy in the oxidative pathway. Oxidation takes place within the mitochondria of the cell (figures 4.4 and 4.5). In addition to glucose-derived pyruvate, the oxidative pathway can use fats or lipids as a fuel. This oxidative pathway consumes O_2 to complete the metabolic processing of either glucose or lipids. In this sense, the glycolytic pathway is a first step in the oxidative production of ATP from glucose.

The Control of Glucose Entry Into the Cell

The non-oxidative (anaerobic) formation of ATP begins with the entry of blood glucose into the cell. The pancreatic hormone, insulin, facilitates glucose entry. Recently it has been recognized that glucose entry into the cell can take place without insulin, particularly at rest. Several transporters carry glucose into the cell, including the proteins Glut-1 and Glut-4. Glut-1 is not sensitive to insulin and is the primary transporter that is active at rest. Glut-4 is sensitive to insulin and is the primary transporter during exercise. Under conditions of high blood glucose, high insulin levels, or muscle contraction,

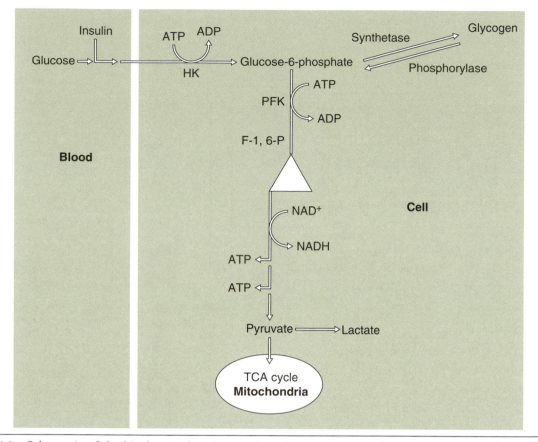

Figure 4.3 Schematic of the biochemical pathways for adenosine triphosphate production in the muscle. The tricarboxylic cycle is discussed on page 46.

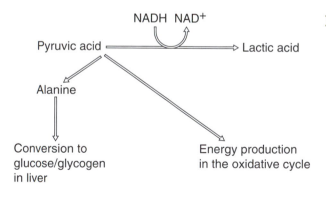

Figure 4.4 Pyruvic acid being processed through oxidation. This process can produce energy or convert pyruvic acid to lactic acid or alanine.

Glut-4 is translocated from sites near the sarcoplasmic reticulum of the muscle cell to the cell surface. The Glut-4 then transports the glucose into the cell.

➤ Sarcoplasmic reticulum is a series of conducting tubules within the muscle that allow communication from the outer membrane of the muscle cell into the interior (discussed in detail in chapter 8).

Glycogen Synthesis and Breakdown

The storage of glycogen is catalyzed by the enzyme **synthetase**. Synthetase is activated by insulin, which is released by the pancreas to promote the reduction of high blood glucose levels. The stimulation of synthetase by insulin opens a storage mechanism for excess blood glucose in both the muscle and the liver. There, glucose is stored as glycogen, a series of long-chain and branched glucose molecules. The breakdown of glycogen is promoted by a complex **phosphorylase** enzyme system, described in figure 4.6. The enzyme phosphorylase is found in an inactive "b" form and an active "a" form. Several factors can directly stimulate the conversion of the b to the a form, including

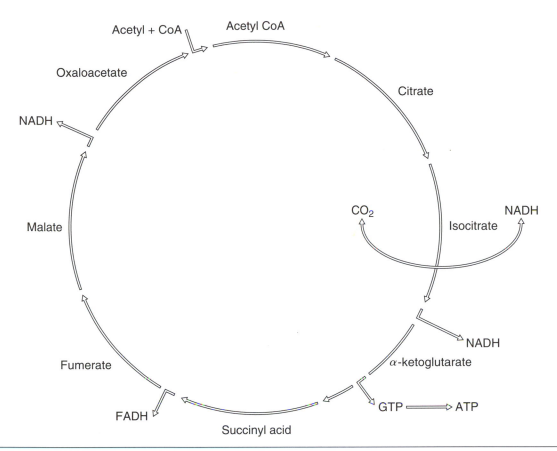

Figure 4.5   The tricarboxylic acid (or Krebs) cycle showing the intermediate steps and by-products. Note that adenosine triphosphate and electrons are by-products of this cycle. CoA = coenzyme A; GTP = guanosine triphosphate.

Ca^{++}. The conversion also is promoted indirectly through stimulation of a sequence of reactions by the catecholamine **epinephrine** (adrenaline). As you will see later, each of these factors become important controllers of energy production under different energy-demand situations.

Glycolysis and Its Control

Once inside the cell, the six-carbon glucose molecule is transformed through a process designed to increase its energy level so that energy can be siphoned off to form high-energy ATP. The first step in this transformation process is the addition of a single phosphate to one end of the glucose (figure 4.3). The phosphate comes from ATP in a reaction catalyzed by an enzyme called **hexose kinase (HK)**. Hexose kinase is one of those important enzymes described as control valves in figure 4.2. The phosphate is placed on the sixth carbon to form glucose 6-phosphate (G-6-P).

At this point the G-6-P can be processed to be stored as **glycogen** or can be processed in the glycolytic pathway to produce ATP (figure 4.6). During times of high energy demand, the stored glycogen is broken down into G-6-P to form ATP through glycolysis. When the breakdown of glycogen occurs under these conditions, the initial use of an ATP as the glucose enters the cell is "spared," with the result that more energy is provided at less cost at that point in time.

The breakdown of G-6-P to form ATP involves a series of 12 reactions designed to manipulate the energy level of the glucose molecule so that some of that energy can be used for ATP production. To grasp the essentials of exercise metabolism, you need to understand only a few of the important control reactions. The first is the reaction catalyzed by the enzyme **phosphofructokinase (PFK)**. Phosphofructokinase adds a second phosphate to the G-6-P to form fructose 1,6-phosphate, which is a six-carbon molecule with a phosphate on both

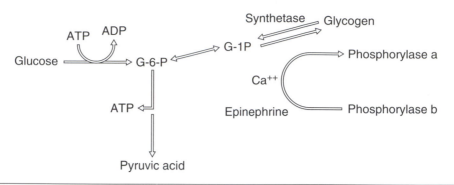

Figure 4.6   Glucose processing in the cell. When glucose is transported into the cell it is converted into glucose 6-phosphate with the addition of a phosphate from adenosine triphosphate. Glucose 6-phosphate can be further processed to pyruvic acid with the production of adenosine triphosphate or can be stored in the form of glycogen. G-1-P = glucose 1-phosphate.

ends. The next reaction takes this six-carbon molecule and splits it into two mirror-image three-carbon molecules. These two molecules can be interconverted, such that there are now two three-carbon units that are further processed in glycolysis. At the next critical step, a hydrogen and its associated electron are removed from each of these units. The removal is accomplished with the help of an electron carrier, **NAD⁺**, to form **NADH** (figure 4.3). This electron carrier, which appears at many steps in the metabolic process, is an important component of energy production. Within the next several steps of glycolysis, ATP is formed using the increased energy in the two three-carbon units. For each three-carbon unit, two ATP are produced, for a gross of four and a net of two ATP for each glucose molecule processed. The end product of glycolytic metabolism is pyruvic acid, a three-carbon pyruvate base with a hydrogen ion. It is important to recognize that the end product of glycolysis is not lactic acid, but pyruvic acid.

Important factors that control the flow of molecules through glycolysis include

- the level of O_2 delivery;
- the level of the metabolic by-products, ADP and Pi (inorganic phosphate); and
- the entry of Ca⁺⁺ into the cell.

As ATP is used, more ADP and Pi are found in the cell. These levels of ADP and Pi are important signals of a shortage of ATP and are examples of the exquisite design of the system to ensure that ATP levels in the cells do not decrease significantly. An increase in ADP and Pi stimulates the important control enzyme PFK, ensuring that the energy-producing system is "turned on" as soon as the need for energy is increased. A second example of this exquisite design is the use of the Ca⁺⁺ as a controller. Calcium is released into the muscle cell to initiate muscle contraction (chapter 8). This same Ca⁺⁺ initiates the conversion of phosphorylase b to phosphorylase a, thus making G-6-P available from the breakdown of glycogen. The Ca⁺⁺-regulated step assures that when muscle contraction takes place, fuel (glucose/glycogen) for energy production is available.

To summarize at this point, a six-carbon glucose has been processed by the glycolytic system to produce ATP. This process involves the use of ATP as the initiating step, the activation of the control enzymes HK and PFK, and the formation of NADH. The final products of this pathway are two three-carbon pyruvic acid molecules. In addition to the glucose from blood, glucose stored in the form of glycogen is also used as a substrate for energy production through glycolysis.

> In addition to the glucose from blood, glucose stored in the form of glycogen is also used as a substrate for energy production through glycolysis.

The pyruvic acid formed through glycolysis can be further processed in the oxidative cycle or can be removed to form **lactic acid** or alanine (figure 4.4). Alanine enters the blood and is transported to the liver, where it is converted to glycogen for storage or to glucose, which enters the blood and becomes available as substrate.

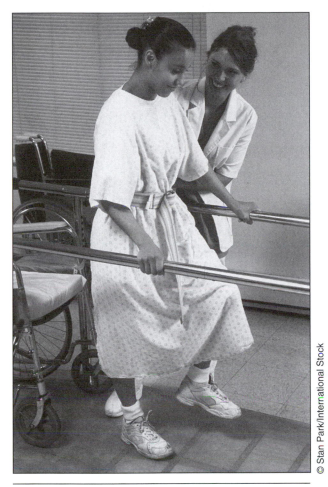

© Stan Park/International Stock

Patient in rehabilitation. Activity lasting from 30 seconds to 3 or 4 minutes relies heavily on glycolytic energy production.

Lactic Acid and Energy Production

The formation of lactic acid from pyruvic acid at rest and during exercise serves several important metabolic functions. Lactic acid formation, with the removal of an electron from NADH, prevents the buildup of pyruvic acid that would inhibit further energy production through glycolysis; additionally, it regenerates NAD^+, which is also required in order for glycolysis to continue (figure 4.3). Under conditions of high energy demand, the continued production of energy through glycolysis is critical; and this formation of lactic acid ensures that glycolysis will operate to meet the demand. Lactic acid

- is used as a substrate in the muscle within which it has been formed;

- can be transported to other muscles, including the heart, where it can be used as a substrate; or
- is transported to the liver, where it is converted to glucose or glycogen.

Thus the formation of lactic acid is a critical metabolic process that ensures continued energy production when oxidation is unable to accept the pyruvic acid being produced, and lactic acid can serve as a metabolic substrate.

> The formation of lactic acid is a critical metabolic process that ensures continued energy production when oxidation is unable to accept the pyruvic acid being produced.

While these positive effects of lactic acid formation need to be recognized, it is also necessary to acknowledge the negative consequences of a buildup of an acid and the consequent lowering of cell pH. The biochemical reactions in the metabolic cycles operate best within such a narrow range of pH that many of these reactions are inhibited under conditions of high acid buildup.

Lactic acid is continually formed in small amounts under resting conditions. Larger quantities are formed during exercise under conditions in which the production of pyruvic acid exceeds the ability of the mitochondria to process those amounts. Mitochondrial processing of substrate, including pyruvic acid, requires adequate availability of O_2. Lactic acid is formed in large amounts when O_2 delivery is insufficient to meet the demands for energy production through the oxidative cycle. Inadequate O_2 delivery occurs at the onset of exercise, when the demand for energy increases instantly but increased blood flow to the working muscles is delayed, and during periods of high energy demand such as highly intense exercise.

> Lactic acid is formed in large amounts when O_2 delivery is insufficient to meet the demands for energy production through the oxidative cycle.

Case Study

During treatment for an unrelated injury, your client complains about muscle soreness related to lactic acid buildup in the muscles. The activity that brought on this complaint was a long, slow run of 16 km (10 miles) three days earlier. What is the possibility that lactic acid was, in fact, responsible for the soreness?

Your reply should dispel the myth that all muscle soreness is a result of lactic acid buildup. The activity described was too low in intensity to result in the metabolic production of high levels of lactic acid. Depending on the level of sophistication of the client, you also might explain that lactic acid is formed only under conditions in which the delivery of O_2 lags behind the demand, or when the exercise intensity is so high that O_2 delivery cannot meet the demand for energy production through O_2 consumption. Only in these two cases does lactic acid form. The muscle soreness must be related to other causes (see chapter 9).

Tricarboxylic Acid, or Krebs, Cycle

The aerobic, or oxidative, production of energy occurs within the **mitochondria**. It is here that

- pyruvic acid is further processed,
- lipids are metabolized, and
- the electrons attached to NAD$^+$ are cycled in the **electron transport chain** (see figure 4.7) to produce ATP.

Both glucose, through the formation of pyruvic acid, and lipids are oxidized in the **tricarboxylic acid (TCA) cycle** within the mitochondria. Other terms for this cycle include the **Krebs cycle**, after the person who first described it,

and the **citric acid cycle**, for the first product in the cycle.

Acetyl Entry Into the Tricarboxylic Cycle

Entry into the TCA cycle is via a two-carbon acetyl molecule. The conversion of pyruvic acid to an acetyl involves several steps catalyzed by the enzyme complex pyruvate dehydrogenase (PDH) (figure 4.8). During this conversion, one carbon is removed to form a carbon dioxide, and one electron is removed to form NADH$^+$. Long-chain lipids are broken into the same two-carbon acetyl molecules, through a process called **β-oxidation** (page 48), to enter the TCA cycle.

The Tricarboxylic Cycle

The TCA cycle is a repeating sequence of reactions in which the final product, oxaloacetic acid (OAA), forms the beginning of the next cycle (figure 4.5). The cycle begins with the combination of OAA, the acetyl group, and a complex called coenzyme A (CoA) to form **citric acid**. As was seen with glucose in glycolysis, the citric acid in the TCA cycle is transformed through a series of steps to increase the energy level so that ATP can be formed. In a process similar to glycolysis, several electrons are removed to form NADH and another carrier, FADH. CO_2 is also a by-product of the TCA cycle.

Oxidation

The final source of ATP through the oxidative process is derived from the electrons accumu-

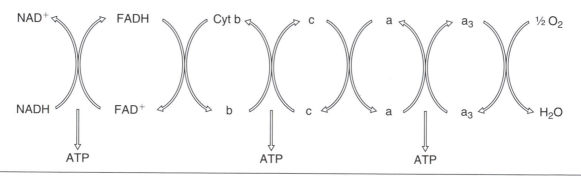

Figure 4.7   The electron transport chain (ETC). The ETC passes electrons accumulated from glycolysis and the tricarboxylic acid cycle to form adenosine triphosphate and water. This transport chain is located in the mitochondrion and is the site of much of the adenosine triphosphate production in the muscle.

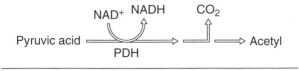

Figure 4.8   The conversion of the 3-carbon pyruvic acid to the 2-carbon acetyl group. This conversion is catalyzed by the enzyme pyruvate dehydrogenase, and the resulting acetyl group is further processed in the oxidative cycle.

lated on NADH and FADH. These electrons are passed through the electron transport chain (ETC), and this transfer of electrons provides the energy to form ATP from ADP and Pi (figure 4.7). Note that processing the electrons from NADH results in the formation of three ATP, while the electrons from FADH form two ATP. The final product of the ETC is water, which is formed by the combination of hydrogen and its associated electron with O_2. The O_2 consumption measured in the laboratory is simply the amount of O_2 consumed within the mitochondria of the body.

Lipid Oxidation

We have seen how glucose is used as a fuel in the glycolytic and the oxidative pathways. In this section we will learn

- how lipids are **mobilized** from their storage sites to the muscle cell and
- how those lipids are **utilized** in the mitochondria to form ATP.

As exercise is performed over time, increased lipid utilization is accompanied by decreased carbohydrate (glucose) utilization. As you read, consider how this balance between carbohydrate and lipid utilization might occur.

Lipid Mobilization

To become available as fuel for energy production, lipids must first be mobilized from storage sites. Lipids are stored in specialized fat cells and within other tissue, including skeletal muscle. As discussed in chapter 2, lipids are stored in the form of **triglycerides**, which must be separated into their component parts—glycerol and free fatty acids—in order to become available for metabolism. In general, the stimulation of the enzyme **lipase** liberates free fatty acids. Lipase is found in two forms:

- A lipoprotein lipase, which promotes storage of lipids
- A hormone-sensitive form (hormone-sensitive lipase, HSL), which promotes the release of lipids in adipose tissue

Insulin and glucose stimulate lipoprotein lipase and therefore lipid storage. The mobilization of lipids for use as a substrate during exercise occurs primarily through the stimulation of HSL by the catecholamines epinephrine and norepinephrine, but growth hormone also plays a role in stimulating the release of lipids from adipose tissue. The exercise-induced release of norepinephrine locally, or of epinephrine from the adrenal gland, results in the cleavage of free fatty acids from the glycerol and the release of these products into the bloodstream. The transport of the free fatty acids to the working muscle was described in detail in chapter 2. In general, specialized lipid carriers, chylomicrons, and other lipoproteins transport fatty acids in the blood. Once the fatty acids have been mobilized and reach the muscle, utilization begins.

Lipid Utilization

Lipid utilization begins with the process of transporting the fatty acids across the muscle membrane. In general, about 50% of the lipid passing through the muscle is transported into the muscle. This principle is important for understanding metabolic integration, since it means that lipid uptake and utilization occur at all times (including during rest) and that the increased blood flow at the onset of exercise automatically increases fatty acid delivery, and therefore uptake and utilization, to the muscles. In this sense, lipid metabolism increases immediately at the onset of exercise.

The two aspects of lipid utilization are transport and β-oxidation:

- Transport of fatty acids across the cell membrane is rapid and is facilitated by a fatty acid binding protein. Once inside the cell, the fatty acid is combined with a carrier, **carnitine**, for transport into the mitochondria. There the lipid molecule is altered so that electrons become available for NADH and FADH and so that two-carbon acetyl units from β-oxidation can enter the TCA cycle.

C-C-C-C-C-C-C-C-C-C-C-C-C-C-C-C   Saturated fatty acid

C-C=C-C-C-C-C-C-C=C-C-C-C-C-C-C   Formation of double bond

H^+C-C   C-C-C-C-C-C-C-C-C-C-C-C-C-C   Formation of acetyl
group and H^+

FADH and NADH   Formation of FADH and
NADH with each cycle

Figure 4.9 The breakdown of fatty acids. Fatty acids are broken down into two-carbon acetyl units in a process called β-oxidation. This process releases electrons that are carried by FAD^+ and NAD^+.

• **β-oxidation** is the process by which long-chain fatty acids are broken down into two-carbon acetyl groups that can enter the TCA cycle (figure 4.5) for the production of energy. Once inside the mitochondria, the fatty acid is transformed in a series of repeated steps so that the second and third carbons are joined by a double bond (figure 4.9). The advantages to this transformation are twofold: the double bond is easier to break than a single bond, leaving an acetyl group, and the electrons removed are used to form NADH and FADH. Each time a double bond is formed and broken, NADH and FADH are produced. These electrons can enter the ETC to supply the energy to produce more ATP. The process of β-oxidation and subsequent oxidation of fatty acid-derived acetyl groups in the TCA cycle has the potential to produce a large quantity of ATP. For instance, if one palmitate fatty acid (16 carbons) is broken down through β-oxidation, and the acetyl groups are passed through the TCA cycle, and additionally the NADH and FADH are oxidized in the ETC, a total of 129 ATP are produced. Contrast this with a total of 38 ATP for the metabolism of a six-carbon glucose.

The process of β-oxidation and oxidation of fatty acid-derived acetyl groups in the TCA cycle has the potential to produce a large quantity of ATP.

Long distance runners. Exercise lasting longer than 10 to 15 minutes relies mainly on the oxidative system for energy production.

Metabolic Integration

It is important to recognize that the consumption of a particular substrate or the use of a particular pathway to produce energy during exercise does not "just happen." Integration of the metabolic pathways is accomplished by both external and internal control factors (table 4.1). Metabolism is controlled externally via hormones and internally via chemical influences from metabolic by-products (Pi, ADP), substrates (glucose, glycogen, G-6-P, pyruvic acid, etc.), and the autonomic nervous system. Understanding the integrated role of each of these factors in metabolic control will help you understand energy production under a variety of circum-

stances. Once you appreciate the integrative nature of metabolic processes, you should be able to apply these concepts in a variety of clinical situations.

Metabolic Control

Those external factors that control metabolism are dynamic, with the contribution of each factor altered by energy demand and substrate availability. One way to learn the contribution of the control factors to exercise energy production is to examine the time course of changes in their levels during exercise. Figure 4.10 depicts changes, during moderate exercise lasting 90 min, in

- blood glucose,
- insulin,
- glucagon,
- epinephrine, and
- free fatty acids.

Before reading the following paragraphs, which explain the causes for these changes, see if you can come up with as many causes as possible yourself, based on the preceding discussion of exercise metabolism.

Integration of Glucose and Lipid Mobilization

You will note that glucose levels initially increase, then begin a slow decrease as exercise progresses. The small increase in blood glucose at the start of exercise reflects the release of glucose from the liver. This early release is likely stimulated by neural mechanisms. The later fall in blood glucose reflects uptake exceeding release. Of course the decrease in glucose availability for energy production is countered by an increase in free fatty acid (FFA) availability. This is the first example of the integrative nature of the metabolic process. The initial rise in blood insulin levels is stimulated by the initial increase in glucose. The later decrease in glucose has the following results:

- A lowering of insulin, decreasing insulin's inhibitory effect on glucose release from the liver
- The secretion of glucagon from the pancreas, which stimulates the breakdown of protein, fat, and liver glycogen into glucose

Table 4.1 Factors That Exert Control on Metabolism

	Inhibits	Stimulates
ADP		PFK
Pi		PFK
Ca^{++}		Phosphorylase
ATP, CP	PFK	
G–6–P	HK	
Citric acid	PFK	
O$_2$	PFK	
Epinephrine	Glycogen storage	Glycogen breakdown, lipid mobilization
Insulin	Fat storage, glucose release from liver	HK, PFK, glycogen storage
Glucagon		Glycogen breakdown, lipid mobilization
Growth		Glycogen breakdown, lipid mobilization

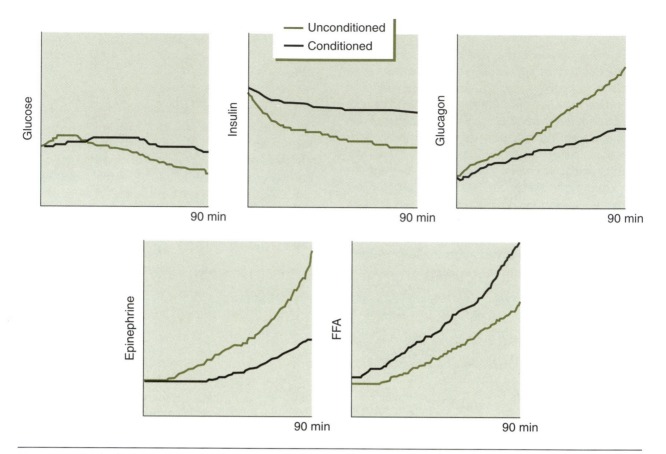

Figure 4.10 Idealized depiction of the changes in blood glucose, insulin, glucagon, epinephrine, and free fatty acids over 90 min of moderate exercise in an unconditioned and a conditioned person.

Thus, the decrease in insulin and the increase in glucagon work to maintain blood glucose levels.

Once understood, the process of energy production "makes sense." That is, the regulation of the system ensures that energy will be made available in the most effective manner possible. You should stop here and be sure you understand how the system "makes sense" so far.

Regulation of the energy production system ensures that energy will be made available in the most effective manner possible.

The decrease in insulin also encourages the release of lipid from fat cells. As exercise continues beyond 5 to 10 min, epinephrine secreted by the adrenal gland begins to stimulate the breakdown of lipids in the fat cells. The increase in glucagon and epinephrine stimulates the release of FFA, which becomes the preferred substrate as exercise continues. You should recall from chapter 2 that the increased utilization of fat at this

point is important because of the high amount of energy available from fat compared to carbohydrate (9 kcal/g vs. 4.5 kcal/g). The increases in epinephrine and glucagon also permit continued glucose release from the liver so that blood glucose levels can be preserved for protection of the metabolic needs of nervous tissue. At this point you should appreciate the way the body is regulated to ensure availability of metabolic substrates during exercise. As one substrate (glucose) decreases in availability, control factors ensure that another substrate (e.g., lipids) becomes available.

Regular exercise, or exercise conditioning, results in enhancement of both the glycolytic and oxidative energy-producing pathways. Chapter 9 discusses in more detail the changes that take place in the muscle as a result of exercise conditioning. Increased use of either pathway stimulates increases in the enzymes necessary to produce energy through that pathway. Repeated exercise of high intensity stimulates increases in the capacity of the glycolytic pathway to produce

ATP, while high volumes of long-duration (aerobic) activity stimulate increases in the oxidative capacity of the muscle. The latter change takes place through increases in the number and size of the mitochondria. Since the glycolytic pathway is always active, aerobic activity will also stimulate increases in glycolytic capacity to produce ATP.

Conditioning results in a tighter regulation of blood glucose so that there is less variability. This tighter regulation also is reflected in the smaller fluctuations in blood hormone levels in the conditioned person as compared to someone who is less well conditioned. Finally, the higher FFA levels in the conditioned person reflect the greater mobilization, and utilization, of lipids as a result of the increased oxidative potential.

Conditioning results in a tighter regulation of blood glucose, which is reflected in the smaller fluctuations in blood hormone levels in the conditioned person as compared to someone who is less well conditioned.

Integration of Metabolic Pathways

Energy production during exercise involves the integration not only of glucose and lipid mobilization, but also of the glycolytic and oxidative metabolic pathways (i.e., glycolysis, the TCA cycle, β-oxidation, and the ETC) using internal factors to control flux through each pathway. One way to conceptualize this integration is to examine metabolic events chronologically from the onset of exercise to the point of steady state. The primary reactions and control points are depicted in figure 4.11; the individual steps in the process are identified by circled numbers in that figure.

Several events occur immediately at the onset of exercise. Adenosine triphosphate is utilized (releasing ADP, Pi, and energy) to provide the energy for muscle contraction. At the same time, the Ca^{++}, which was released to initiate contraction, stimulates the breakdown of glycogen (step 7, figure 4.11), and glycolysis is stimulated by the increased appearance of ADP and Pi at step 2. The transport of glucose into the cell is enhanced by the translocation of Glut-4 from its position within

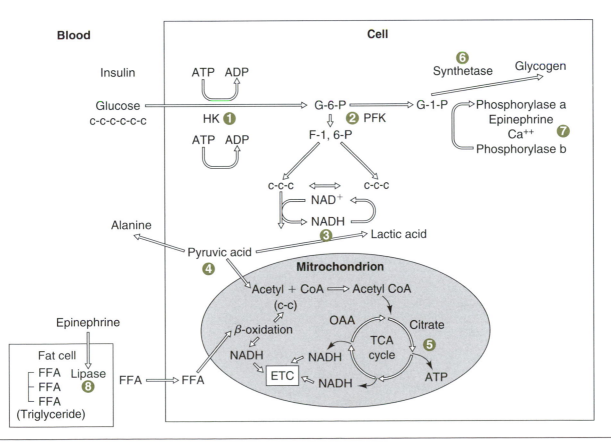

Figure 4.11   Schematic diagram of the primary reactions and control points (circled numbers) in the systems responsible for producing energy during exercise.

the cell to the cell membrane (step 1). Very soon thereafter, blood flow is increased so that the delivery, and therefore availability, of lipids is increased within the cell.

Thus within the first few moments after the onset of exercise the carbohydrate and lipid fuels are enhanced within the muscle, and the processes by which these fuels are broken down to produce energy—glycolysis, β-oxidation, and the TCA cycle—are stimulated.

Within the first few moments after the onset of exercise, carbohydrate and lipid fuels are enhanced within the muscle, and the processes by which these fuels are broken down to produce energy are stimulated.

During the transition period before the oxidative process is stimulated to full activity, the glycolytic-produced pyruvic acid that cannot be further oxidized must be converted to lactic acid to ensure continued energy production through glycolysis (step 3). Under conditions of mild to moderate activity, this lactic acid is readily used for energy within the cell and is transported to other cells for further metabolic energy production. As O₂ delivery is increased to meet the demand, the TCA cycle is able to produce energy through the oxidation of both pyruvic acid (steps 4 and 5) and fatty acids (steps 8 and 5). Epinephrine-induced release of fatty acids makes this substrate available to the working muscles. Of course, the increased transport leads to increased utilization of these fatty acids, resulting in higher levels of fatty acid-derived acetyl and citric acid—which in turn inhibits PFK (step 2) and therefore glycolysis (table 4.1). This citric acid inhibition of glycolysis ensures that glycolytic activity is reduced while the TCA cycle is providing an increased amount of energy using fat as a substrate. As exercise continues toward a steady state, oxidation contributes as much as 80% of the energy production while glycolysis provides the greatest proportion of the remaining 20% (figure 4.12). Thus a balance is maintained between energy production through glycolysis and oxidation. Again, these mechanisms ensure that, when O₂ delivery is meeting demand, the oxidative system becomes the primary source of ATP. Because fats provide twice the energy per gram that carbohydrates do (9 kcal/g vs. 4.5 kcal/g), they are the preferred substrate when O₂ availability is not limiting oxidation.

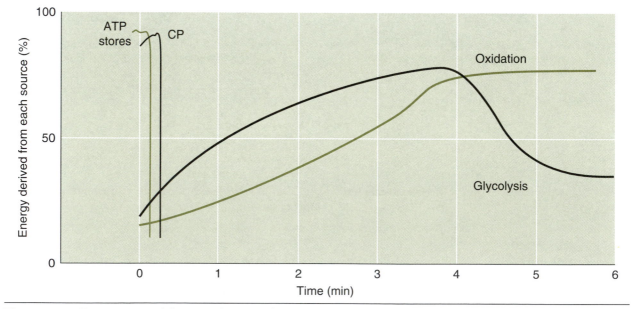

Figure 4.12   Time course of the contribution of various sources to the production of energy to meet increased demands at the onset of exercise to steady state. The ordinate represents the percent contribution of each energy source. The sum of all of these sources must add up to 100%. Thus at the onset of exercise, nearly 95% of the energy is derived from stored adenosine triphosphate and creatine phosphate, while the remaining 5% is derived from a combination of glycolysis and oxidation.

Adapted from Keul, Doll, and Keppler 1972.

When O_2 delivery is meeting demand, the oxidative system becomes the primary source of ATP.

Once a steady state is achieved, the same release of epinephrine that has stimulated the breakdown of lipids also stimulates the breakdown of glycogen (step 2, figure 4.11), ensuring continued energy availability through glycolysis. This continued glycogen breakdown via the stimulation of epinephrine during long-duration exercise creates a paradox. The maintenance of muscle glycogen above a critical level is important in each muscle fiber's ability to continue to contribute to movement. As glycogen is used during exercise, levels inside the cell decrease. Once these levels fall below the critical level, continued contraction of the muscle depends on recruitment of other fibers with higher levels of glycogen. Why is epinephrine used to stimulate both the lipid mobilization, necessary for long-term work, and the breakdown of glycogen, which limits the performance of long-term work? Of course the reason for this paradox cannot be known for certain. One way to understand it, however, is to think of the maintenance of a reserve of glycogen as a mechanism to ensure that the getaway car (glycolysis), required for emergency energy, is left running while the rest of the system (oxidation) is "robbing the bank." It would be foolish to turn off the emergency engine and take the keys into the bank!

Exercise performed at an intensity approaching 100% of aerobic power derives energy primarily from non-oxidative, glycolytic, sources, since this exercise can be maintained for only 4 or 5 min. This time is too short to allow the mobilization and utilization of lipids to increase to the point where inhibition of glycolysis can occur. At these intensities, the metabolic system operates best by extracting the maximum amount of energy from the available O_2. The consumption of carbohydrates produces 5.0 kcal of energy per liter of O_2, whereas the consumption of fat produces only 4.8 kcal/L O_2. This makes carbohydrates the preferred substrate and the glycolytic system the preferred pathway during high-intensity exercise.

The metabolic energy sources for intermittent (repetitive) exercise provide another example of metabolic integration. During the first interval of exercise of 1 min, for example, the metabolic response is as we have already described. If the next exercise interval is preceded by a pause interval of 0.5 to 1.0 min, the metabolic response to the subsequent exercise interval will be similar to that of the first. However, some stimulation of oxidation has likely occurred during the first interval such that somewhat more of the energy for the second interval will be derived from oxidation. In subsequent exercise intervals the metabolic response will tend more and more toward oxidation, but will derive energy less from oxidation than if the exercise had been continuous.

Case Study

An 800 m runner has come to you for advice. She explains that she runs excellent practice times but falls far short of these times during competition. Further questioning about her habits reveals that she has a typical warm-up lasting 5 to 10 min, consisting of stretching followed by running at progressively increasing intensities. She also reveals that she follows this regimen before competition, but that for logistical reasons her competition warm-up takes place almost 15 min before her event. What is your assessment, and what would you advise?

One of the functions of warm-up is to increase metabolic activity to the point where oxidation can contribute maximally to the competitive situation. During practice, her warm-up and even her prerace routine stimulate oxidative metabolism. During her practice runs, which are of high intensity, more of the pyruvic acid produced during the activity can be oxidized in the TCA cycle than if she hadn't stimulated the cycle prior to that run. The time between warm-up and competition, however, is too long to allow her system to take advantage of the stimulated oxidative system, and more of the pyruvic acid must be converted to lactic acid to allow glycolysis to continue. This increased muscle lactic acid inhibits her performance. You should advise the athlete to try to shorten the time between warm-up and competition. Since she appears to perform better at the end of practice, she also seems to require a considerable warm-up period. You should work with her to optimize this routine so that she can warm up before competition in a manner that she knows will optimize her performance.

All of the metabolic systems contribute to the total energy production at all times, but in different proportions dependent on the intensity, and thereby the duration, of the exercise. The metabolic systems are controlled to ensure that ATP requirements are met through use of the pathways that provide energy in the most efficient manner possible.

All of the metabolic systems contribute to the total energy production at all times.

Glycogen Utilization

Glycogen plays a critical metabolic role at the onset of exercise, during periods of high-intensity exercise, and during long-duration exercise. Glycogen is the primary fuel during the first few critical minutes at the onset of exercise. Because it is stored inside the cell, glycogen does not require mobilization from a distant site. It becomes available as soon as contraction is started, due to the Ca^{++} activation of the phosphorylase enzyme (step 7, figure 4.11). The use of glycogen stored in the cell produces energy through glycolysis without consuming the ATP that had been used to store the glycogen (steps 1 and 6). This ATP had been used at some time previously. The utilization of

muscle glycogen thus provides an energy advantage under conditions of high energy demand and at the onset of exercise.

In light of the dependence of glycolytic metabolism on glycogen, it is somewhat surprising that long-term exercise also is dependent on this substrate. As already mentioned, the "reasons" for this multiple dependency are not known. The dependency of long-term work on glycogen provides opportunities to enhance performance through manipulation of glycogen storage. Figure 4.13 shows the change in muscle glycogen during moderate exercise lasting for a prolonged period. Note that the greatest rate of utilization (the fastest drop in muscle levels) occurs during the early stages of exercise, reflecting the considerable reliance on glycogen at the onset of exercise. The lowest trace shows the rate of glycogen utilization under normal dietary conditions in a sedentary person. The person terminates the exercise when the glycogen reaches critically low levels in the recruited fibers. The middle trace shows what might happen as a result of regular exercise conditioning. The initial level of muscle glycogen is increased such that, even if no other changes in glycogen utilization occurred, the person would be able to continue exercise longer. In addition to the effect of conditioning on the initial glycogen

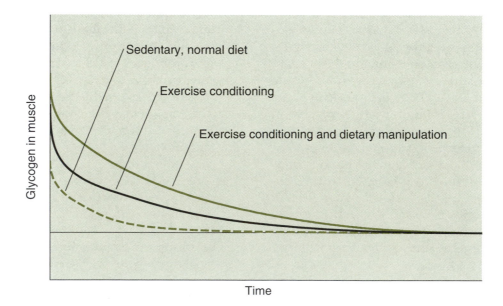

Figure 4.13 Glycogen levels in working muscle over the duration of exercise in three individuals. Note that the initial level is different and that the progressive decrease in glycogen is slightly different in each case. The horizontal dashed line indicates the glycogen level below which there is insufficient glycogen to support contraction.

levels, the enhancement of oxidative metabolic potential through conditioning reduces the rate of glycogen utilization, thus "sparing" the glycogen. The use of oxidative metabolism instead of glycolysis allows prolongation of the work because less glycogen is used. Finally, the highest curve shows the situation that might occur as a result of conditioning and dietary manipulation to enhance muscle glycogen storage.

Case Study

A high-level athlete has approached you to see if you could provide a conditioning edge. An examination of his conditioning program shows little room for improvement, but you do discover that he is having difficulty completing a marathon. He is able to lead the pack for the first 35 or 37 km (22 or 23 miles), but then seems to lose energy. He is able to lead for most of the race through two strategies. He stops at the drinking stations only infrequently for water and never for a carbohydrate drink. In addition, he throws in bursts of sub-5 min mile "sprints" several times during the first 30 km (18 miles). What is your assessment?

Your understanding of metabolic processes allows you to recognize several problems. You explain that his sprints are primarily nonoxidative/glycolytic and tend to use his glycogen stores so that they are unavailable later during the race. You also recognize that his refusal to ingest a glucose solution means that less blood glucose will be available for glycolysis, resulting in a greater utilization of glycogen. Both of these strategies cause an early depletion of muscle glycogen and exhaustion. You recommend that he ingest a carbohydrate drink early in the race and that he continue doing so throughout the race so that glucose will be available, thus sparing glycogen. While you may have no control over his tendency to sprint at specified times during the race, you might recommend he rethink this strategy so that he has glycogen available later in the race. Finally, you advise him to attempt to enhance muscle glycogen levels through dietary manipulation.

You tell the athlete that there are several important enzymes responsible for storing glycogen. You suggest that about a week before the marathon, he exercise hard enough to decrease glycogen levels significantly because this will stimulate those enzymes. If he repeats this regime for a second day, the enzymes that stimulate glycogen storage will be stimulated even further. Finally you suggest that the third day be one of regular conditioning. For the fifth and sixth days you prescribe a high-carbohydrate diet. The muscle cells, which have increased their enzyme activity, now have considerably more glycogen stored than usual. This increased muscle glycogen should enhance his performance during the marathon.

Respiratory Quotient

Understanding metabolic processes can be difficult. It is important not to get too involved in the "chemistry," but to concentrate on the general concepts of metabolic control and the way in which this control contributes to integrating the supply of appropriate amounts of ATP during all kinds of activity. Examining the changes in the respiratory quotient or the respiratory exchange ratio will help you understand metabolic integration.

The **respiratory quotient** (**RQ**, also called the respiratory exchange ratio, or **RER**) is the ratio between the amount of CO_2 produced from metabolic reactions and the amount of O_2 consumed ($RQ = \dot{V}CO_2 / \dot{V}O_2$). This ratio is dependent on the mixture of fuels that are being used to produce the energy. For instance, if the only fuel you were using were carbohydrates, you would produce one CO_2 for each O_2 consumed, giving you an RQ of 1.0. On the other hand, if you were consuming only lipids, the RQ would be closer to 0.72. Since we never consume only one fuel, the ratio falls between these two numbers during steady-state exercise. When you are sitting at rest, the ratio is close to 0.85, indicating that you are producing about 50% of your energy from each of these fuels (the total possible change is 1.0 − 0.72, or 0.28. Half of this is 0.14, and 0.72 + 0.14 = 0.86).

Figure 4.14 is a schematic showing how the RQ might change over the course of an exercise bout of about 35 min. During phase A, the RQ is increasing. This increase reflects the increased use of carbohydrates at the start of exercise. These carbohydrates are initially stored glycogen and then blood glucose. The primary but not exclusive metabolic pathway, then, must be glycolysis. You should be able to recreate the steps involved

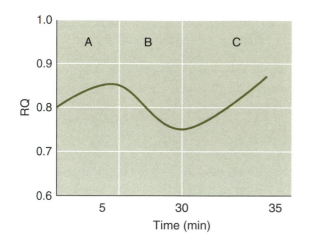

Figure 4.14 Changes in the respiratory quotient over time during exercise.

in the activation of the glycolytic pathway. In phase B, the RQ decreases, reflecting a shift to predominately lipids as the fuel. This should suggest to you that the exercise intensity is low to moderate and that the metabolic pathway is shifting from glycolytic to oxidative. You should be able to recreate the steps that have stimulated the shift toward lipid metabolism, activated oxidation, and slowed down glycolysis. Finally, during phase C the RQ is increasing again. Since this increase represents a shift back toward the use of carbohydrates, you might interpret it as an increase in intensity such as might occur at the end of a race. You should be able to describe the factors involved in metabolic control that allow this shift back to carbohydrate metabolism through the glycolytic pathway.

What You Need to Know From Chapter 4

Key Terms

adenosine diphosphate (ADP)
adenosine triphosphate (ATP)
aerobic metabolism
- citric acid cycle
- Krebs cycle
- tricarboxylic acid (TCA) cycle

anaerobic
β-oxidation
carnitine
citric acid
creatine phosphate (CP) or phospho-creatine (PC)
cytoplasm
electron transport chain
epinephrine
Glut-1 and Glut-4
glycogen
glycolysis
hexose kinase (HK)
insulin
lactic acid
lipase
lipid
- mobilization
- utilization

mitochondria
$NAD^+/NADH^+$
oxidation
phosphofructokinase
phosphorylase
pyruvic acid
respiratory quotient (RQ), respiratory exchange ratio (RER)
substrate
synthetase
triglyceride

Key Concepts

1. You should understand the three sources of ATP (stores, glycolysis, and oxidation) and should know how metabolism is controlled to ensure that ATP levels in the muscle do not decrease.

2. You should know the substrates used for glycolysis (glucose and glycogen) and the end products of glycolysis (pyruvic acid and ATP).

3. You should know the reasons for the production of lactic acid and understand why this is a desirable process.

4. You should know the factors that control the mobilization and utilization of lipids.

5. You should understand the source of electrons and know how electrons are used to produce ATP though the metabolic pathways.

6. You should understand the role of glycogen in energy production and its importance for the performance of long-term work.

7. You should be able to describe the substrate used for a variety of exercise intensities and the ways in which metabolism is controlled to ensure that the most effective substrate is used for each of these intensities.

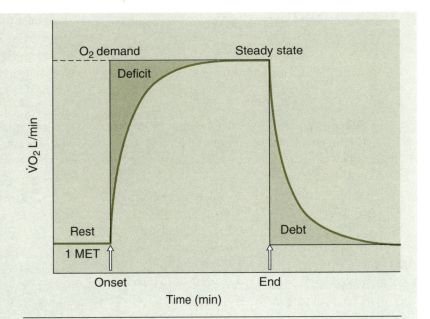

Figure 4.15 The change in $\dot{V}O_2$ at the start and end of exercise.

Review Questions

1. Examine figure 4.15 and then answer the questions that follow.

a. Demand for energy increases immediately, but a few minutes elapse until oxidation can meet that demand. How is this deficit for energy met?

b. What is the primary fuel during the first 3 to 5 min of exercise?

c. Describe the various factors that are controlling metabolism during the transition to the steady-state level.

d. Does all of the energy come from fats when the oxidative steady state is reached?

2. If you can answer the following questions based on figure 4.16, you have achieved an acceptable understanding of the metabolic process.

a. Describe the exercise intensity in each phase.

b. Name the primary metabolic pathway in each phase.

c. Describe the events that determine which pathway is being used in each phase and the events that determine a shift from one pathway to another.

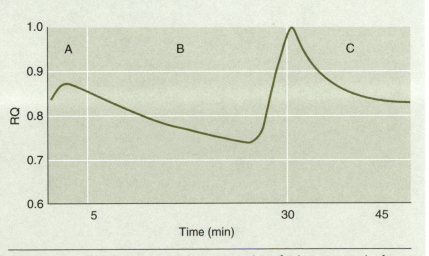

Figure 4.16 The change in respiratory quotient during an exercise bout.

Bibliography

Brooks GA and Mercier J. Balance of carbohydrate and lipid utilization during exercise: The "crossover" concept. *J Appl Physiol* 1994, 76:2253–2261.

Coggan AR and Coyle EF. Carbohydrate ingestion during prolonged exercise: Effects on metabolism and performance. *Exerc and Sport Sci Rev* 1991, 19:1–40.

Gollnick PD, Piehl K, and Saltin B. Selective glycogen depletion pattern in human muscle fibers after exercise of varying intensity and various pedaling rates. *J Physiol (London)* 1974, 241:45–57.

Katz A and Sahlin K. Role of oxygen in regulation of glycolysis and lactate production in human skeletal muscle. *Exerc and Sport Sci Rev* 1990, 18:1–28.

Lewis S and Haller R. The pathophysiology of McArdle's disease: Clues to regulation in exercise and fatigue. *J Appl Physiol* 1986, 61:391–401.

McAllister RM and Terjung RL. Training-induced muscle adaptations: Increased performance and oxygen consumption. *J Appl Physiol* 1991, 70:1569–1574.

O'Brien MJ, Viguie CA, Mazzeo CS, and Brooks GA. Carbohydrate dependence during marathon running. *Med Sci Sports Exerc* 1993, 25:1009–1017.

Wright D, Sherman W, and Dernbach A. Carbohydrate feedings before, during, or in combination improve cycling endurance performance. *J Appl Physiol* 1991, 71:1082–1088.

M etabolic Diseases

This chapter will discuss

➤ the effects of diabetes on metabolism and the ways that diet and exercise can be used as part of the treatment of this disease; and

➤ the effects of McArdle's disease, which is rarer than diabetes but helps us understand exercise metabolism.

There are several disease states that disrupt normal metabolic energy production. The most common of these is diabetes, which this chapter covers in some detail. We briefly discuss a less common pathology, McArdle's disease, because it allows some insight into metabolic regulation that can be applied to other situations. As you read about these two diseases, try to picture the sites in the metabolic scheme (figure 4.11) that are affected and the consequences on energy production.

Diabetes

Diabetes occurs in two forms (table 5.1). Although both forms are characterized by high blood glucose levels, the elevated glucose levels have different etiologies in the two forms.

Type 1 diabetes, also known as **insulin-dependent diabetes mellitus (IDDM)** or childhood or juvenile [onset] diabetes, usually appears before the age of 20 years but can appear as late as 40 to 45 years of age. Type 1 diabetes is caused by a reduction or termination of insulin production by the pancreas. It may surprise some to learn that type 1 accounts for only 10% of the total diabetic population. There is evidence of a genetic component in many groups of patients with

© JimWestPhoto.com

A hiker with diabetes adjusting the setting on her insulin pump. The use of an insulin pump enables a person with diabetes to regulate her blood sugar levels more precisely and easily than other forms of control, resulting in higher quality of life and fewer complications from fluctuating blood sugar levels than can otherwise be achieved.

Table 5.1	Characteristics of the Two Major Categories of Diabetes	
Type I	**Type 2**
Juvenile onset	Adult onset
Reduced insulin production	Reduced insulin receptor sensitivity
Insulin-dependent diabetes mellitus (IDDM)	Non-insulin-dependent diabetes mellitus (NIDDM)
Genetic component	Possible genetic component

type 1 diabetes. Onset of type 1 diabetes is generally rapid and is accompanied by symptoms such as weakness, frequent urination, rapid weight loss, reduced exercise capacity, and in severe cases ketoacidotic coma. Both symptoms and medical emergencies occur more often in type 1 than in **type 2 diabetes**, also known as **non-insulin-dependent diabetes mellitus (NIDDM)** or adult (onset) diabetes. Type 1 diabetes requires the replacement of **insulin** by regular injection or infusion.

Type 2 diabetes appears in adulthood and is slow to develop, with symptoms appearing over a period of months or years. Increasing scientific evidence suggests that there is a genetic component to type 2 diabetes. Insulin resistance, a reduced sensitivity to insulin, can be overcome by increased insulin secretion, but may progress to type 2 diabetes. Insulin resistance is associated with visceral obesity and coronary heart disease and often occurs during pregnancy (gestational diabetes). This disorder is usually treatable through implementation of dietary recommendations given to people with type 1 or type 2 diabetes (see page 63). Obesity is a primary risk factor for type 2 because it reduces the sensitivity of the insulin receptors to insulin. Inactivity, in addition to its contribution to obesity, may be an independent risk factor for the development of type 2 diabetes. Contrary to popular belief, excess sugar intake is not thought to lead to diabetes unless it is associated with excess caloric intake and subsequent obesity. Weight loss and physical activity together or independently reduce, or may even eliminate, type 2 diabetes. Type 2 diabetes can be treated by exercise and careful control of the diet or the use of oral hypoglycemic agents or insulin-like skin patches.

Case Study

A friend of yours has just been told that she is pregnant and that she has diabetes. She is pretty distraught over the possibility of having to use insulin for the rest of her life. She has come to you for advice.

This relatively common occurrence is called gestational diabetes. You should tell her that gestational diabetes usually does not require insulin shots and resolves after the baby is born, but these women are at increased risk for future type 2 diabetes.

Complications

Diabetes, particularly if not well controlled, can lead to several acute and chronic complications. Acute increases (hyperglycemia) and decreases (hypoglycemia) in blood glucose levels are hallmarks of diabetes. If poor, control of diabetes is chronic, and vascular and neural disorders will occur.

Acute Complications

Acute complications include hyper- and hypoglycemia and skin infections and abscesses.

• Hyper- or hypoglycemia. Insulin, of course, promotes the uptake of glucose from the blood into the liver or muscle. Patients who are not receiving enough insulin develop high blood glucose (**hyperglycemia**) because glucose is not leaving the blood. Symptoms of hyperglycemia include fatigue/weakness, thirst, and frequent urination. In cases in which a patient is receiving too much insulin, blood glucose levels will decrease, leading to **hypoglycemia**. Symptoms of hypoglycemia include weakness, hunger, profuse sweating, headaches, shaking, and confusion. Exercise, by enhancing the effect of insulin, can lead to hypoglycemia in the person with diabetes. It is important to have the patient with presumed hypoglycemia ingest a simple, fast-acting sugar, such as three to four glucose tablets.

• Ketoacidosis. Although relatively uncommon today, **ketoacidosis** is a potential complication of poorly controlled diabetes, or it may result from the onset of any unrelated illness—even when the person normally achieves tight control. When glucose cannot be metabolized, cells rely on increased lipid metabolism. Accumulation of **ketone bodies** (acetoacetate, beta-hydroxybutyrate, acetone), a by-product of this lipid metabolism, results in metabolic **ketoacidosis,** a potentially fatal complication of diabetes.

• Surface infections and abscesses. Finally, acute infections of the skin, gums, and other body surfaces may result from poor control of diabetes. Poor disease-related circulation promotes the development of ischemic ulcers that may go unnoticed and untreated because of reduced peripheral sensations.

➤ Ischemia means little or no blood flow.

Chronic Complications

Chronic complications of diabetes include vascular diseases and neuropathies.

• Vascular diseases. The most common vascular abnormality is associated with the retina. Microaneurysms in the area of the retina eventually lead to impaired vision and potential blindness. Likewise, diabetic nephropathy leads to renal dysfunction and potential failure. Lesions within the kidney contribute to increases in blood pressure in patients with diabetes. On a larger scale, diabetes leads to an increased risk of cardiovascular diseases, including atherosclerosis. In addition to diabetes-related hypertension, increased very low-density lipoprotein (VLDL) with a reciprocal reduction in high-density lipoprotein (HDL), elevated blood lipid levels, and obesity contribute to an increased risk of cardiovascular diseases. In extreme cases, poor circulation may lead to tissue necrosis and the eventual need for limb amputation.

➤ Low-density lipoprotein and high-density lipoprotein are lipoproteins that carry lipid in the bloodstream.

• Neuropathies. The mechanism for diabetic neuropathy is not known. Diabetes can affect sensory or motor nerves, as well as the central nervous system. Diabetic neuropathy results in numbness, pain, muscle atrophy, anomalous motor control, and abnormal cardiovascular responses to exercise. The possibility of neuropathy with abnormal circulation leads to increased risk of infection due to open wounds from abrasions and cuts. For this reason, particular attention should be given to proper fit of footwear for exercise. Poorly fitting footwear can cause sores on the foot that, if not treated immediately and properly, can lead to infection, gangrene, and possible amputation.

Treatment

Treatment of diabetes should include

• insulin-replacement therapy,
• dietary manipulation, and
• exercise.

Each of these is a critical element, necessary to attain optimal control of the disease. Your background in nutrition and its interaction with exercise metabolism will help you understand how these three treatment elements can work together.

Insulin

For type 1 diabetes, insulin injection or infusion using portable pumps is an important part of treatment. In type 1 diabetes, or other types that require periods of insulin replacement, the patient has access to four types of insulin: short-acting, rapid-acting, intermediate-acting, and long-acting insulins (table 5.2). Blood glucose is controlled via insulin replacement through injection, in appropriate combinations, of short-acting (regular) or rapid-acting (Lispro) with intermediate-acting (NPH [Neutral Protamine Hagedorn] or Lente) and/or long-acting (Ultralente) insulin. The short- and rapid-acting insulins stimulate glucose uptake within 30 min after injection and are effective for up to 8 hr, while intermediate- or long-acting insulins are effective for up to 24 hr. The selection of insulin type, amount, and timing is dependent on the severity of the disease, long-term dietary and activity patterns, blood glucose level at the time of the injection, the timing of meals, the type of meal, level of emotional stress, and immediate exercise history.

In general, patients attempt to control their blood glucose through the use of a background of one of the longer-acting solutions along with short-term adjustments provided by one of the shorter-acting solutions. The intermediate- and long-acting insulins (NPH, L; U) are used when the patient is not eating (i.e., insulins used to maintain an individual's basal rate). Approximately 50 to 60%

Table 5.2   Comparison of the Different Types of Treatments Available for Diabetes

Type 1 (insulin-dependent)				
Insulin name	Action	Initial onset*	Peak time*	Duration*
Insulin lispro (LP; e.g., Humalog)	Rapid-acting insulin	5-15 min	1-2 hr	4-5 hr
Regular (R)	Short-acting insulin	30 min	2-4 hr	6-8 hr
NPH (N)	Intermediate-acting	1-3 hr	5-7 hr	13-18 hr
Lente (L)	Intermediate-acting	1-3 hr	4-8 hr	13-20 hr
Ultralente (U)	Long-acting insulin	2-4 hr	8-14 hr	20-24 hr
Premixed 70/30 NPH and Regular	Short-acting and intermediate-acting insulin mix	30 min	7-12 hr	16-24 hr

*Onset, peak time, and duration are given as ranges because individuals differ in their responses to the various insulins available for treating type I diabetes.

Type 2 (non-insulin-dependent)	
Medication	Action
Sulfonylureas	Stimulates pancreatic production of insulin
Biguanides	Decreases liver production of glucose
Alpha-glucosidase inhibitors	Slows absorption of dietary starches
Thiazolidinediones	Increases insulin sensitivity
Meglitinides	Stimulates pancreatic production of insulin

of the total insulin should be given as NPH, L, or U, with the remainder as Lispro (Humalog) or R. The precise combination and schedule, whether using injections or an insulin pump, should be worked out with the patient's physician. While type 2 diabetes may require insulin replacement there are alternative therapies that are effective (table 5.2). The medications listed for treatment of type 2 diabetes are taken in pill form. Again, the exact medication and dosage should be worked out with the treating physician. Insulin pumps are used to infuse insulin (usually Lispro [Humalog]) subcutaneously, intravenously, or intraperito-

neally. The object of all treatment is to maintain pre-meal blood glucose levels within a range of 75 to 150 mg/dl. Measuring **HbA₁c**, a glycosylated hemoglobin that reflects the patient's serum glucose levels over the past three to four months, monitors effectiveness of treatment.

The interaction between available insulin and exercise is critical to recognize. Exercise increases the sensitivity of the insulin receptors. If injected or infused insulin is high, exercise enhances the action of that insulin, thereby promoting greater glucose uptake than if exercise had not been performed. In fact, patients who exercise regularly can decrease their insulin intake because of the exercise-induced sensitization of the insulin receptors.

> Patients who exercise regularly can decrease their insulin intake because of the exercise-induced sensitization of the insulin receptors.

Diet

The diet of the patient with diabetes must take into consideration the total caloric intake as well as the exact composition of the diet. If the caloric intake of the diet is sufficient to meet, but not exceed, caloric expenditure, the patient can include more carbohydrates in the diet than if the caloric intake exceeds expenditure. In addition, patients who exercise regularly should include greater amounts of healthy complex carbohydrates in their diets than if they were not exercising. In addition to controlling the total caloric intake, the timing and regularity of the meals should receive special attention.

It is beyond the scope of this book to describe specific dietary recommendations, because each diet must be individualized on the basis of each patient's status. The diet prescribed will require a specific balance of carbohydrates, fats, and proteins. A system of "exchanges" or carbohydrate counting is used to give the patient flexibility in the selection of foods (table 5.3). An exchange consists of the grams of nutrients and the calories assigned to a particular portion of a food. For instance, a bread exchange consists of 15 g of carbohydrate and 2 g of protein with a total of 70 kcal. One exchange would be equivalent to one slice of bread, a half bagel, or a half cup of cooked pasta. One exchange of lean meat would consist of 7 g of protein and 3 g of fat with a total of 55 kcal. In general, one exchange is equivalent to about 1 oz of meat. A complete diet is composed of milk, vegetable, fruit, bread, meat, and fat exchanges.

Table 5.3   List of Food Exchanges for Diet Planning

Food group	Example of one exchange	Nutrition from one exchange	Energy from one exchange (kcal)
Breads, cereals, pasta (6–11 servings)	1 slice bread 1/3 c cooked rice 1/2 c cooked pasta 1/3 c cooked beans 3/4 c cereal 1 small potato 1/2 bagel or muffin	15 g carbohydrate 3 g protein trace fat	80 kcal
Vegetables (3–5 servings)	1/2 c cooked greens 1 c raw carrots 1/2 c cooked carrots 1 tomato	5 g carbohydrate 2 g protein	25 kcal
Fruits (2–4 servings)	1 small apple 1/2 grapefruit 1/2 fruit juice	15 g carbohydrate	60 kcal
Dairy (2–3 servings)	1 c milk 1 c yogurt, plain	12 g carbohydrate 8 g protein fat dependent on fat content	90 kcal for nonfat 120 kcal for low fat 150 kcal for whole
Meat (2–3 servings)	1 oz meat 1 egg	7 g protein fat dependent on fat content	55 kcal for lean 75 kcal for medium

Case Study

Ann is an 18-year-old patient with diabetes. She is 165 cm (5 ft 6 in.) tall and weighs about 59 kg (130 lb). She has been extremely conscientious about her treatment. She monitors blood glucose frequently throughout the day and makes adjustments in her insulin and diet based on these readings. Her HbA$_1$c levels affirm that she has maintained good glycemic control. In addition to her other activities, she has recently become an avid tennis player and has joined the high school tennis team. She has noticed that her blood glucose levels have been fluctuating more in recent days than in the past. She has been working with her physician to adjust her insulin, and her physician has advised that she consult a sports nutritionist, a professional specializing in the interactions between sport activities and diet.

You find out that Ann has been on a 2000 kcal/day diet. Your assessment, based on the amount of tennis she plays, is that she needs to increase this to over 2500 kcal. How would you work out her diet?

• Going back to the food pyramid (chapter 2), you start with the basic fact that the average person needs 6 to 11 servings of bread, cereals, pasta, and rice; 2 to 4 servings of fruits; 3 to 5 servings of vegetables; 2 to 3 servings of meat; and 2 to 3 servings of dairy products.

• 2500 Cal requires 12 to 14 exchanges from group 1 (starches/breads); 5 to 7 exchanges each from groups 2 and 3 (vegetables and fruits, respectively); 3 to 5 exchanges from group 4 (dairy); 3 to 4 exchanges from group 5 (meats); and 5 to 10 exchanges from the fat group. Table 5.4 provides an example of a typical day's meals that would be appropriate for Ann.

The exchange approach gives Ann the maximum amount of flexibility while ensuring a balanced diet that meets her caloric needs.

Table 5.4   Typical Meal Plan for Ann in Case Study

	Breakfast	Lunch	Dinner	Snack	Total
Grain	2 toast 2 cereal	2 bread 2 pasta salad with croutons	2 rice 1 bean	2 bagel	13
Vegetable		1 tomato 1 carrot	1 salad 2 vegetable servings		5
Fruit	2 orange juice	2 apple (large)	1 peach	2 banana	7
Dairy	1 skim milk	1 skim milk	1 skim milk	1 yogurt 1/2 cream cheese	4.5
Meat		2 chicken	2 fish		4
Fat	1 margarine	2 nuts	2 oil	1 cream cheese	6

Case Study

James is a 12-year-old patient who was diagnosed as having diabetes at age 6. His treatment includes dietary controls and insulin injections. He injects Humalog and NPH before breakfast, Humalog before lunch and dinner, and NPH at bedtime. He has recently joined a local soccer club that practices and plays games in the early evening. James has been eager to participate in athletics because he has always felt that his diabetes made him an "outsider." In spite of his best efforts, however, he seems to run out of energy after about an hour of exercise and must sit on the sidelines. Fortunately, James and his parents have increased the number of times they test his blood sugars to four to six times a day. One day during practice when he begins to complain of weakness, they decide to test his blood glucose. From a finger prick they obtain a small amount of blood for testing. To their surprise, his blood glucose level is < 50 mg/dl and they have asked you what changes they need to make to anticipate the effects of exercise on his blood glucose levels.

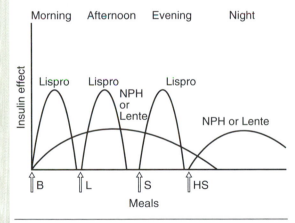

Figure 5.1 Graphic representation of James' insulin regimen.
Reprinted from Skyler 1998.

If you are familiar with the biochemistry of exercise from chapter 4 and with the earlier description of diabetes, you will be able to tell immediately what has happened. The Humalog injection before dinner has, as usual, increased James' ability to remove the glucose (carbohydrate) ingested from the meal. The exercise has increased the effectiveness of the insulin, further promoting glucose removal.

You tell James and his parents that they have three choices.

1. They can decrease his Humalog at dinner.
2. They can have him ingest some carbohydrate before exercise or at the first signs of tiredness.
3. They can use some combination of the first two options.

In consultation with you, James and his parents decide to reduce his pre-dinner insulin on those days when James will be playing soccer. As always, they will continue to make sure James has a source of carbohydrate available for times when he feels hypoglycemic. Upon follow-up three weeks later, you find that James is no longer feeling abnormally tired and that he has been able to decrease the total amount of daily insulin as his activity level has increased.

James's parents are now concerned that the site of insulin injection might affect how the insulin is absorbed during exercise, and they want your advice.

You need to tell them that although the evidence is somewhat contradictory, it appears that the site of injection does affect absorption of the insulin during exercise. You recommend that they use a standard rotation of sites. A standard rotation may involve using the same area of the body for a particular injection; that is, the morning shot should always be given in the arm, the lunch shot in the leg, the supper shot in the abdomen, the bedtime shot in the buttock. You also emphasize that the legs and arms should not be used for an injection that is given immediately before activity involving that extremity since the muscle activity can increase the rate of absorption. Studies have shown that the more consistent you are in what site you use, and the more accurate the injection technique, the higher the predictability of insulin absorption. In fact, current studies suggest that only one anatomic region of the body should be used, and that the rotation of shots should be confined within that part of the body, with the abdomen usually recommended. If the same part of the anatomy is used routinely, then a shot should not be given within an inch of a previous shot with a lag time of at least three days before reusing the site. You emphasize that James' parents should consult his physician or diabetes educator regarding the type of standard rotation they should use so it can be individualized for James. You finally suggest that when they speak with the medical professional, they ask about a visual chart designed especially for children to help them track injection sites.

Exercise

With few exceptions, regular exercise should be part of the treatment plan for every patient with diabetes. The primary exception is the patient

who is not under good glucose control (see discussion later in this chapter).

The use of exercise as treatment must be understood in the context of the acute response to exercise (figure 5.2).

Regular exercise should be part of the treatment plan for every patient with diabetes. The primary exception is the patient who is not under good glucose control.

Acute Response

A non-diabetic person may show a small decrease in blood glucose during prolonged exercise. This decrease in blood glucose is accompanied by a decrease in the level of circulating insulin. The person with well-controlled diabetes may start exercise with elevated blood glucose levels, and because of insulin injected, show a decrease in blood glucose over the course of this

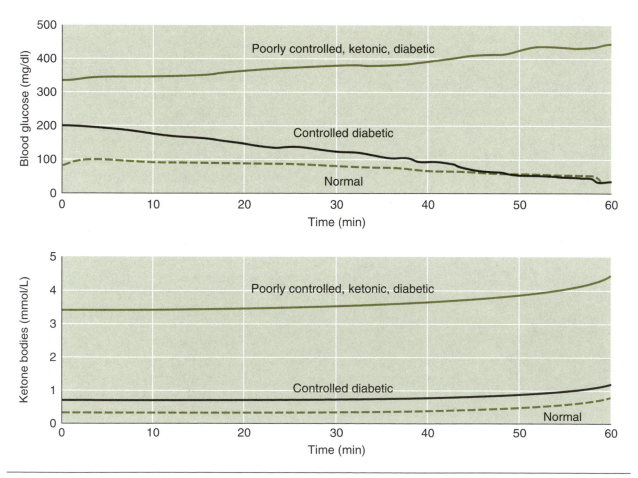

Figure 5.2   The blood glucose and ketone body responses to moderate exercise in a non-diabetic person, a patient with well-controlled diabetes, and patient with poorly controlled diabetes.
Adapted from Berger et al. 1977.

exercise. The rate of decline is greater in the person with diabetes than in others because there is no way to decrease insulin levels as in the non-diabetic person. Acute or chronic hypoglycemia results in an increase in lipid utilization. As a result of increased lipid utilization during prolonged, moderate exercise, ketone bodies will increase over time. The person with poorly controlled diabetes has a significantly higher blood lipid level at rest and during exercise and shows elevated resting ketone bodies that will increase during exercise. This tendency to increase both glucose and ketones during exercise in the person with poorly controlled diabetes is the reason exercise is contraindicated in these patients. Exercise should be postponed until glucose is under better control.

Glucagon and **growth hormone** are counterregulatory hormones; that is, they oppose the actions of insulin (see chapter 4). These hormones, along with epinephrine, promote lipid mobilization and utilization. The changes in these hormones during moderate exercise are shown in figure 5.3. The normal glucagon and growth hormone responses to moderate exercise are characterized by a slow increase to a peak at about 45 min of exercise. This increase is part of the regulatory mechanism, which promotes lipid utilization during prolonged exercise. In the patient with well-controlled diabetes these responses are exaggerated, reflecting the increased tendency toward lipid utilization. The response in the patient with poorly controlled diabetes is even more exaggerated, reflecting still greater lipid utilization.

Response to Regular Exercise

Patients with diabetes respond to regular exercise conditioning in the same way as non-diabetics. Many of the cardiovascular disease risk factors associated with diabetes such as obesity, atherosclerosis, hypertension, and increased low-density

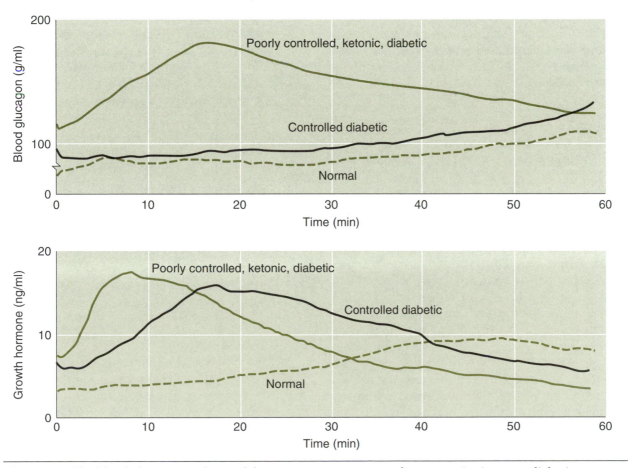

Figure 5.3   The blood glucagon and growth hormone responses to moderate exercise in a non-diabetic person, a patient with well-controlled diabetes, and a patient with poorly controlled diabetes.
Adapted from Berger et al. 1977.

lipoprotein (LDL) are ameliorated with regular exercise. In addition, regular exercise in the patient with diabetes results in increased insulin sensitivity and thus a decreased insulin-replacement requirement. The change in insulin sensitivity, however, is an acute response; no increase in sensitivity appears to be retained for more than 48 to 72 hr after the last bout of exercise. The short-term nature of this effect emphasizes the need for regular exercise in the treatment of diabetes.

The most effective exercise for improvement in diabetes control is exercise lasting 30 to 60 min performed at a moderate intensity. This activity, performed four to six days a week, may be sufficient to reduce insulin requirement and stabilize blood glucose levels in patients with diabetes. These patients can safely perform resistance training, but this type of exercise is effective as treatment only when combined with more prolonged exercise or as part of a circuit weight-training program. Intense conditioning is not contraindicated for athletes at any level who have diabetes. In fact, most athletes with diabetes decrease insulin requirements significantly with intense conditioning or training. Good diabetic control for these athletes requires regular monitoring of blood glucose levels and adjustments in insulin and diet to minimize disturbances in control.

Case Study

Latisha is a competitive skier who has diabetes. She monitors her blood glucose four or five times a day depending on her training and work schedules, but doesn't vary her insulin schedule significantly. Although her diabetes generally is under good control, she has noticed that during the season her glucose levels fluctuate greatly; she has performed poorly on days when she knows that her glucose levels are either quite high or quite low. During an interview, you find that she has the most trouble on race days when her schedule is unpredictable. On some days she may compete early in the morning but on others later in the morning, and she always competes again in the afternoon. She has been reluctant to alter her insulin schedule based on these contingencies because she has been afraid of upsetting her balance. Now she realizes that this strategy may not be working and that she needs advice.

Working with Latisha's physician, you devise an insulin and diet schedule based on race-day contingencies that are most likely. Under this plan, if she is to race early, she reduces her morning Humalog insulin and eats a breakfast higher in complex carbohydrates than on other days. If she is to race later in the morning, she has her usual amount of Humalog insulin and eats a moderate breakfast with a good mix of carbohydrates and fats. You devise other plans based on other scenarios. In addition, you teach Latisha how to alter her diet in the short term on the basis of feedback from her blood glucose monitor. For instance, if she measures a blood glucose of 50 mg/dl, she knows that she must ingest both simple carbohydrates to increase glucose levels and complex carbohydrates to maintain glucose levels over a longer term of several hours. Fruit, because of its high level of fructose, is an excellent source of carbohydrate in these situations. Whole fruits do not induce extremely high levels of blood glucose but instead provide carbohydrates that can be converted to glucose for use as a fuel during the period of competition. Fruit juices are more likely to result in high blood glucose levels because of the concentration of sugars.

An "insulin response" or hypoglycemic reaction 12 to 15 hr postexercise is not uncommon. Typically, these reactions take place after a particularly strenuous exercise session in which glycogen depletion is likely. The glycogen depletion enhances glycogen synthetase activity and enhances glucose transport into the cell. This enhanced glucose transport, along with injected insulin and the usual enhanced postexercise increase in insulin sensitivity, may result in a precipitous decrease in blood glucose levels.

Case Study

Robert had a 3 hr afternoon basketball practice today. Since this was one of the hardest practices early in the season, he was feeling quite fatigued. As usual, he took his Regular insulin injection before dinner, had a big meal, and studied before injecting his nighttime NPH and retiring at 10:30. At about 3 A.M., Bob awoke feeling dizzy, sweating, and very anxious. He recognized the signs of insulin shock and immediately ingested some simple carbohydrate. After an hour, he was feeling better and returned to bed.

Robert approaches you, his therapist/trainer, before practice on the following day and expresses concern whether he should consider quitting the team because he felt he was at some risk. After hearing his account, you realize what happened. You describe to him the events that led to his episode. The heavy exercise had lowered his muscle sugar, which stimulated a normal response to increase glucose removal from the blood into the muscle. In addition, the exercise had enhanced the sensitivity of the insulin receptors. Finally, Robert had injected his usual insulin before dinner and retiring, which was more than he needed under the circumstances, and had loaded the system with carbohydrates from dinner. The simple solution is to reduce or eliminate his pre-dinner Regular insulin on the days when he has had a prolonged, hard exercise session and reduce his nighttime dose of NPH. In addition to making these postexercise recommendations, you educate Robert regarding the use of carbohydrate-supplemented drinks during prolonged, highly intense exercise sessions. The strategy of minimizing glycogen depletion and reducing insulin injections appears to be a good solution to Robert's problem and allows him to participate normally in athletics.

Regular exercise can prevent the onset of type 2 diabetes. Since a lack of exercise and associated obesity are potent risk factors for the development of type 2 diabetes, it would be surprising if regular exercise did not prevent this form of diabetes. The prevention of type 2 diabetes by regular exercise is independent of the effect of the exercise on obesity, so that the effect of exercise is enhanced as obesity is reduced.

Since a lack of exercise and associated obesity are potent risk factors for the development of type 2 diabetes, it would be surprising if regular exercise did not prevent this form of diabetes.

McArdle's Disease

McArdle's disease is an uncommon disorder that interferes with normal metabolic production of energy. Patients with McArdle's disease present with symptoms of sudden onset of muscle fatigue, pain, and cramping during exercise. Muscle biopsies from these patients show that they are deficient in the enzyme phosphorylase (figure 5.4). The defect results in little energy production through glycolysis. A lack of pyruvic acid production ensures that there will be a low capacity to produce lactic acid. The low rate of glycolytic energy production is reflected in some reduction in oxidative capacity due to minimal movement of glycolytic products into oxidation. Increasing the level of blood glucose, to compensate for the low breakdown of glycogen, enhances glycolytic

Reprinted from Sternberg 1992.

Figure 5.4 (*a*) Normal enzyme, and (*b*) absent enzyme. A histochemical stain has been used to identify a deficiency of the enzyme phosphorylase.

energy production and the movement of glycolytic end products into the oxidative cycle.

Case Study

Paul, a thoracic surgeon in your facility, has sought your advice regarding a recently appearing problem. He has been relatively inactive for most of his life, but after becoming concerned about his weight he decided to begin a modest exercise program. He has been attempting to avoid using the elevators by climbing the stairs. His complaint is that after only two flights of stairs he has so much pain in his muscles that he must stop to rest. In addition, he has been parking his car at the far end of the parking lot and walking a little farther to get to the hospital. His complaint here is that he gets very tired on this walk, and the tiredness doesn't seem to be getting any better.

Your first thoughts are cramps due to an electrolyte imbalance, but tests reveal that this is unlikely to be the cause of Paul's complaints. A blood lactic acid measurement at the end of a standard graded progressive exercise test is not different from his resting blood lactic acid. You are aware of McArdle's disease, and this finding seems to point in that direction. A muscle biopsy confirms the lack of phosphorylase and McArdle's disease.

You suggest that Paul become more aware of his glucose intake before exercising and try to avoid sudden, strenuous exercise such as stair climbing, since this activity places the maximum amount of stress on the glycolytic pathway. You should tell him to be extremely careful to start any activity at a very low intensity in order to optimize blood flow to the working muscles and optimize the level of energy from oxidative metabolism.

What You Need to Know From Chapter 5

Key Terms

diabetes
- type 1, insulin-dependent diabetes mellitus (IDDM)
- type 2, non-insulin-dependent diabetes mellitus (NIDDM)

glucagon

growth hormone

HbA$_1$c

hyperglycemia

hypoglycemia

insulin

ketoacidosis

ketone bodies

Key Concepts

1. You should understand the causes of type 1 and type 2 diabetes.

2. You should be familiar with diabetic complications such as cardiovascular and renal (nephropathic) diseases.

3. You should be familiar enough with the role of diet, insulin, and exercise in the treatment of diabetes to make recommendations for patients with this disease.

4. With an understanding of any particular metabolic derangement such as McArdle's disease, you should be able to determine the effects of that derangement on exercise metabolism.

Review Questions

1. Type 1 and type 2 diabetes differ in their causes, yet exercise is an important part of the treatment for each. Why?

2. Why is exercise contraindicated for patients with diabetes that is not under good control?

3. What are common strategies for preventing diabetes-related hypoglycemia during exercise?

4. Why do patients with McArdle's disease produce less pyruvic acid than normal?

Bibliography

American Diabetes Association, American College of Sports Medicine. Diabetes mellitus and exercise joint position paper. *Med Sci Sports Exerc* 1997, 29(12):vi.

Berger M, Berchtold P, Cuppers H, Drost H, Kley HK, Muller WA, Wiegelmann W, Simmerman-Telschow H, Gries FA, Kruskemper HL, Zimmerman H. Metabolic and hormonal effects of muscular exercise in juvenile type diabetes. *Diabetologia* 1997, 13:355–365.

Bogardus C, Ravussin E, Robbins D, Wolfe R, Horton ES, and Sims EH. Effects of physical training and diet therapy on carbohydrate metabolism in patients with glucose intolerance and non-insulin-dependent diabetes mellitus. *Diabetes* 1984, 33:311–318.

Ekoe JM. Overview of diabetes mellitus and exercise. *Med Sci Sports Exerc* 1989, 21:353–355.

Eriksson KF, Lindgarde F. Prevention of Type II (non-insulin dependent) diabetes mellitus by diet and physical exercise. *Diabetologia* 1991, 34:891–198.

Ford ES, Herman WH. Leisure-time physical activity patterns in the US diabetic population. *Diabetes Care* 1995, 18:27–33.

Helmrich SP, Ragland DR, Paffenbarger RS. Prevention of non-insulin dependent diabetes mellitus with physical activity. *Med Sci Sports Exerc* 1994, 26:824–830.

Kelley DE, Simoneau J-A. Impaired FFA utilization by skeletal muscle in NIDDM. *J Clin Invest* 1994, 94:2349–2356.

Kriska AM, Blair SN, Pereira MA. The potential role of physical activity in the prevention of non-insulin dependent diabetes mellitus: The epidemiological evidence. *Exerc Sport Sci Rev* 1994, 22:121–143.

Lewis SF, Haller RG. The pathophysiology of McArdle's disease: clues to regulation in exercise and fatigue. *J Appl Physiol* 1986, 61:391–401.

Rogers MA. Acute effects of exercise on glucose tolerance in non-insulin-dependent diabetes. *Med Sci Sports Exerc* 1989, 21:362–368.

Wallberg-Henriksson H. Acute exercise: fuel homeostasis and glucose transport in insulin-dependent diabetes mellitus. *Med Sci Sports Exerc* 1989, 21:356–361.

Neural Control of Movement

This chapter will discuss

➤ how neural information is transmitted;

➤ how action potentials are generated and propagated;

➤ the general organization of the sensory and motor systems;

➤ the functional organization of motor units, motor unit heterogeneity including the interaction of the three major functional systems, the role of proprioceptive input in controlling movement; and

➤ patterns of motor unit recruitment.

The neural regulation of movement is perhaps the most complex function in humans. Successful completion of any motor task depends on intricate, precise, and simultaneous control of many different complementary (agonistic) and opposing (antagonistic) muscles or muscle groups. At the same time, the basic principle underlying movement is quite simple: completion of motor tasks depends solely on the degree of skeletal muscle activation. Another requisite for precise, coordinated movement is the continual supply of information from external and internal environments. This constant flow of sensory input provides information about the progress of the motor task and is vital for its successful completion. For example, according to Teasdale et al., individuals with impaired proprioception tend to overcompensate for deviations in motor tasks, and consequently their

movement patterns are greatly exaggerated. Sensory input from many sources is filtered and integrated (or centrally processed); then adjustments, if needed, are made by the motor system on the basis of this input. All these processes are multilayered, involving a complex array of pathways and structures within the **central nervous system (CNS)**. Rather than covering the structures and pathways of the CNS in depth, however, we will focus on the means by which neural information is generated and transmitted within the system and the ways that nerves and muscle fibers communicate with one another. You must have a good grasp of these basic principles to make informed decisions involving a range of applications, from exercise conditioning to the use of electrical stimulation for preventing loss of contractile protein or restoring mass under conditions of disuse or reduced muscle use.

Completion of motor tasks depends solely on the degree of skeletal muscle activation.

Transmission of Neural Information

A key principle governing movement is the ability of cells to successfully communicate among one another. Indeed, loss of a particular communication pathway through injury or disease inevitably leads to an identifiable functional deficit. First, let's examine how signals are generated.

The Action Potential

All information within the nervous system, and between nerves and muscle cells, is transmitted by electrical signals called **action potentials**. Action potentials are produced in excitable cells, such as nerves and muscle fibers, when the resting membrane potential is altered, or **depolarized**. The resting membrane potential is dependent on the separation of charged ions across the cell membrane, which results in the presence of more positive charges outside the cell than inside the cell, and more negative charges inside the cell than outside. Thus, in a cell at rest, the inside charge is negative compared to the outside charge. Thus the cell membrane is polarized.

All information within the nervous system, and between nerves and muscle cells, is transmitted by electrical signals called action potentials.

There are many charged ions, amino acids, and proteins in extra- and intracellular fluid; but the membrane potential is determined largely by the separation of three major ions—sodium and chloride (higher concentration outside the cell) and potassium (higher concentration inside). How do these ions get positioned on opposite sides of the cell membrane? This charge separation is a product of

- the membrane permeability characteristics of these three ions and
- ion pumps that transport K^+ ions into the cell and Na^+ ions out.

Membrane potential is determined largely by the separation of three major ions—sodium, chloride, and potassium.

Membranes of most excitable cells are more permeable to K^+ than to Na^+ ions; therein lies the mechanism to generate an electrical potential. The number of positively charged potassium ions that diffuse out of the cell (down its concentration gradient) is greater than the number of sodium ions that diffuse in. Thus, the inside of the cell becomes slightly negative. Now, because K^+ ions are being attracted back into the cell by the increasing negative charge inside, an **electrical potential** is also created. So, we have an interaction of two opposing forces:

- A concentration gradient, which tends to move K^+ out of the cell
- An electrical potential difference, which attracts the potassium ions back in

The concentration gradient is a very powerful force, so more and more K^+ ions diffuse out of the cell, and the inside becomes more and more negative. Eventually, however, there is no net movement of K^+, and an **equilibrium potential** is reached. However, the membrane equilibrium potential is a function of equilibrium potentials for both K^+ (approximately –90 mV) and Na^+ (approximately +60 mV). Because the membrane is much more permeable to K^+, the membrane potential is closer to the equilibrium potential for K^+ than to that for Na^+. A typical resting membrane potential for both nerve and skeletal muscle cells is –70 mV.

If the equilibrium potential depended solely on the different rates of Na^+ and K^+ diffusion across the membrane, though, the ionic gradient would eventually disappear. Equilibration of Na^+ and K^+ concentrations across the cell membrane, and thus dissolution of the equilibrium potential, is prevented by **active transport** of Na^+ out of the cell and of K^+ into the cell. Active transport is accomplished by the adenosine triphosphate (ATP)-dependent Na-K pump (figure 6.1). To maintain the resting membrane potential, then, active and passive fluxes for K^+ and Na^+ must be balanced, so no net movement of either occurs.

But shouldn't Cl^- ions have a central role in determining the membrane potential? Nerve and muscle cell membranes are, indeed, permeable to

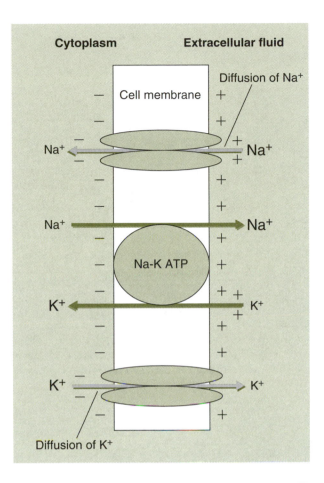

Figure 6.1 Polarization of cell membranes. Sodium and potassium ions are separated across excitable membranes as a result of (1) greater membrane permeability to potassium and (2) the adenosine triphosphate-dependent Na-K pump. The tendencies of K^+ to diffuse out and Na^+ to diffuse into the cell are balanced by the Na-K pump. When an action potential is generated, Na^+ rushes into the cell and K^+ out. Because only a small number of ions actually move across the membrane, the Na^+-K^+ pump can quickly restore membrane polarization. In the figure, the smaller Na^+ and K^+ represent lower concentrations than the larger Na^+ and K^+.
Adapted from Vander et al. 1994.

Cl^- ions. But whereas separation of K^+ and Na^+ is dependent on both active and passive forces, only passive forces determine distribution of Cl^- across the cell membrane. Thus, the equilibrium potential for Cl^- at rest equals the membrane potential. Chloride, then, does not contribute to the magnitude of the membrane potential. (For a discussion of the importance of chloride, see page 79).

A key to understanding the development and maintenance of the membrane potential is the fact that very few K^+ and Na^+ ions need to be separated or to move across the membrane to generate a potential difference or produce an action potential. This process is basically a membrane event: charge separation exists in a very narrow band on either side of the membrane, and the overall concentrations of Na^+ and K^+ inside and outside the cell change very little. It's pretty amazing to consider that all signaling among excitable cells, and thus the essence of being alive (and being able to move), is rooted in this simple fact—separation of electric charges across a cell membrane. To begin the signaling process, the polarization of charges is rapidly altered and the cell membrane becomes depolarized. This is the first step in generating an electrical current.

Generation and Conductance of Action Potentials

Now that we have established the basic requisites for maintaining a separation of electric charges across the cell membrane, let's examine how action potentials are generated and conducted.

Action Potential Generation

An action potential is triggered when the membrane permeability to Na^+ (P_{Na}) increases and Na^+ rushes into the cell, driving the membrane potential toward the equilibrium potential of sodium. After a slight delay, K^+ channels are opened (P_K), causing K^+ to exit the cell, and Na^+ channels are inactivated. This sequence of events initially drives the membrane potential in the positive direction and then, with the efflux of K^+ ions, back toward zero (figure 6.2). When K^+ channels are inactivated, the Na^+-K^+ pump reestablishes a separation of charges across the membrane; the cell is **repolarized** and once again is primed to generate another electrical signal. Action potentials are normally followed by a short **afterhyperpolarization**—a period during which the membrane is **hyperpolarized** (figure 6.2). The length of afterhyperpolarization is shorter in neurons of fast versus slow motor units, thus affecting the rate of action potential discharge.

An important point to remember is that the electrical current is elicited by movement of relatively few ions across the cell membrane. If large

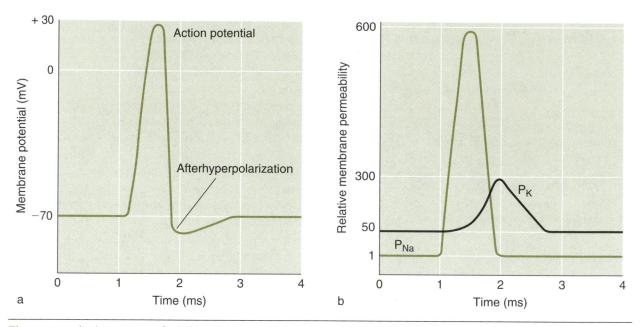

Figure 6.2 Action potentials. When the threshold voltage is reached, an action potential is triggered in an all-or-none response (*a*). Upon repolarization, the membrane potential "overshoots" the resting potential, resulting in a short afterhyperpolarization. Voltage-gated Na^+ channels are the first to open, resulting in a rapid increase in membrane permeability (P_{Na}) (*b*). K^+ channels (P_K) are opened as permeability to Na^+ is declining. The length of the afterhyperpolarization determines when the next potential can be generated.
Adapted from Vander et al. 1994.

numbers of ions moved across the membrane with each action potential, repolarization or recharging of the cell would take too long, and the cell would be **refractory** (unable to generate an action potential) during this extended period of repolarization. Because only small numbers of ions move back and forth, many thousands of action potentials can be generated before the overall concentration differences of Na^+ and K^+ change enough to compromise the electrical potential across the membrane. This characteristic is essential for proper functioning of the neuromuscular system, because neural information is transmitted by **trains**, or groups of action potentials. Rapid recharging of the cell membrane enables action potentials to be generated at a very high frequency. One of the basic principles of communication among excitable tissues depends on this property. The intensity of an output signal is dependent on the number of action potentials delivered per unit time (or action potential frequency). The higher the frequency of action potentials, the higher the output signal and the greater the response of the target or **effector** cell and vice versa. For example, a weak input signal would generate very few action potentials, and the corresponding output signal (e.g., muscle contraction) would be weak.

Excitable cells can also be hyperpolarized by stimuli that drive the membrane potential in a negative direction. The more negative the inside of the cell becomes, the more difficult it is to generate an action potential, and so hyperpolarization inhibits cell excitation. Most movement behavior is a culmination of inhibitory and excitatory inputs. Turning specific motor neuron pools on and off in particular patterns (such as in locomotion) is essential in order for smooth and precise movements to occur.

The intensity of an output signal is dependent on the number of action potentials delivered per unit time (or frequency).

Action Potential Propagation

In order for action potentials to signal the next excitable cell in the pathway, they must be propagated, or conducted, along the nerve fiber. **Myelination** of the axons of motor nerves facilitates high-speed transmission of electrical signals. During development, **Schwann cells** form a special lipid-rich tissue around the axon that is called myelin. Each Schwann cell covers about 1 mm of the axon, and there is a 1 to 2 μm gap between cells. These unmyelinated regions, called **nodes of Ranvier**, are areas of low resistance along the axon, as opposed to the high membrane resistance that myelin imparts on the membrane. The electrical current jumps from node to node (**saltatory conduction**), as opposed to the local spread of the current between neighboring membrane sites that occurs in **unmyelinated** axons. Thus, the transmission of the nerve impulse is relatively slow in unmyelinated fibers because the entire length of the cell membrane has to be depolarized, whereas myelinated fibers transmit the signal very fast as the impulse jumps from node to node.

The utility of nerve axon myelination becomes apparent when one considers the need to conduct signals, often a very long way, with as much speed as possible. If we did not transmit sensory and motor signals quickly, we would not be able to exert timely and precise control over our movements. Consider the state you would be in if it took several seconds between the time you issued a command (such as "Get out of the way of this onrushing rhino") and the time you effected the movement (i.e., actually got out of the way)!

Another property of nerve fibers that affects the speed of impulse transmission is the diameter of the axon. Just as large-diameter electrical cables conduct electricity faster than small-diameter cables, large-diameter nerve axons transmit electric currents faster than axons with small diameters. Thus, the **cable properties** of nerve axons have a direct effect on the speed of signal transmission. Myelinization of motor axons, together with cable properties, gives motor axons the highest conduction velocity of all nerves in the body.

A good understanding of these basic principles is vital if one is to understand how muscles are voluntarily and involuntarily activated as well as how muscles (and nerves) respond to artificial electrical stimulation. For example, the best results for restoration of functional properties of denervated muscle occur when electrical stimulation resembles the normal motor neuron firing patterns of that muscle. As we will see later, this task is not nearly as simple as it might seem.

Myelination of motor axons, together with cable properties, gives motor axons the highest conduction velocity of all nerves in the body.

Input, Conduction, and Output Signals

Generation of action potentials is an all-or-none response. That is, once the depolarizing threshold has been reached, the membrane potential rises to its maximum value over the same time course each time, giving each action potential the same amplitude and duration. In fact, unless the nerve is fatigued or altered by a pathologic state, all action potentials generated by various nerves throughout the nervous system are essentially the same. How, then, do we elicit such different behavioral responses to different stimuli? An obvious partial answer is that different signals have different ports of entry (different sensory nerves), different destinations (different pathways and structures), and different embarkation points from the CNS (to different effector cells or organs). But, in order for us to interpret different inputs within a particular pathway, a mechanism must exist that enables us to sense and react to stimuli of various intensities. This is achieved in a very simple manner: the magnitude of the behavioral response is dependent on the number of signals that are generated per unit time and the total number of signals delivered. Communication between various nerves and between nerves and muscles, then, is accomplished by a basic but very simple process. Trains of impulses (action potentials) are generated in excitable cells at a frequency and train duration proportional to the strength and duration of an input stimulus. The output signal, and therefore the outcome in the effector (target) cell, are in turn proportional to the total number of action potentials that arrive per unit time. To facilitate this process, the initial stimulus, or input signal to the nerve, creates a graded response whereby the number of action potentials generated is proportional to the magnitude of the input signal.

The output signal, and therefore the outcome in the effector (target) cell, is in proportion to the total number of action potentials that arrive per unit time.

Graded Potentials

Graded potentials can be generated either in sensory or motor nerves and begin the process of information transmittal.

• **In sensory nerves.** Stimulus inputs to sensory neurons (stretch, mechanical deformation, chemical substances, etc.) are transformed into graded electrical signals called **receptor potentials**. The sensory nerve ending is depolarized in a graded fashion; that is, the magnitude of the depolarization is directly proportional to the strength and duration of the input signal. The signal is passively transmitted along the neuron until it reaches the **trigger zone** where action potentials are generated (figure 6.3). The train of action potentials travels the length of the axon, and the signal is passed to the next neuron. If the magnitude of the graded potential falls below the action potential threshold, it does not escape from the trigger zone and ends without causing further activity. The difference between graded and action potentials, then, is obvious—the magnitude of the graded potential is variable, whereas that of the action potential is not.

The magnitude of the graded potential is variable; that of the action potential is not.

• **In motor nerves.** Inhibitory and excitatory inputs to the cell body or motor nerve **dendrites** (tree-like processes extending from neuron cell body) result in graded potentials, called **synaptic potentials**, that pass to the integrative zone of the motor axon (**axon hillock**). If the summed input is sufficiently excitatory, then a train of action potentials is triggered and the message is carried, without fail, to the next cells in the pathway—the muscle fibers. The outcome, or level of muscle activation, then, is directly proportional to the input. Signaling between cells across the synaptic cleft is achieved by the release and diffusion of a chemical substance. This chemical substance or **neurotransmitter** represents the **output signal** for the **presynaptic** cell (the cell delivering the stimulus), and at the same time is the input signal for the **postsynaptic** cell (the cell receiving the stimulus).

Output Signals

A train of action potentials arriving at an axon terminal causes neurotransmitter release in a concentration exactly proportional to the strength of the arriving signal. (You should understand by now that the strength of the signal is determined by the action potential frequency and train duration.) The neurotransmitter diffuses from the presynaptic terminal across a narrow gap and binds to receptors on the postsynaptic membrane, altering its membrane permeability. Whether the signal is excitatory or inhibitory depends on the specific neurotransmitter that is released and in turn on which ion channels are opened. For example, what if the neurotransmitter caused an increase in K^+ permeability in the postsynaptic

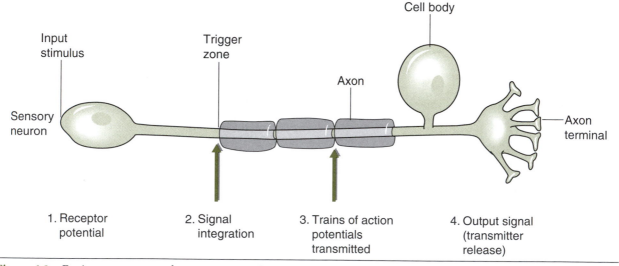

1. Receptor potential   2. Signal integration   3. Trains of action potentials transmitted   4. Output signal (transmitter release)

Figure 6.3 Basic components of sensory nerve. Sensory nerve endings generate graded receptor potentials (1) that travel passively to the trigger zone (2), generating a train of action potentials. The signal is then propagated to the axon terminal (3), where neurotransmitter is released (4) in proportion to the frequency and total number of action potentials. Thus, the identical signal is transmitted from cell to cell.
Adapted from Kandel et al. 1991.

cell? Would you expect a depolarization (excitatory) or hyperpolarization (inhibitory) response? It helps to remember that the potassium equilibrium potential is –90 mV but the resting membrane potential of the cell is about –70 mV. If you guessed hyperpolarization, you were right. If more potassium channels are opened, potassium diffuses out of the cell, making the inside more negative—and less likely to be depolarized.

The utility of grasping this principle may not be evident now, but this basic knowledge is useful for understanding related mechanisms. For example, bradycardia (decreased heart rate) is caused by increased input to the **sinoatrial (SA) node** from the **parasympathetic nervous system (PNS)**. The PNS releases a neurotransmitter, acetylcholine (ACh), which hyperpolarizes the pacemaker cells of the SA node, making it harder to generate a spontaneous action potential and thus slowing the heart. As another example, increasing concentrations of H^+ (a by-product of non-oxidative metabolism) cause an increased permeability of muscle cell membranes to Cl^- ions. What is the outcome? The muscle fiber becomes less excitable (because negatively charged chloride ions are diffusing into the cell), and force production decreases. This is one of the many mechanisms of muscle fatigue to be examined in chapter 8.

Case Study

A friend of yours, Karen, jogs regularly and logs about 32 to 40 km (20-25 miles) a week. A group of classmates talked her into running a marathon, even though she hadn't really been training that intensively. Karen usually runs in the early morning when the temperature is cool. On the day of the race, which begins at 10:00 A.M., the temperature is 75° F and relative humidity is 85%. As the race progresses, Karen notices that she is sweating profusely, and by the end of the race she notices that she has accumulated what she assumes to be salt on her body and running clothes. During the race she consumed water frequently at the drinking stations and estimates that she drank about 5 L (about 5 quarts) of water during the 4.5 hr it took to complete the race. After the race, Karen describes uncontrollable muscle twitching and generalized muscular hypertonia. A friend finds her tendon reflex to be high. Do you have any explanation for the nerve/muscle changes that took place?

Your initial evaluation is an electrolyte imbalance. The apparent extreme sweat loss (sweat has a high concentration of sodium chloride, with smaller amounts of potassium) indicates that Karen experienced a change in the excitability of either motor nerves or muscles, or both. In drinking large volumes of water, she may have exacerbated the problem by diluting the concentrations of both electrolytes. Excessive sweat loss may also have upset the calcium homeostasis within the muscle, but this is only conjecture. No evidence exists that totally explains this phenomenon. Karen should drink a solution containing small amounts of sodium and potassium.

Summing Postsynaptic Inputs

A postsynaptic cell typically has a number (maybe hundreds) of presynaptic cells affecting it at any one time. As already mentioned, it is the sum of the excitatory and inhibitory inputs that dictates the final outcome—generation of action potentials. An excitatory input generates an **excitatory postsynaptic potential (EPSP)**, and an inhibitory signal an **inhibitory postsynaptic potential (IPSP)**. The total effects of all EPSPs and IPSPs are summed in two ways:

- When a signal arrives from a single presynaptic cell before the effect of the previous potential has died away, the two potentials are summed. This is called **temporal summation**.
- When two or more signals arrive at a postsynaptic cell from different sources (and at different locations) simultaneously, these inputs also sum. This is referred to as **spatial summation** (figure 6.4).

An excitatory input generates an excitatory postsynaptic potential (EPSP), and an inhibitory signal an inhibitory postsynaptic potential (IPSP).

An excitatory and an inhibitory signal arriving simultaneously at different locations on the postsynaptic cell usually cancel each other out. However, if the sum of the excitatory inputs exceeds the sum of inhibitory inputs by a large enough margin, the cell will reach the necessary threshold to trigger action potentials. This process of excitation and inhibition of motor nerves

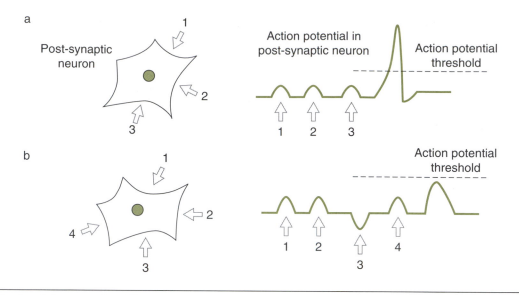

Figure 6.4 Spatial summation. Three excitatory action potentials arrive at a postsynaptic cell simultaneously and are summed (*a*). The magnitude of the resulting receptor potential is high enough to surpass the threshold for generating an action potential in the postsynaptic cell. Three excitatory signals and one inhibitory signal are delivered to the postsynaptic cell. The resulting receptor potential is weak, and no action potential is generated (*b*).

is essential for precise regulation of all movement behaviors. For example, the simultaneous excitation and inhibition of leg and hip extensors and flexors achieves locomotion. For successful completion of this movement behavior, a given muscle group (let's say the quadriceps) is turned on and off in a precise timing pattern. More complex movement patterns obviously require control at a more sophisticated level, and the cooperation of a number of muscle groups throughout the body is necessary.

The Stretch (Myotactic) Reflex

We now have the basics in place to better understand movement control. Let's emphasize this material by putting all of these basic concepts into a very simple model—the stretch, or **myotactic reflex**. This type of reflex—a **monosynaptic** reflex—is the most basic of movements and is regulated entirely at the level of the spinal cord. We will learn later that the most stereotypical, or simplest, movements are controlled at the lower levels of the CNS, and the more complex movements at the higher levels. The knee-jerk reflex is an example of a monosynaptic reflex; that is, the sensory input and the motor outcome (or muscle contraction) are separated by a single connection, or synapse.

The knee-jerk reflex is an example of a monosynaptic reflex; that is, the sensory input and the motor outcome (or muscle contraction) are separated by a single connection, or synapse.

The stretch reflex is mediated by encapsulated structures (**muscle spindles**) containing modified muscle (intrafusal or spindle) fibers that lie parallel to the skeletomotor (extrafusal) fibers. Sensory nerve fibers wrap around the spindle fibers and increase their firing rate in proportion to the **amount** and **rate** of muscle stretch (figure 6.5). The signal travels passively to the trigger zone, generating a train of action potentials. The action potentials are conducted the length of the axon and cause a release of neurotransmitter in the axon terminal in a concentration that is, again, proportional to the initial input signal. Since this is a monosynaptic reflex, the sensory nerves from the muscle spindles synapse directly on motor nerves that innervate the stretched muscle. The synaptic potential generated in the motor nerve triggers a series of action potentials that travel to the muscle fibers and elicit a contraction, whose intensity reflects the magnitude of the original input or stretch. Remember that action potentials themselves do not change in amplitude or duration; rather the signal inten-

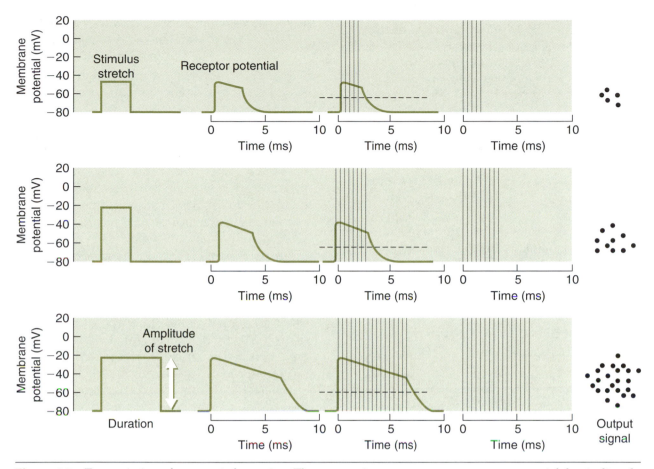

Figure 6.5   Transmission of sensory information. The sensory input generates a receptor potential that is directly proportional to the intensity and duration of the input. The output signal (chemical neurotransmitter) is also directly proportional to the original input signal. In the example shown, the input stimulus is the stretching of a muscle spindle. If either the amplitude or the duration of the stimulus changes, the corresponding frequency and total number of action potentials change. Accordingly, the response of the muscle will be greater or less. Adapted from Kandel et al. 1991.

sity is reflected in the frequency and total number of action potentials.

How would the signaling process change if the stretch of the muscle spindles were very large? The answer is simple: the receptor potential would be greater in magnitude, more action potentials would be generated, more neurotransmitter would be released, and the muscle contraction would be more forceful. We can use this simple model to understand one of the basic principles of muscle fiber recruitment during voluntary movements, which are initiated from higher brain centers: a higher stimulation frequency of motor nerves results in a greater force production by the muscle. Of course this is true only to a point, because muscle fibers, depending on their basic contractile properties, will sum forces only until the maximum force potential (**tetanic force**) has been

achieved. Stimulation of a fiber at a higher frequency, once it has reached maximum tetanic force, will not result in an increase in force production. Slow fibers sum forces and reach maximum force potential at lower frequencies than fast fibers.

A higher stimulation frequency by motor nerves results in a greater force production by the muscle.

Motor nerves receive inputs from many sources, and it is the sum total (temporal and spatial) of all of these inputs (excitatory and inhibitory) that determines whether the nerve will fire or not. Once an action potential is triggered in the motor neuron, it is transmitted without fail to the axon terminal where it will synapse with the final link in the pathway, the muscle fiber.

Motor nerves receive inputs from many sources, and it is the sum total (temporal and spatial) of all of these inputs (excitatory and inhibitory) that determines whether the nerve will fire or not.

Organization of Sensory and Motor Systems

So far we have examined how input signals are transformed into electrical signals and how this information is transmitted and interpreted to eventually produce output signals resulting in movement behavior. Now let's consider how the sensory and motor systems are linked and how they function together to produce precise and coordinated movements.

Divergence and Convergence

The principles of divergence and convergence of information play central roles in our understanding of the way neural information is distributed and processed in the CNS, also providing insight into the functional organization of the brain. The circuitry of the sensory and motor systems is organized in a fashion that

- enables a single piece of information to be distributed simultaneously to many centers, a process called **divergence**; and
- enables diverse information from many sources to be delivered to a single target, a process that is termed **convergence**

As mentioned in the previous section, a single postsynaptic nerve cell, such as a motor neuron, may have many hundreds of cells synapsing on it. At any given moment the motor neuron may be bombarded with information from many different sources; the cells that are affecting it, in turn, receive multiple input as well. This principle of progressive convergence is important in providing flexibility to the system by allowing integration of diverse information from different levels of the entire system (figure 6.6). Thus, a great deal of information processing occurs before a motor neuron is activated or inhibited. This multilayered processing of information that precedes motor neuron activation prompted Sherrington, one of the pioneers in the area of neural control of movement, to designate the motor nerve as the "final common pathway."

Sensory nerves, on the other hand, tend to distribute their information by divergence. This

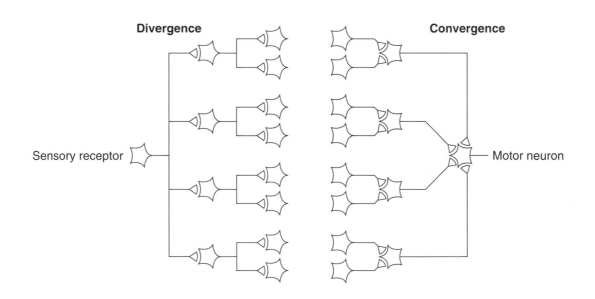

Figure 6.6 Divergence and convergence of neural information. Sensory input tends to be distributed by divergence, the passing on of information to several target cells simultaneously. Each motor neuron receives input from many presynaptic cells, which in turn receive multiple input (convergence). This allows integration of diverse information from different levels of the central nervous system, enabling precise control of movements.
Adapted from Kandel et al. 1991.

process, whereby a single sensory neuron passes its information on to many target cells, has the advantage of distributing discrete inputs widely to various processing units. These chains of connections are organized in series to enable information to be filtered or integrated at many levels. At the same time, this functional organization facilitates **parallel processing**, a means by which similar information is delivered to, or from, a brain center via parallel pathways. This process increases the likelihood that a sensory perception will be properly interpreted or that a motor command will be effectively delivered. As we will see later in this chapter, parallel processing reduces the likelihood that a lesion to a single neural pathway will completely eradicate a specific movement behavior.

Structural and Functional Hierarchy of Motor Control

As mentioned in the previous section, input from sensory neurons tends to diverge and impact on many target cells. The first cells targeted are located in the spinal cord. We have already examined the myotactic, or stretch, reflex and learned that primitive movements of this kind are regulated with little or no input from the higher brain centers. This example illustrates the arrangement of the sensory and motor systems in both a structural and a functional hierarchy. That is, simple movements that require little processing are carried out at the lower levels of the CNS, and more complex movements that require more integration and processing are regulated at the higher centers. As the complexity of the movement increases, successively higher centers become involved, first the brainstem, then the diencephalon, then the highest and most sophisticated center—the cerebrum.

Central Nervous System and Motor System Organization

The CNS consists of the spinal cord (the most caudal structure in the system) and the brain, which is subdivided into three major regions (figure 6.7):

- Brainstem
- Diencephelon
- Cerebrum

The last major structure, the cerebellum, is located dorsal to the brainstem and is also important in controlling movement. Let's examine the basic structural and functional characteristics of these regions separately.

The Spinal Cord and the Brainstem

The spinal cord receives sensory input through the **dorsal horn**, which contains the cell bodies or dorsal root ganglia of sensory nerves. Sensory information is then delivered into the spinal cord; there it either synapses directly with motor neurons (or with motor neurons through **interneurons**) or is forwarded to specific processing areas in the higher brain centers. The **ventral horn** of the spinal cord houses the two types of motor neurons. These are

- **alpha motor neurons**, which innervate skeletal muscle (or extrafusal) fibers; and
- **gamma motor neurons**, which excite smaller muscle fibers (called intrafusal) that are encapsulated in muscle spindles.

The most numerous neurons in the spinal cord are interneurons, which are responsible for much of the final refinement of information before it reaches the motor neurons.

As mentioned previously, some of the simpler movements are coordinated in the spinal cord at the level of the interneurons. For example, the repetitive activation and inhibition of flexors and extensors in the lower extremities that enable us to locomote, once initiated by the higher centers, are regulated entirely at the level of spinal cord interneurons. Changes in direction, speed, and the like are controlled at the higher levels, but the act of locomotion itself is largely a spinal event. This is another example of the hierarchical organization of the motor system in that simple, or automatic, movements are controlled in the most caudal structures. Another function of interneurons involves the process of **gating** (figure 6.8), a feature of motor control whereby descending input from higher centers enhances or inhibits the action of specific interneurons and thus alters the level of activation of a particular motor neuron. For example, a sensory input that depolarizes a spinal cord interneuron can be nullified by a hyperpolarizing signal from higher brain centers. We will learn later that gating of sensory input provides an important neural adaptation to resistance training that enhances gains in muscle strength.

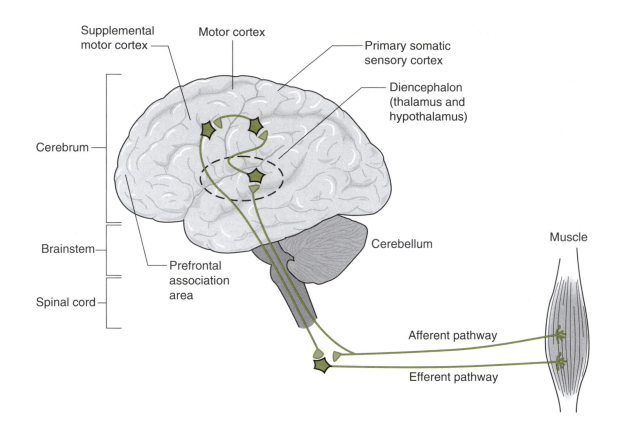

Figure 6.7 Hierarchical organization of the CNS. The central nervous system has four major regions, which are organized in a structural and functional hierarchy: the spinal cord, brainstem, diencephalon, and cerebrum. Simple movements such as reflexes are regulated at the lower levels of the central nervous system; more complex movements are processed at the higher levels. The cerebellum coordinates motor activity by comparing motor output with sensory input.
Adapted from Kandel et al. 1991.

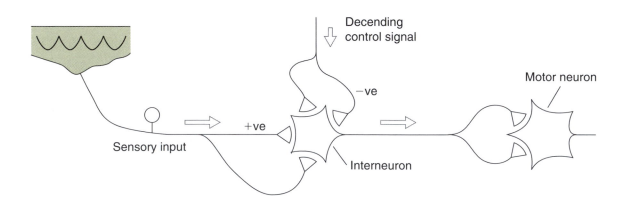

Figure 6.8 Gating by interneurons. Sensory input can be modified at the level of the spinal cord as illustrated here. Sensory input that would normally depolarize the interneuron is nullified by a hyperpolarizing signal from higher levels of the central nervous system. −ve = negative; +ve = positive.
Adapted from Kandel et al. 1991.

Immediately rostral to the spinal cord is the **brainstem**, consisting of the medulla oblongata, the pons, and the midbrain. Information passed between the spinal cord and the higher centers of the brain travels through the brainstem in one of two ways:

- Direct pathways, which carry information between higher centers and motor neurons with very little synapsing—which means that the information is largely unaltered as it travels from its origin to its destination
- Indirect pathways, which carry information that is refined along the way by integration of new information via input from multiple sources

The Diencephalon

The **diencephalon** is next in the functional and structural hierarchy of the motor system. It contains the **thalamus**, an important processing and relay station for most inputs to the cerebral cortex, and the **hypothalamus**, which coordinates many neural and endocrine functions essential for life. The hypothalamus has many interconnections with, and is intimately related to, the **limbic system**, a circuitous structure that encircles the upper brainstem and has myriad functions related to behavior, emotion, and motivation. We rely greatly on the interplay between the limbic system, in concert with the hypothalamus, and the motor cortex for the quality of motor performance. A person who is highly motivated obviously will perform certain motor tasks, especially gross movements, more effectively than one who is not motivated or is bored. On the other hand, high levels of excitement or emotion may actually hinder highly complex and precise motor tasks, such as a difficult gymnastics movement.

> We rely greatly on the interplay between the limbic system, the hypothalamus, and the motor cortex for the quality of motor performance.

The Cerebrum

The largest structure in the brain, and the highest in the functional hierarchy, is the **cerebrum**, made up of the **cerebral cortex** and the **basal ganglia**. The cerebral cortex is the most complex processing center of the brain; it is responsible for transforming basic incoming signals into meaningful conceptual impressions. The motor and premotor areas of the cerebral cortex have ultimate control over most movements. The motor cortex primarily regulates movements involving the use of single muscle groups; the premotor area exerts control over more complicated patterns of movement such as alteration of force or velocity, coordination of groups of muscles to enable performance of specific tasks, and changing over from one task to another. The premotor area also serves as a relay station for passing information from the prefrontal and posterior parietal association areas and from the **supplementary motor cortex** to the motor cortex (figure 6.7). These areas represent the highest level in the functional hierarchy of the CNS and are responsible for planning and initiating movements.

The basal ganglia are collections of neurons lying deep within the cerebrum that, in conjunction with the motor and premotor regions, play an important role in the execution of complex movements. Basal ganglia, via connections with the association areas of the cerebral cortex, also help organize thought processes to determine which patterns of movement will be used together and in what sequence. Lesions to the basal ganglia, then, result in loss of smooth coordination of motor tasks and are reflected by uncoordinated and clumsy movements.

The Cerebellum

The last region important for motor control is the **cerebellum**, which lies dorsal to, and receives and relays information through, the pons, which is part of the brainstem (figure 6.7). Although the cerebellum does not initiate movement, it is vital in movement control because it coordinates motor activity by comparing afferent sensory information to efferent information received from the motor and premotor cortexes. In essence the cerebellum compares the intended to the actual movement and makes adjustments in the output as needed. In addition, the cerebellum helps the motor cortex plan the next movement in a sequence by anticipating, based on the backdrop of peripheral sensory input, what the next movement should be. Together, then, the cerebellum and the basal ganglia act to provide precision and smooth coordination and to ensure proper sequencing and patterning of movements. As mentioned, it is vital that a constant flow of sensory information be provided to these and other structures to ensure successful completion of a particular movement. Especially important for precise movement control is the sense called **proprioception**.

Proprioception

The precise control of movements, necessarily, has a high priority in human behavior. Indeed, the ancient Greeks thought of the ability to move as the essence of being alive. As you may have guessed, it is important to have a supply of sensory input about the position of body parts and movement of the limbs. This information comes from an array of different receptors located at all sites where movement can be detected: the muscles, tendons and joints, and even the skin. In the early 1900s, Sherrington first used the term proprioception to refer to all of the senses that originate in the musculoskeletal system. The definition of proprioception has been broadened since that time and now includes

- the sense of stationary limb position (static limb sense);
- the sense of limb movement, or **kinesthesia**; and
- the senses affecting balance, which is largely regulated by the vestibular apparatus.

Examination of this last sense is beyond the scope of this book. We will concentrate solely on the position and kinesthetic senses.

Proprioception, as already noted, is largely a function of receptors located in joints, skin, and muscles. Receptors in and around joints are sensitive to the extremes of joint angulation and contribute (in conjunction with muscle spindles) information about the rate of change of limb position. Cutaneous mechanoreceptors also contribute to position sense when the skin is stretched by movement at the joints or by contraction of underlying muscles. Skin receptors are especially dense in the fingers and thus provide important feedback in movement of fingers. Perhaps the most important proprioceptors, though, are located within the muscle itself.

The two primary muscle receptors responsible for "muscle sense" are

- muscle spindles, which are stretch-sensitive mechanoreceptors; and

- **Golgi tendon organs,** which provide information about the magnitude of muscle force production.

Muscle Spindles

Muscle spindles are embedded within the body of the muscle and are in parallel with the "ordinary" or **extrafusal** muscle fibers, anchored in place by attachments to connective tissue. The highest density of muscle spindles is found in muscles dedicated to fine motor control, such as those in the hand, and in muscles whose primary purpose is control of posture such as those in the neck (feedback about head position and movement is vital for the maintenance of posture). Spindles contain basically two types of muscle fibers: bag (1 and 2) fibers, and chain fibers.

These **intrafusal** fibers, as described by Proske, are much smaller than extrafusal fibers and are also structurally different in that they have contractile elements at both ends, separated by a central region containing the cell nuclei. You should note that this structural orientation prevents the muscle spindle from contributing to the overall force production of the muscle itself. The central region of an intrafusal fiber is innervated by large sensory afferent nerve fibers, and the polar regions (containing the contractile elements) are innervated by two types of gamma motor neurons, static and dynamic. One end of bag 2 fibers and chain fibers is also innervated by smaller afferent nerve fibers.

Thus, each muscle spindle has two types of sensory nerves sending information to the CNS and two types of gamma motor nerves innervating the contractile portion of the fibers. You may be asking, if the muscle spindle simply responds to stretch of a muscle, why is it so complicated? The answer lies in the complexity of our interaction with the environment. Interaction with the physical environment depends on the simultaneous occurrence of so many different types of movements and changes in direction, force production, velocity, and so on, that precise regulation can be exerted by the central controllers only if sufficient and exact information is received from the periphery. To simplify this complicated feedback mechanism, think of muscle spindles as basically responding to changes in muscle length and in turn helping to regulate muscle length.

They supply information not only about the absolute change in length but also the rate of change in length. Bag 2 fibers are responsible for providing feedback about passive stretch, and bag 1 fibers are more important during dynamic activity, when both alpha (innervating extrafusal fibers) and gamma (innervating intrafusal fibers) motor neurons are activated.

Muscle spindles respond to changes in muscle length and in turn help to regulate muscle length.

So, what is the functional significance of having extrafusal and intrafusal fibers innervated by separate, but parallel, motor nerves? The answer is simple. When the muscle is shortening during a contraction, the intrafusal muscle fibers would fall slack if an equivalent input were not delivered to the contractile elements in the polar regions to maintain stretch on the central region (where the afferent nerve endings are located). Valuable input about muscle length would be lost if this dual innervation (alpha and gamma) did not exist. So the muscle spindle fibers are **coactivated** along with the extrafusal fibers during shortening contractions to maintain a constant supply of sensory information (figure 6.9). Bag 1 and chain fibers are responsible for providing feedback during concentric (shortening) contractions; bag 1 fibers are responsible in the case of the slower-velocity contractions, and the chain fibers (which can contract rapidly) contribute most to those of high velocity.

This complicated scheme will perhaps become clearer with an example. Suppose you are approaching a table to lift an object of unknown

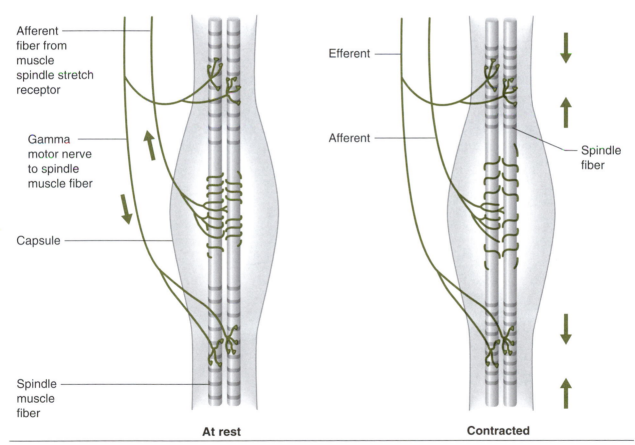

Afferent fiber from muscle spindle stretch receptor

Gamma motor nerve to spindle muscle fiber

Capsule

Spindle muscle fiber

At rest

Efferent

Afferent

Spindle fiber

Contracted

Figure 6.9 Co-activation of alpha and gamma motoneurons. To avoid unloading of muscle spindles during a concentric contraction, gamma motor neurons are coactivated with alpha neurons to maintain a constant stretch on the sensory afferents. When you lift an object that is heavier than you anticipated, an imbalance between the length of the extrafusal fibers (which has not changed) and the stretch of the muscle spindle through gamma motor neuron activation results in increased recruitment and an increased strength of the contraction.
Adapted from Vander et al. 1994.

weight. The premotor and supplementary areas of the brain enact a program that will recruit the appropriate muscles and **motor units** within those muscles to carry out the task. You bend over to pick up the object, but it is heavier than you anticipated. What happens? Does the object remain on the table until you consciously decide that the darn thing is not moving so you'd better exert a little more force? No. In fact, you are able to lift the object "automatically"—largely as a result of the input you receive from the muscle spindles. Remember, alpha and gamma motor neurons are coactivated (both motor neuron pools are located in the same vicinity in the spinal cord); so as soon as you lift, and the object doesn't move, an immediate imbalance between the extrafusal and intrafusal fibers is sensed. The extrafusal fibers have not shortened (because the object did not move); but contraction of the polar regions of the intrafusal fibers has increased the stretch on the central region, where the sensory afferents are located (figure 6.9). This increased stretch results in additional motor unit recruitment through the stretch-reflex mechanism. Thus, the object is moved without any conscious effort on your part. Incidentally, as figure 6.10 shows, the stretch reflex works so well because the same sensory input that activates the muscle being stretched

- inhibits antagonist muscles (or muscles that oppose the movement),
- activates synergist muscles, and
- informs the brain about the whole process.

The stretch reflex works so well because the same sensory input that activates the muscle being stretched inhibits antagonist muscles, activates synergist muscles, and informs the brain about the whole process.

Case Study

As you are treating a patient he develops severe, painful cramps in the gastrocnemius muscle. Your initial response is to massage the muscle, digging into the belly with your thumbs. This doesn't relieve the cramp, so you get the patient to stand up and stretch the muscle by keeping his heel on the ground while leaning forward. After 30 s the cramp is relieved. Describe mechanistically why the static stretch worked whereas the massage did not.

The gentle static stretch is an effective mechanism to relieve such a cramp because it avoids stimulation of the muscle spindle that

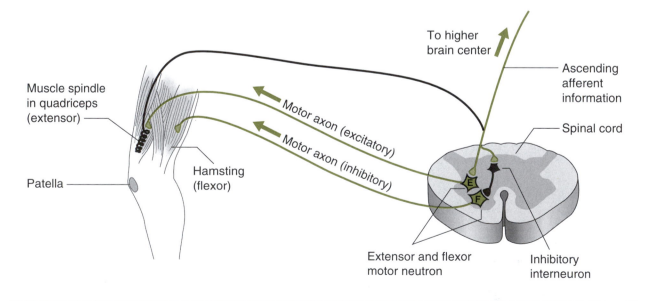

Figure 6.10 Muscle spindle. Muscle spindles are sensitive to the magnitude and the rate of change in the length of a muscle when it is stretched. The sensory afferents synapse directly on the motor neurons of the stretched muscle, causing it to contract. Antagonist muscles are inhibited, and the higher levels are "informed" of the movement through ascending pathways.
Adapted from Vander et al. 1994.

would accompany a fast stretch. This stimulation would cause a reflex contraction of the muscle. Holding the stretch for 30 s or more causes the muscle spindle to remain in equilibrium, minimizing the stretch reflex while providing sufficient tension to stretch the muscle elements. The deep massage may reduce the pain through accommodation, but it is unlikely to have much effect on the cramp.

Golgi Tendon Organ

The **Golgi tendon organ** (GTO) is the other muscle receptor important in proprioception. This encapsulated structure is located at the myotendinous junction and is sensitive to changes in muscle tension. Because it is arranged in series with the muscle fibers, whereas the muscle spindle lies in parallel, the GTO is much more sensitive to active force than to passive stretch of the muscle. When a muscle is activated and begins to generate force, the GTO increases its firing rate in proportion to the magnitude of the contraction. This information is sent to the spinal cord where, through connections with interneurons, it inhibits the motor neurons of the contracting muscle and its synergists and excites the antagonistic muscle (figure 6.11). This action is the exact opposite of the input

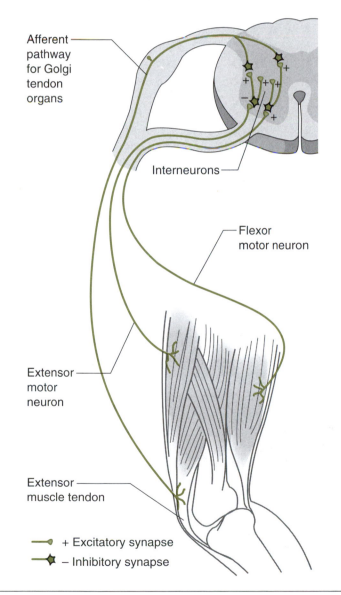

Afferent pathway for Golgi tendon organs

Interneurons

Flexor motor neuron

Extensor motor neuron

Extensor muscle tendon

— + Excitatory synapse

— – Inhibitory synapse

Figure 6.11   Golgi tendon organ. As muscle force increases, the Golgi tendon organ increases its output; this inhibits the motor neurons of the contracting muscle and excites neurons of the antagonists.
Adapted from Vander et al. 1994.

from the muscle spindle. Why do we need a force-monitoring sensor that is located in the muscle itself? The reason is that the output (or force) of the muscle is not always equivalent to the input signal. Force, per given neural input, varies with factors such as fatigue, fiber damage, initial length, and type of contraction. Thus the brain needs constant input about actual force being generated so that appropriate adjustments can be made under different conditions or states of the muscle.

The GTO is much more sensitive to active force than to passive stretch of the muscle, because it is arranged in series with the muscle fibers.

One could also surmise that another major function of the GTO is to protect the muscle from injury. As more and more force is generated, the GTO is sending a stronger and stronger signal to inhibit the contraction, preventing the person from exerting maximum voluntary strength of a particular muscle or muscle group. This seems to be especially true in untrained people. But some evidence suggests that in the early stages of a resistance-training program, inhibition of the inhibitory influence of the GTO, that is, **disinhibition** by the higher brain centers, is a neural adaptation that contributes to early strength gains. This adaptation, among others, enables the weight trainer to get "stronger" without an increase in the size of the muscle. Chapter 9 addresses this principle of disinhibition in conjunction with neural adaptations to strength training.

Major Functional Systems

Up to now we have examined some of the major principles that underlie neural control of movement. We have seen that precise and coordinated movements rely on myriad sensory inputs that are processed centrally, then on a motor system that must choose the correct muscles in a specific sequence, and finally on motor units within those muscles to carry out the movement task. But what about the quality of the movement? That is, does the level of interest of the person (e.g., excitement or boredom) have any bearing on successful completion of the movement behavior?

The answer, of course, is yes. Motivation and emotion are modulated by the limbic system, a combination of structures that encircle the upper brainstem. The limbic system exercises control over movement through direct connections with various structures of the motor system as well as the hypothalamus. The hypothalamus, in turn, is the main controller of the **autonomic nervous system (ANS)**, which modulates the body's state of readiness. Among other things, the ANS helps regulate heart rate, blood flow, sweating, and hormone release—all important factors in the exercise response. The sympathetic arm of the ANS helps prepare us for action, while the parasympathetic system is more important in returning the body to a resting state. Unlike motor nerves, which synapse only with muscle fibers once they exit the spinal cord, nerves of the ANS synapse in ganglia before they reach target organs or glands. The ganglia for sympathetic nerves lie close to the spinal cord, whereas parasympathetic ganglia are in close proximity to or are actually within target organs. Thus, sympathetic activation elicits a more extensive response throughout the body, and parasympathetic stimulation is more local, affecting only a specific gland or organ.

The sympathetic arm of the autonomic nervous system helps prepare us for action, while the parasympathetic system is more important in returning the body to a resting state.

The motivational system—through connections with both the motor system (which has a direct effect on the quality of the movement behavior) and the ANS (which modulates visceral responses)—plays a major role in the initiation and successful completion of motor tasks.

Let's examine how the three major functional nervous systems—motor, sensory, and motivational—work together to produce movement using the example of catching a batted baseball. First, as Kandel and Schwartz point out, auditory, visual, proprioceptive, and tactile senses play central roles in your judging where the ball will land, knowing where your various body parts are and what they are doing as you run to meet the ball, and finally feeling the ball when it hits the glove. The motor system must choose the correct muscles and coordinate their activation or inhibition to carry out the task successfully. The **motivational** system will most definitely impact the quality of this task—a professional baseball player whose contract is up for renewal may have a different level of motivation than you or I playing

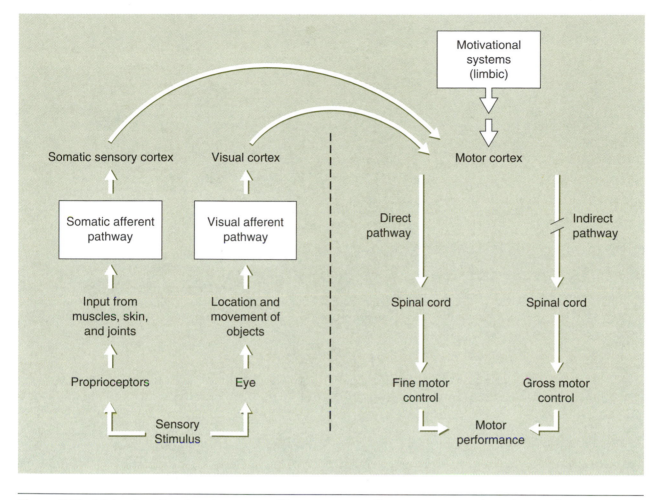

Figure 6.12 Motor functional systems controlling movement. Sensory, motor, and motivational systems must work together to produce efficient movements. Continual sensory input is needed in order for a person to arrive where a batted ball has been hit and to catch it. Muscle groups are activated in a specific sequence based on this sensory input. Motivation affects the movement by acting directly on the motor cortex and by coordinating the sympathetic and motor systems.
Adapted from Kandel et al. 1991.

a pickup game in the park. The motivational system, then, impacts the quality of movement behavior by coordinating the excitement and emotion with motor output (figure 6.12).

Motor Units

This chapter has stressed the importance of selecting the appropriate muscles or muscle groups in a specific timing pattern. But the motor system obviously does not just select, or reject, an entire muscle to enact a particular movement; rather it chooses the appropriate motor units within a muscle to exert an even finer control.

Components of a Motor Unit

A motor unit has four basic components (figure 6.13):

1. Motor neuron cell body (or soma)

2. Axon

3. Neuromuscular junction

4. Muscle fiber

Let's examine how the first three of these components function together to initiate, deliver, receive, and interpret an electrical signal. We will consider the last component, the muscle fiber, in chapter 8.

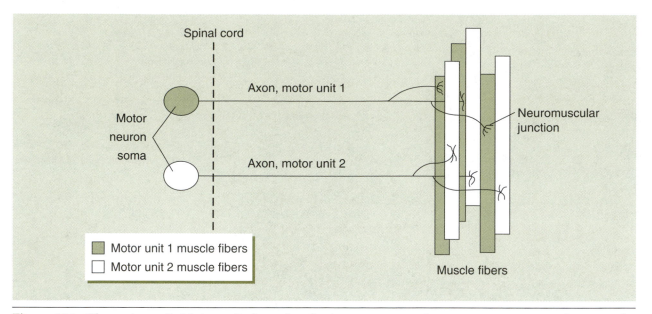

Figure 6.13 The motor unit. Motor units have four basic components: the motor neuron soma, the axon, the neuromuscular junction, and the muscle fiber.

Each motor unit innervates many muscle fibers, but each muscle fiber is innervated by only one motor nerve.

Motor Neuron Cell Body

We have already learned that the soma of the alpha motor neuron lies in the ventral horn of the spinal cord and receives input from many sources, some excitatory and some inhibitory. A major feature of the motor neuron cell body, one that largely determines the ease with which action potentials are generated, is its size. According to Stuart and Enoka, smaller somas have a lower membrane resistance and are therefore more easily excited; larger motor neurons have a higher membrane resistance so that it takes a stronger input to depolarize them. The larger motor neurons are associated with muscle fibers that are fast contracting; conversely, slower muscle fibers are innervated by motor neurons with smaller-diameter cell bodies. Remember from our earlier discussion that a weak input is synonymous with a low frequency of action potentials and that strong inputs are achieved by high frequencies of stimulation. Because it takes a weaker input to activate a motor neuron belonging to a slow motor unit, it follows that slow motor units are recruited most frequently and fast units least frequently. Taking this a step further, we can say that slow motor units are recruited mostly for activities of daily living such

as posture, locomotion, and so on, while fast motor units are used for activities that require high force or power (combination of speed and strength) outputs. It follows, then, that many people (in fact, the majority of the population) do not recruit fast-type motor units very much at all. This information gives us some insight into the **specificity** of muscle adaptation. One parameter that determines the way that a muscle fiber adapts to an external stimulus is the frequency (or number of times per day or week) at which a motor unit is recruited. Slow motor units that are recruited daily for posture and locomotion adapt specifically to these activities, whereas faster motor units, if not recruited for higher-intensity activities, adapt very little, or not at all. Specificity is a key principle in muscle conditioning that we will examine in more detail later.

Motor neurons with large somas are associated with muscle fibers that are fast contracting; conversely, muscle fibers are innervated by motor neurons with smaller-diameter somas.

Slow motor units are recruited mostly for activities of daily living such as posture and locomotion while fast motor units are used for activities that require high force or power outputs.

The last feature of the motor neuron cell body that we will consider is afterhyperpolarization. According to Stuart and Enoka, this is the period immediately following depolarization of a membrane; its duration influences the **discharge rate**, or the rate at which action potentials can be generated. Large-diameter motor neurons, which innervate fast fibers, have a shorter afterhyperpolarization than slow motor neurons and are therefore capable of generating action potentials at higher frequencies. Conversely, action potentials are generated at a lower frequency in slow motor units. Maximum stimulation frequencies in human skeletal muscle range from approximately 5 to 30 Hz for slow motor units and from 30 to 65 Hz for most fast motor units. Although most human motor units normally discharge over a range of approximately 5 to 40 Hz, the fastest motor units in the body (e.g., those that control saccadic movements of the eye) have recruitment frequencies as high as 100 Hz. Normally, motor units are recruited at frequencies below that which produces **fused tetanus** (or maximum force potential). Unfused tetanus produces a "ripple" effect in the force output, that is, peaks and valleys in force production between the individual twitches. To negate this and produce smooth movements, especially during sustained contractions, motor units are activated **asynchronously**. In this manner, various motor units are turning on and off at different times during the movement, in effect canceling the ripple seen with unfused tetanus of individual motor units.

➤ Saccadic refers to a quick jump of the eye from one fixation point to another, such as occurs in reading.

Motor neurons with large cell bodies, characteristic of fast motor units, are less excitable than smaller motor neurons, need a strong stimulus to be activated, and are capable of generating action potentials at high frequencies. Neurons of slow motor units have small cell bodies, are excited by weak inputs, and are limited in the rate at which they can generate action potentials. These properties form the basis of motor unit recruitment patterns and give us our first insight into the principle of specificity of training.

The properties of large and small motor neuron cell bodies form the basis of motor unit recruitment patterns and give us our first insight into the principle of specificity of training.

Axon

The axon of a motor nerve originates as the axon hillock, which extends into the **initial segment**. It is here that the excitatory and inhibitory inputs are summed, and here—if threshold is reached—that the signal to the muscle fiber originates. The axon exits the spinal cord through the ventral horn and joins the "companion" sensory nerve within an **epineurial sheath**, (connective tissue that encapsulates nerve fibers). The motor and sensory axons actually share the same peripheral path to (and from) the muscle. When the motor axon enters the muscle it divides into many branches, each innervating a single muscle fiber at a specialized region called the **motor end plate**. The number of axon branches corresponds to the innervation ratio of the motor unit. As the axon nears the end plate, it loses its myelin sheath and divides again into several fine branches that enter the **synaptic cleft** (the space between an axon terminal and postsynaptic surface) of the muscle fiber to form a **neuromuscular junction** (NMJ). A single motor nerve may innervate as many as 2000 muscle fibers in larger muscles (high innervation ratio) and as few as 10 to 15 fibers in smaller, more finely controlled muscles (low innervation ratio). But as already mentioned, even though a single motor nerve synapses with many muscle fibers, each muscle fiber is innervated by only one motor nerve.

Just as fast motor units have large-diameter motor neuron cell bodies, they also have large-diameter axons. We learned earlier that cable properties of nerve axons dictate the speeds with which electrical signals are conducted along the length of the axon. Thus, fast motor units have high axonal conduction velocities, and slow ones have low speeds of conduction. In general, though, motor axons have the largest diameter, and thus the highest conduction velocities, of any axons in the body. This high conduction velocity underlines the importance of transmitting motor signals from central generators to peripheral effectors with as little delay as possible.

Fast motor units have high axonal conduction velocities, and slow ones have low speeds of conduction.

Neuromuscular Junction

Each of the fine branches of the motor axon terminates in a sac-like structure called a bouton, embedded in the junctional folds of the motor end plate. These presynaptic terminals contain vesicles loaded with the chemical transmitter acetylcholine (Ach). An action potential arriving at the axon terminal depolarizes the membrane, opening voltage-sensitive calcium channels, which allows Ca^{++} to enter the terminal. Calcium triggers the fusion of the synaptic vesicles to the inner surface of the terminal membrane, which leads to the release of ACh into the synaptic cleft. Calcium is important here, then, to couple the electrical signal (action potential) with the chemical signal (the neurotransmitter ACh). We will learn later that Ca^{++} has several other important roles as a messenger within the muscle cell.

Acetylcholine released from the axon terminal binds with its receptor in the membrane of the muscle fiber, causing a change in permeability to both Na^+ and K^+. The resulting **end-plate potential**, which has an amplitude of about –70 mV, is generated in much the same way as an action potential in a nerve. The end-plate potential triggers an action potential that is subsequently propagated along the surface of the muscle fiber (figure 6.14).

Just as a nerve axon must be "recharged" without much delay to be able to generate a series of action potentials, so must the muscle fiber membrane be repolarized quickly to receive another signal from the motor nerve. This is done through a two-step process:

1. Acetylcholine is removed from the site by
   • simple diffusion of ACh out of the synaptic cleft and
   • inactivation by the enzyme **acetylcholinesterase**, located on the surface of the motor end plate.
2. The resting membrane potential is restored by the NA^+-K^+ pump, which moves Na^+ out of the cell and K^+ in.

In this manner, the muscle fiber membrane repolarizes quickly and is able to generate as many action potentials as it receives from the motor nerve. (Recall from our earlier discussion that if large numbers of ions needed to be moved back and forth across the membrane, the cell would become refractory for long periods of time and be unable to generate action potentials in quick succession.) Thus, the original signal, whether it originated from a sensory organ in the muscle itself or from the motor cortex in the brain, is transmitted in its entirety to its final destination. We need to note here that signaling between the motor axon and the muscle fiber is only excitatory, unlike signaling between nerves, which can either be excitatory or inhibitory.

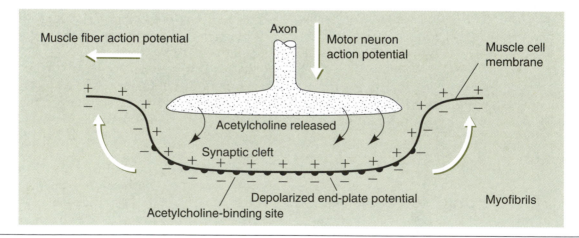

Figure 6.14 Neuromuscular junction. Action potentials arriving at the motor nerve axon terminal trigger the release of acetylcholine, which diffuses across the synaptic cleft and generates an end-plate potential in the muscle fiber. The resulting muscle fiber action potential is propagated the length of the fiber.
Adapted from Vander et al. 1994.

Neuromuscular Junction Fatigue

It is worthwhile at this point to touch on a couple of mechanisms associated with repeated muscle activation that "feed back" into the area of the NMJ and alter the responsiveness of the muscle. Anything that occurs to alter the sequence of events between excitation of the muscle and force production is manifested as muscle fatigue. This is a condition in which the force production for a given neural input is lower than in the nonfatigued state. Of course a continuous input to the axon terminal will eventually deplete the stores of ACh; but this rarely, if ever, occurs during volitional muscle recruitment. Two substances associated with repeated muscle activation, however, interfere with transmission of the action potential from the presynaptic terminal to the motor end plate: lactic acid and adenosine.

These two substances represent the first of many checks and balances that the body uses to reduce the amount of force that a muscle is able to generate under various conditions. This makes perfect sense, since most of us need these internal checks to stop us from injuring ourselves. Unabated, continuous high-force contractions would quickly injure involved structures such as tendon, ligament, bone, and even the muscle itself.

Lactic Acid

The first of the two substances, lactic acid, is produced and accumulates within the muscle under conditions of high-intensity exercise. Actually, it is the H^+ that interferes with the normal function of the NMJ. Lactic acid is a fairly labile compound and is easily dissociated into lactate and H^+ (see chapter 4). As more and more H^+ is produced and is not effectively buffered, it leaks out of the muscle cell and competes with ACh for its binding site on the motor end plate. Eventually, not enough Na^+ and K^+ channels are opened to generate an end-plate potential of sufficient magnitude to trigger an action potential. Under this condition, then, the number of action potentials generated in the muscle fiber is less than the number delivered by the motor nerve, and the result is a lower force output than expected.

Adenosine

The second substance that affects NMJ function is adenosine. To refresh your memory, adenosine is the nucleoside base of ATP and is the end result of ATP breakdown (refer to chapter 4). Adenosine concentration builds up in the muscle over a longer time course than H^+, so its effect is associated with prolonged exercise. Adenosine interferes with the release of ACh from the presynaptic terminal. Again, the effect is to decrease the response of the muscle to a given neural input.

Here are two mechanisms, then, whereby substances produced by repeated muscle contractions provide feedback to the NMJ about the condition or state of the muscle. As the concentration of these substances increases, activation of the muscle is compromised. These are only two of many mechanisms that protect the muscle from overwork or produce muscle fatigue. Chapter 8 covers these other mechanisms.

Characteristics of Motor Units

The precise control of movement or gradations of force production are dependent on specific patterns of excitation and/or inhibition. Each motor unit is distinct from the others in a muscle, largely due to differences in contractile properties and oxidative or glycolytic potential. In essence, each motor unit is constructed to perform a specific task.

Motor Unit Heterogeneity

We often think of motor units (or single muscles) as fast or slow (according to contractile characteristics) or as oxidative or glycolytic (according to metabolic properties). On the basis of these properties in combination, motor units are commonly divided into three (or four) categories:

- Type I or slow oxidative
- Type IIA or fast oxidative, glycolytic
- Type IIB or fast glycolytic

Recently a fourth type of motor unit, type IID, has been characterized; it is intermediate in properties between types IIA and IIB. But placing motor units into three or four discrete categories is a gross oversimplification that does not accurately convey the complexity of motor unit function. In humans, with very few exceptions, skeletal muscles contain many kinds of motor units, all with different characteristics. For example, a muscle contains not just one slow oxidative motor unit but many, all with different characteristics. Thus, according to Pette and Vrbova, motor units form a functional continuum, rather than discrete entities, within each muscle. This **motor**

unit heterogeneity enables single muscles or muscle groups to perform a wide variety of tasks that require wide ranges in force production, speed of contraction, and time to fatigue.

In humans, with very few exceptions, skeletal muscles contain many kinds of motor units, all with different characteristics.

Motor units themselves are homogeneous in that all muscle fibers innervated by a single motor nerve have the same characteristics. That is, diameter, enzyme activity, contractile properties, and so on are identical in fibers that belong to the same motor unit. A motor unit can have very few muscle fibers associated with it or many hundreds of fibers. Muscles that need to exert fine control over a movement (e.g., ocular muscles) have a low innervation ratio, whereas muscles that are more important for gross movement skills (such as the gastrocnemius) have a high innervation ratio. Regardless of the number of fibers associated with a single motor neuron, all fibers within a motor unit receive the same stimulus when the motor neuron fires, causing the fibers to contract as a unit.

Motor units are homogeneous in that all muscle fibers innervated by a single motor nerve have the same characteristics.

Although muscle fibers of the same unit tend to occupy the same region of a particular muscle, the fibers are not contiguous; that is, the fibers of one unit are interspersed with fibers from other units. Thus, a fiber from one motor unit is surrounded by fibers from other motor units; the surrounding fibers have different contractile characteristics and various diameters. This arrangement of motor units gives human muscle fibers, in cross section, a mosaic appearance (figure 6.15). If muscle fibers from a single motor unit are separated from one another within this mosaic, how can they have identical structural, contractile, and metabolic characteristics? The answer lies in the type of neural stimulation that each fiber receives.

Muscle–Nerve Interaction

The motor nerve has the potential to influence structural, metabolic, and contractile properties within a muscle cell, giving a fiber its particular

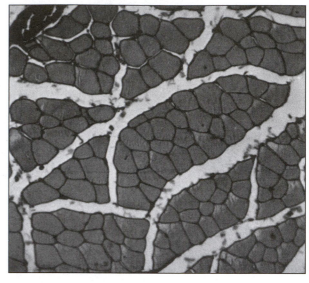

Dept. of PTES, University at Buffalo, SUNY

Figure 6.15   Motor unit heterogeneity. This cross section of normal muscle shows the diversity of muscle fibers. Note the different diameters of fibers. The large-diameter fibers are fast glycolytic (high force, low oxidative potential); the smaller fibers are slow oxidative (low force, high oxidative capacity). But many fibers of other units lie between these two extremes, having various capacities for force development and oxidative power. This motor unit heterogeneity gives muscles the ability to perform a wide variety of tasks.

characteristics in two ways. It has been proposed that (a) the nerve (through the presynaptic terminal) can supply the fiber with trophic factors that regulate cell differentiation or (b) muscle cell characteristics can be determined by the pattern of nerve stimulation. Several investigators, including Pette and Vrbova, have shown experimentally that fast muscles can be converted to slow muscles, and vice versa, by cross-innervation. That is, a slow muscle will adopt characteristics of a fast muscle over time when a fast nerve is attached to it. Remember that different motor neurons have different membrane resistances, among other characteristics, that influence the rate at which they generate action potentials. Lomo et al. demonstrated that it is not the nerve attachment per se that causes the conversion from one type to another; rather the pattern of nerve activity imposed by the nerve is the most important determinant of the specialization of the fiber. These investigators showed that the same changes in fiber type occurred whether the muscle was stimulated directly or through the motor nerve. It appears, then, that nerve stimulation patterns rather than trophic factors released from the motor nerve play the major role in determining fiber type.

Fiber specialization occurs early in life, beginning in the fetal stage even before nerve attachment. Once the nerve does attach in the early stages of muscle development, the pattern of nerve impulse activity, aided by hormones such as thyroid hormone, directs the differentiation process. Fiber-type distribution within the various muscles of each person, then, is established early in life. The question then becomes, are we able to convert fiber types from fast to slow and vice versa through specific types of training? As we will see, this question is not easy to answer.

Motor Unit Plasticity

As reported by Donovan and Faulkner, the plasticity of skeletal muscle, or the ability to adapt to specific stimuli, is demonstrated by transplantation experiments in which muscles, transferred from one site in the body to another and attached to the nerve in the recipient site, eventually adopt the characteristics of the muscle originally in that site. For example, as demonstrated by Freilinger, the condition known as Bell's palsy (unilateral facial palsy) can be corrected by transferring a muscle from another site in the body, like the gracilis muscle of the leg, to replace atrophied and dysfunctional facial muscles. When teased apart into different functional units and attached to the facial nerve, the gracilis muscle will adapt to its new site and eventually function as a muscle indigenous to that site. This is further evidence that the motor nerve directs the expression of a muscle fiber's phenotype.

These observations provide valuable insight into fiber-type conversion that results from physical conditioning. Are we able now to answer the question regarding the ability to transform motor units from slow to fast or fast to slow with training? When we examine all of the evidence presented here, it is apparent that

- muscles are plastic and can express fast or slow, or oxidative or glycolytic phenotypes;
- muscle phenotype is regulated by pattern of neural stimulation; and
- characteristics of the motor nerve are related to such properties as cell size and axon diameter.

It seems that fiber characteristics of a motor unit can in fact be altered through changes in the pattern of activity, that is, by the way the motor unit is recruited and the number of times per day or week it is recruited—but, to truly transform a muscle from one type to another, we would need to alter the characteristics of the nerve. Nobody has yet documented unequivocally that this occurs with training. Views differ widely on whether one can change the fiber-type makeup of skeletal muscle through training.

Patterns of Motor Unit Recruitment

We have already discussed the concepts underlying the size principle of motor unit recruitment. That is, motor units, in general, are recruited in order during physical activity. The slower motor units, because they have small neuron cell bodies and because they are recruited with the weakest input, are recruited first; the faster motor units, requiring a stronger input, are recruited last. This phenomenon is often referred to as the Henneman principle of motor unit recruitment after the first investigator to describe it.

> Slower motor units are used for tasks requiring low force production and are recruited first; faster motor units, needed for high forces, are recruited last.

Fiber Recruitment and Specificity of Training

Motor neurons that innervate a particular muscle are grouped together in a motor neuron pool. Within the pool, neurons form a continuum of cell body sizes as previously described. During a particular movement, motor units are recruited as needed. Low-force contractions are accomplished with slow motor units; as more and more force is needed, the faster units (whose fibers have larger diameters) are activated. This information helps us grasp the principle of specificity of training. Because slow motor units are designed specifically for low-intensity, long-duration activities (the fibers have small diameters and are highly oxidative), one would expect a primarily low-intensity, long-duration training regimen to produce adaptations in those motor units recruited during this type of activity. This of course is true; only those fibers used to perform the activity will receive the necessary stimulus for adaptation to occur. On the other hand, fibers of fast motor units have a larger diameter and use primarily non-oxidative sources for energy. Consequently, fast motor units are used primarily for activities that require a high force production over a short period, or for movements that are very powerful (that combine speed and force). Again, one would expect that the motor units used primarily to produce force during these types of activity (fast motor units) would be the ones that adapt. Does this mean that slow motor units are derecruited (or deactivated) during movements that require high power outputs? No. As shown by Faulkner et al., the slow motor units are still being activated; but because of the small cross-sectional area of the fibers, they contribute little to the movement. It is useful to note here that fast motor units generally have more muscle fibers per unit than slow motor units, so greater increments in force are achieved as fast motor units are recruited. These key principles of motor unit recruitment patterns are illustrated in figure 6.16.

Preferential Recruitment of Fast Fibers

Although the Henneman principle explains motor unit recruitment for most types of movements, some evidence suggests that under certain cir-

cumstances the orderly recruitment of motor neurons does not occur. Various investigators have indicated that under conditions of explosive powerful movements (such as sprinting), fast motor units may be preferentially recruited. At any rate, even if the slow motor units are in fact recruited for explosive powerful movements, they probably contribute little if anything to the overall movement because of the short time the movement lasts. In addition, a painful stimulus, such as touching a hot object, initiates a reflex withdrawal that bypasses the slow motor units and selectively recruits fast units.

Fast motor units may be preferentially recruited under conditions of explosive powerful movements (such as sprinting).

What Determines Motor Unit Recruitment?

So far we have established that slow motor units are recruited primarily for low-intensity long-lasting activities, such as posture and low-speed locomotion, that are part of everyday living. Fast units are normally used less because they are recruited for higher-force activities, which most people perform infrequently. The classification of motor units as slow or fast can be confusing

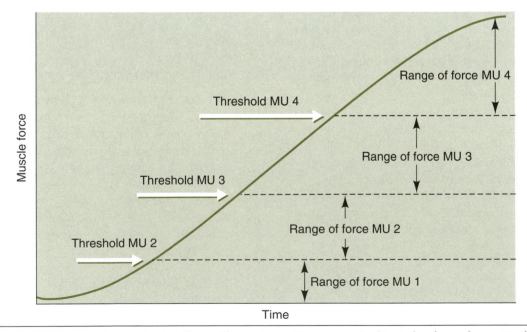

Figure 6.16   Motor unit recruitment. Generally, motor units are recruited in order, from slowest to fastest. In this example, motor unit (MU) 1 has a low threshold and is recruited for tasks that require low force development. As force requirements increase, more units are recruited to meet the requirements. Each unit has a range of forces over which it responds. The force that each unit produces varies as the frequency of stimulation changes.
Adapted from Astrand and Rodahl 1986.

because recruitment of specific types of motor units is governed more by force requirements than by speed of contraction. Skeletal muscle fibers were originally classified as slow or fast on the basis of **twitch** characteristics, that is, the time it takes to reach maximum force and then relax when a single electrical impulse is delivered. But a light load, which requires activation of only slow motor units, can be moved through a range of motion much more quickly than a heavy load. For example, think of yourself pedaling a stationary bicycle. If the resistance you are pedaling against is set very high, you can move the pedal only very slowly with a great deal of effort. Which types of motor units are the primary contributors to this task? The answer, because of the high force needed to move the pedal, is fast. Now, if the resistance is slowly reduced, you can move the pedal faster and faster. Which fibers are being derecruited? Even though the speed you can turn the pedal at is increasing, the motor units dropping out are the faster ones. When the resistance is reduced to a very low level, you can turn the pedal at a very fast rate, but the primary contributors are the slow motor units. Thus, for most types of movements, it is easier to think of the activity in terms of its force requirements (intensity) than in terms of its speed if we want to determine which motor units are the major contributors. Of course, as we saw earlier, in movements that require explosive power, fast motor

units are the primary contributors. Again, a good grasp of these basic principles governing motor unit recruitment is essential for understanding the complexities of human movement—but more importantly, for making informed decisions on conditioning, whether for rehabilitation, athletics, or recreation.

Primary Mechanisms for Increments in Force

Coupled with our discussion of orderly recruitment of motor units, the principle of motor nerve discharge rate (action potential frequency) provides us with two primary mechanisms that increase force production:

- Increased frequency of action potentials in a single motor unit
- Recruitment of additional motor units

Not every muscle uses the same strategy to increase force, however. Smaller muscles (e.g., muscles of the hand) tend to increase force by both strategies up to approximately 50% of maximum voluntary contraction (MVC); then, any further increases in force occur because of increases in discharge rate. Larger muscles (e.g., biceps brachii) use both strategies to increase force up to approximately 85% of MVC; then, any further increase in force occurs because of increased discharge rate only.

What You Need to Know From Chapter 6

Key Terms

acetylcholine (ACh)

acetylcholinesterase

action potentials

active transport

afterhyperpolarization

alpha motor neuron

asynchronously

autonomic nervous system (ANS)

axon hillock

basal ganglia

cable properties

central nervous system (CNS)

cerebellum

cerebral cortex

cerebrum

coactivated

concentric contraction

convergence

dendrites

depolarized

diencephalon

discharge rate

disinhibition

(continued)

divergence

dorsal horn

effector

electrical potential

end-plate potential

epineurial sheath

equilibrium potential

excitatory postsynaptic potential (EPSP)

extrafusal

fiber type

fused tetanus

gamma motor neuron

gating

Golgi tendon organ

Henneman principle

hyperpolarized

hypothalamus

inhibitory postsynaptic potential (IPSP)

initial segment

interneurons

intrafusal

kinesthesia

limbic system

monosynaptic

motivational

motor end plate

motor neuron pool

motor unit heterogeneity

motor units

muscle spindles

myelination

myotactic reflex

neuromuscular junction

neurotransmitter

nodes of Ranvier

output signal

parallel processing

parasympathetic nervous system (PNS)

postsynaptic

presynaptic

proprioception

receptor potential

refractory

repolarized

saltatory conduction

sinoatrial (SA) node

Schwann cells

spatial summation

supplementary motor cortex

synaptic cleft

synaptic potential

temporal summation

tetanic force

thalamus

trains

trigger zone

twitch

unmyelinated

ventral horn

Key Concepts

1. You should be able to describe how resting membrane potentials are developed and how action potentials are generated.

2. You should be able to differentiate between receptor potentials and action potentials and understand how an input stimulus to a nerve cell is delivered in its entirety from cell to cell.

3. You should be able to describe the basic organization of the CNS.

4. You should be able to describe the function of the muscle spindle and GTO.

5. You should be able to relate the interaction of the three main functional systems of movement behavior—sensory, motor, and motivational.

6. You should be able to describe the features of a motor unit that distinguish between fast and slow units.

7. You should be able to state how muscle fibers obtain their specific characteristics and should know whether exercise training can change fibers from one type to another.

8. You should be able to name the factors that are most important in determining which type of motor unit will be recruited for a specific task.

Review Questions

1. Describe how the resting membrane potential is developed and maintained in excitable cells.

2. How is an excitable cell depolarized? What is the key feature of the resting membrane potential that enables generation of many action potentials in rapid succession, without compromising the electrical potential across the membrane?

3. Why is the length of afterhyperpolarization important in distinguishing among the properties of fast versus slow motor units?

4. What is the reason for differences in the velocity of nerve impulses in myelinated versus unmyelinated nerve axons?

5. What are the similarities and differences in graded potentials between sensory and motor nerves?

6. Use the knee-jerk reflex as a model to describe how neural information is generated (sensory input), transformed into electrical signals, and transmitted along the neural pathway to eventually trigger a response in the effector cells (muscle fibers). How would the input and output signals change if the stimulus (stretch of the muscle spindle) is increased or decreased in magnitude?

7. How do divergence and convergence of neural information optimize movement behavior?

8. How does parallel processing enable a patient to relearn a movement pattern if a particular neural pathway is destroyed or injured?

9. What is the functional significance of the hierarchical arrangement of the major regions of the CNS?

10. How is the quality of a motor task influenced?

11. What two areas of the brain exert control over the motor system by determining which patterns of movement will be used together and in which sequence muscles will be activated?

12. Describe the roles of muscle spindles and GTOs in proprioception.

13. Describe how the three major functional neural systems—motor, sensory, and motivational—interact to ensure successful completion of a movement.

14. Describe how three of the four components of a motor unit (neuron soma, axon, NMJ) function together to deliver an excitatory stimulus to muscle fibers.

15. What are two substances produced by contracting muscle fibers that inhibit generation of muscle fiber action potentials, and what are the mechanisms by which they do so?

16. What is the most important factor determining the type of fibers within a given motor unit? Can fibers be transformed from one type to another? Why or why not?

17. What is the functional significance of the heterogeneous makeup of motor units?

18. Your friend Karen is hooked on running marathons, even though her first experience wasn't that positive (see page 79). For her second marathon, Karen is intent on keeping hydrated and replacing her lost energy stores. She decides that eating lots of citrus fruit before her race and drinking orange juice frequently during the race will accomplish both. Unfortunately, the large amounts of citrus products don't agree with Karen. She throws up twice during the race, and once finished she is vomiting and has severe diarrhea. After a few hours she feels OK but has widespread muscle weakness and is lethargic. She can barely walk up a flight of stairs. You immediately recognize Karen's problem and insist that she drink a solution containing small amounts of potassium. A short time later her muscle weakness is gone. What was the cause of Karen's muscle hypotonia (reduction or loss of muscle tone), and why did her symptoms disappear after you made her drink the potassium solution?

19. You have been treating a young cerebral palsy patient for several years. She experiences severe, painful muscle contractures that are getting progressively worse, making it difficult for her to remain mobile. Her physician recommends dorsal rhizotomy,

(continued)

whereby the spinal nerve roots are transected to relieve pain and contractures. The child's parents are asking you to explain the rationale behind this procedure. You should be able to describe both the anatomical and the physiological basis for relief of contractures after a dorsal rhizotomy.

20.   One of the feedback inputs to the respiratory control centers is from joint receptors that sense changes in muscle force produced by muscles around that joint. The greater the muscle force, the greater the stimulus to increase ventilation in order to increase O_2 supply to the working muscles. A carpenter has observed that if he uses predominately his arms to lift a load, instead of predominately the legs, his ventilation is significantly higher. How can you explain this?

Bibliography

Burke RE, Levine DN, Tsairis P, and Zajac FE. Physiological types and histochemical profiles in motor units of the cat gastrocnemius. *J Appl Physiol* 1973, 234:723–748.

Donovan CM and Faulkner JA. Plasticity of skeletal muscle: Regenerating fibers adapt more rapidly than surviving fibers. *J Appl Physiol* 1987, 62:2507–2511.

Enoka RM. Single joint system activation. In *Neuromechanical basis of kinesiology*, 2nd ed., pp. 193–220. Champaign, IL: Human Kinetics, 1994.

Faulkner JA, Claflin DR, and McCully KK. Power output of fast and slow fibers from human skeletal muscles. In *Human muscle power*, ed. NL Jones, N McCartney, and AJ McComas, pp. 81–94. Champaign, IL: Human Kinetics, 1986.

Freilinger GA. A new technique to correct facial paralysis. *Plastic Reconstruct Surg* 1975, 56:44–48.

Ghez C. The control of movement. In *Principles of neural science*, 3rd ed., ed. ER Kandel and JH Schwartz, pp. 534–547. New York: Elsevier, 1991.

Ghez C and Sainburg R. Proprioceptive control of interjoint coordination. *Can J Physiol Pharmacol* 1995, 73:273–284.

Gordon J and Ghez C. Muscles and muscle receptors. In *Principles of neural science*, 3rd ed., ed. ER Kandel and JH Schwartz, pp. 565–580. New York: Elsevier, 1991.

Henneman E, Clamann HP, Gillies JD, and Skinner RD. Rank order of motoneurons within a pool: law of combination. *J Neurophysiol* 1974, 37:1338–1349.

Kandel ER. Nerve cells and behavior. In *Principles of neural science*, 3rd ed., ed. ER Kandel and JH Schwartz, pp. 18–32. New York: Elsevier, 1991.

Kelly JP and Dodd J. Anatomical organization of the nervous system. In *Principles of neural science*, 3rd ed., ed. ER Kandel and JH Schwartz, pp. 273–282. New York: Elsevier, 1991.

Kukulka CG and Clamann HP. Comparison of the recruitment and discharge properties of motor units in human brachial biceps and adductor pollicis during isometric contractions. *Brain Res* 1981, 219:45–55.

Larsson L, Edstrom L, Lindegren B, Gorza L and Schiaffino S. MHC composition and enzyme-histochemical and physiological properties of a novel fast-twitch motor unit type. *Am J Physiol* 1991, 261:C93–101.

Lomo T, Westgaard RH, and Engebresten L. Different stimulation patterns affect contractile properties of denervated rat soleus muscles. In *Plasticity of muscle*, ed. D Pette, pp. 297–309. Berlin, W. de Gruyter: 1980.

McComas AJ. Motor unit estimation: Methods, results and present status. *Muscle Nerve* 1991, 14:585–597.

Monster AW, Chan NC, and O'Connor D. Activity patterns of human skeletal muscle: Relation to muscle fiber type composition. *Science* 1978, 200:314–317.

Pette D and Vrbova G. Invited review: Neural control of phenotypic expression in mammalian muscle fibers. *Muscle Nerve* 1985, 8:676–689.

Pette D and Vrbova G. Adaptation of mammalian skeletal muscle to chronic electrical stimulation. *Rev Physiol Biochem and Pharmacol* 1992, 120:115–202.

Proske U. The mammalian muscle spindle. *News Physiol Sci* 1997, 12:37–42.

Stuart DG and Enoka RM. Motoneurons, motor units and the size principle. In *The clinical neurosciences,* ed. RN Rosenberg, pp. V471–V517. New York: Churchill Livingstone, 1983.

Teasdale N, Forget R, Bard C, Paillard G, Fleury M, and Lamarre Y. The role of proprioceptive information for the production of isometric forces and for handwriting tasks. *Acta Psychol* 1993, 82:179–191.

Zernicke RF and Smith JL. Biomechanical insights into neural control of movement. In *Handbook of physiology—exercise: Regulation and integration of multiple systems,* ed. LB Rowel and JT Shepard, pp. 293–330. New York: Oxford University Press, 1996.

Neuropathies of the Nervous System

This chapter will discuss

➤ the more common neuropathies of the nervous sytem (including those caused by infections and inflammation of the central nervous system, degenerative diseases of the central nervous system, vascular lesions, and diseases of the peripheral nervous system); and

➤ the etiology of each disease, along with functional consequences and implications for rehabilitation.

Common Neuroanatomical Terms

acetylcholine (ACh)—A chemical that acts as a neurotransmitter. Nerve signals result in the release of ACh, which then results in stimulation of more nerves or muscles.

alpha motor neuron—Neuron with its cell body in the anterior horn of the spinal cord and its axon that extends to the skeletal muscle cell.

anterior horn cells (alpha motor neurons)—Cells with their cell bodies in the anterior horn of the spinal cord.

axon—Limb of the nerve cell that conducts impulses away from the cell body.

basal ganglia—Group of nerve cells (gray matter) at the base of the cerebral hemisphere.

brainstem—"Stem" connecting the upper part of the brain with the spinal column. Consists of the midbrain, the pons, and the medulla oblongata.

caudate—One of the two primary circuits of the basal ganglia. The other is the putamen.

cerebral cortex—Gray matter covering the surface of the cerebral hemisphere.

cerebrospinal fluid—Fluid circulating throughout the spinal column and brain.

corticobulbar fibers—Cortical fibers projecting to motor cranial nerve cells, into the reticular formation or sensory cells.

corticospinal fibers—Cortical fibers projecting through the pyramidal tract to the lower motor neurons (anterior horn cells) in the spinal cord.

cranial nerve—Twelve pairs of nerves originating in the cerebrum and brainstem and projecting to the body to control sensation and movement.

(continued)

fusimotor system—Gamma motor neurons that innervate the intrafusal fibers of the muscle.

gamma-aminobutyric acid (GABA)—A neurotransmitter.

glial cells—Supporting cells of the nervous system, including Schwann cells.

hydrocephalus—Excessive fluid in the cerebral ventricles.

meninges—Fibrous coverings surrounding the brain and the spinal cord.

oligodendrocytes—Cells that produce myelin.

putamen—One of the two primary circuits of the basal ganglia. The other is the caudate.

pyramidal cells—Corticospinal cells.

subarachnoid—The space between the dura mater and the middle meningeal layer of the brain.

substantia nigra—A nucleus within the basal ganglia.

terminal bouton—The ending of the axon.

thalamocortical pathway—Efferent connections of the thalamus and the cerebral cortex.

Neuropathies of the motor system can develop from infection, inflammation, degeneration, ischemia, or trauma of the central or peripheral nervous system. In some disease states the damage to the nervous system is diffuse, impacting all functions of the brain. In other conditions the disease is localized to a specific region of the brain, and impairments (signs and symptoms) are associated with specific dysfunction of that region of the brain (see Common Neuroanatomical Terms on page 105). A common impairment associated with diffuse or localized disorders of the nervous system is weakness, which is the inability to generate normal levels of muscle force. In disorders of the nervous system, weakness may be related to a low rate of motor unit firing or the inability of the motor units to fire. Lack of motor unit activity may result from any of the following conditions:

- Cortical lesions with inadequate corticospinal activation
- Spinal cord lesions with interruption of descending corticospinal impulses
- Disruption of the impulses sent by alpha motor neurons to muscle fibers
- Synaptic dysfunction (e.g., neuromuscular junction)
- Damage within muscle tissue

A common impairment associated with diffuse or localized disorders of the nervous system is weakness, which is the inability to generate normal levels of muscle force.

Muscle weakness also occurs with deconditioning, a process by which the body loses strength and cardiovascular endurance as a result of long-term disuse associated with illness, hospitalization, or chronic neurologic conditions (see chapter 9).

This chapter reviews commonly occurring diseases and disorders of the central and peripheral nervous system. We emphasize the etiology of the disorder, pathophysiologic changes, impairments and disabilities, and impact on exercise capacity in a person with neurologic dysfunction.

Infections and Inflammation of the Central Nervous System

Infections of the central nervous system (CNS) from bacterial or viral agents can involve the white matter (myelinated axons) or the gray matter (primarily cell bodies of the neurons). Unlike other diseases of the nervous system, infections can involve a large region of the brain or spinal cord, resulting in numerous impairments (signs and symptoms). Some infections can be terminal; others result in a full recovery. This section reviews two common infections of the CNS, **meningitis** and **encephalitis** (see table 7.1).

Meningitis

Meningitis is caused by a virus or bacterium that originates not in the meninges but from an infection located elsewhere in the body. As a result of the infection, the meninges become inflamed, with the pia and arachnoid layer becoming con-

Table 7.1   Infections and Inflammation of the Central Nervous System

Disease	Etiology	Symptoms	Limitations	Rehabilitation
Meningitis	Viral or bacterial inflammation of the meninges	• Fever • Headache • Neck stiffness and pain	• Pain with movement • Deafness • Palsy • Progression to seizures and coma possible	• Overcome deconditioning • Resolve residual neural deficits
Encephalitis	Direct viral invasion of central nervous system neurons	• Headache • Nausea • Vomiting	• Paresis • Paralysis • Ataxia • Aphasia	• Overcome deconditioning • Treat neurological dysfunction

gested and opaque. The inflammatory process of the meninges can lead to scarring and compression of cortical veins, resulting in thrombosis. Blockage of cerebrospinal fluid may also occur, causing hydrocephalus or a subarachnoid cyst that can compress underlying neural tissue. The impairments seen immediately with meningitis are fever, headache, and stiffness and pain in the neck. Clinically a test for **Kernig's sign** and/or a **Brudzinski's sign** is performed (figure 7.1). With the patient in the supine position and hands cupped behind his head, he attempts to raise his head to his chest and raise the lower limbs, which causes stretching of the meninges. If the meninges are inflamed, the patient will feel pain (Kernig's sign) or will flex the lower limbs (Brudzinski's sign). Meningitis can progress to seizures and coma; the patient presents with focal neurologic signs, including cranial nerve palsies and deafness.

➤ Thrombosis is the formation or presence of a blood clot within a blood vessel.

➤ Palsies are diseases characterized by paralysis.

The impairments seen immediately with meningitis are fever, headache, and stiffness and pain in the neck.

Immediate medical intervention is needed to manage the infection and symptoms. In the acute phase the patient is placed in isolation and on bed rest, with fluid and electrolyte monitoring. With successful management of the acute phase, the

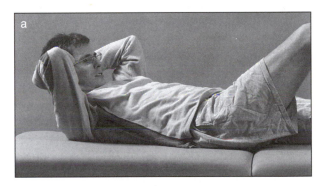

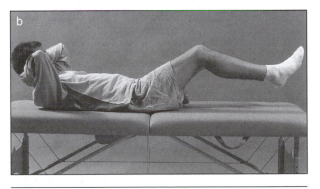

Figure 7.1 Kernig-Brudzinski test for inflamed meninges. When a patient with inflamed meninges places his hands behind his head and attempts to raise his head and unbent leg simultaneously, he will manifest either Kernig's sign (a), that is, pain, or Brudzinski's sign (b), that is, flexing his leg to avoid pain.

prognosis is usually positive with full recovery, especially in adults with viral meningitis. Patients will require rehabilitation to address residual focal neurologic signs and deconditioning resulting from bed rest during the acute phase. Once rehabilitation is complete, people who have had meningitis should be able to resume a normal exercise program with appropriate modifications if permanent motor deficits, such as cranial nerve palsies or balance dysfunction, persist.

Encephalitis

Encephalitis is an acute inflammatory disease of the brain caused by direct viral invasion of the neurons within the CNS. Mosquito or tick bites are the most common form of transmission, as these insects can carry the encephalitis-causing virus. In over 50% of encephalitis cases, the etiology cannot be established. In some patients, encephalitis may be caused by the herpes simplex virus or may be bacterial in origin.

> Encephalitis is an acute inflammatory disease of the brain caused by direct viral invasion of the neurons within the CNS.

Encephalitis is inflammation of and damage to the gray matter of the CNS. Death of neurons in the gray matter produces neurologic dysfunction and can cause cerebral edema as well as damage to the vascular system and meninges. In herpes simplex encephalitis, the temporal lobe is the primary area of inflammation.

Early signs and symptoms of encephalitis depend on the causative agent. Most often the early symptoms include headache, nausea, and vomiting with alterations in consciousness. Since damage occurs to the gray matter, more specific neurologic impairments such as **paresis** (partial paralysis), paralysis, **ataxia** (an inability to coordinate muscles to produce smooth movements), and **aphasia** (impaired ability to communicate) can occur. When paresis or paralysis is present, physiologic changes occur also within the muscle because of lack of corticospinal and alpha motor neuron activity.

> Early signs and symptoms of encephalitis include headache, nausea, and vomiting with alterations in consciousness.

Medical management of encephalitis depends on the infectious agent. Presently antiviral agents are not available to treat encephalitis, with the exception of an agent for treatment of herpes simplex virus. During the acute phase of the illness the patient is monitored closely for symptoms of cerebral edema, which may require surgical decompression. Prognosis for patients with encephalitis depends on the type of infectious agent: some viral agents cause terminal illness; others produce permanent neurologic dysfunction; and in many cases patients experience full recovery.

Degenerative Diseases of the Central Nervous System

Degenerative disease of the CNS is progressive and can affect gray matter, white matter, or both. In some disease states such as Parkinson's disease and multiple sclerosis (MS) the degeneration is a progressive process, occurring over several years with a slow loss of bodily functions. In other disorders such as amyotrophic lateral sclerosis (ALS), the degeneration can occur quickly, within two to five years, and often results in death (see table 7.2).

Amyotrophic Lateral Sclerosis

Amyotrophic lateral sclerosis is the most common adult-onset progressive motor neuron disease of unknown origin. In a small percentage of the cases (less than 10%), there is a familial autosomal trait and the disease is characterized by childhood onset and prolonged survival.

➤ Autosomal means related to a chromosome other than a sex chromosome.

> Amyotrophic lateral sclerosis is the most common adult-onset progressive motor neuron disease of unknown origin.

The pathophysiologic changes in ALS include widespread degeneration of motor neurons located in the cerebral cortex, the brainstem, and the lateral aspect of the spinal cord. Pyramidal cells in the cerebral cortex, which give rise to the corticospinal and corticobulbar tracts, are involved, as are the anterior horn cells (alpha motor neurons) in the spinal cord (see figure 7.2). Since degeneration occurs in the pyramidal cells (upper motor neurons) and anterior horn cells (lower motor neurons), ALS is referred to as a disease of the upper and lower motor neurons with the primary impairment being weakness progressing to paralysis.

Table 7.2   Central Nervous System Degenerative Diseases

Disease	Etiology	Symptoms	Limitations	Rehabilitation
Amyotrophic lateral sclerosis	Unknown	• Progressive weakness starting distally or in brainstem	• Muscle twitching (fasciculations) • Loss of muscle function	• Stretching and strengthening exercise within the patient's limitations
Alzheimer's	Unknown	• Slow decline in memory, visual spatial skills, personality, and cognition	• Inability to comprehend • Loss of memory	• Attempt to maintain independence as long as possible
Huntington's chorea	Hereditary	• Abnormal movement superimposed on normal movement patterns	• Gait abnormalities with other functional limitations	• Maintain functional mobility by strengthening and stretching muscles
Parkinson's	Unkown	• Rigidity • Tremor • Akinesia	• Loss of fine and eventually gross motor function	• Maintain and increase flexibility and strength • Attempt to maintain mobility
Multiple sclerosis	Unknown	• Paresthesia • Diplopia • Optic neuritis • Paresis	• Loss of motor function	• Use flexibility, strengthening, and cardiovascular exercise in a cool environment

ALS is referred to as a disease of the upper and lower motor neurons with the primary impairment being weakness progressing to paralysis.

Weakness in ALS is asymmetrical; it usually starts in distal musculature, with cramping of muscles accompanying active movement. In some cases the disease starts in the brainstem with early symptoms of **dysphasia** (difficulty in swallowing) and **dysarthria** (speech and language dysfunction). As the disease progresses, atrophy of muscles occurs, and there is evidence of **fasciculations** (muscle twitching) as the muscle undergoes denervation. Deep tendon reflexes are altered as a result of inflammation and then degeneration of both upper and lower motor neurons.

At present there is no effective treatment to arrest the progression of ALS. Treatment focuses on relief of symptoms (when possible), maintaining muscle flexibility, monitoring for respiratory problems with intervention as needed (figure 7.3), and physical therapy to address progressive limitation in functional mobility. Death usually occurs within two to five years of onset of symptoms as a consequence of pneumonia or other respiratory complications. Patients with ALS are not candidates for aggressive exercise programs, but they do benefit from stretching and strengthening exercise within their tolerance and according to their level of impairment.

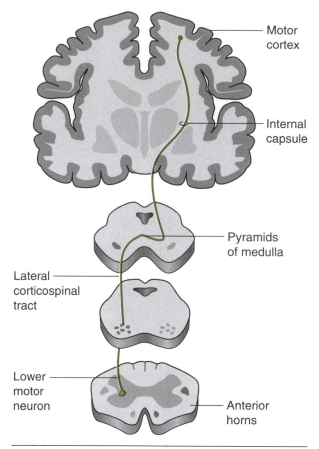

Motor cortex

Internal capsule

Pyramids of medulla

Lateral corticospinal tract

Lower motor neuron

Anterior horns

Figure 7.2 Pyramidal cells and the anterior horn cells.
Reprinted from Martin and Jessell 1992.

Alzheimer's Disease

Alzheimer's disease (AD) is a progressive loss of cognitive and intellectual function, or **dementia**, associated with a slow decline in memory, visual spatial skills, personality, and cognition. The etiology of Alzheimer's is unknown, but the possibility of a genetic predisposition is suspected.

Alzheimer's disease is a progressive loss of cognitive and intellectual function associated with a slow decline in memory, visual spatial skills, personality, and cognition.

The pathophysiologic changes observed in AD are cell death within the cerebral cortex, often involving neurons in the frontal lobe and limbic system. A progressive accumulation of an insoluble fibrous material known as amyloid has been found in these regions, surrounded by fragmented axons, altered glial cells, and cellular debris. Neurofibrillary tangles develop around the pyramidal cells and lead to cell death.

➤ Neurofibrillary tangles are tangled masses of nervous system tissue, including the cell body, dendrites, and axons.

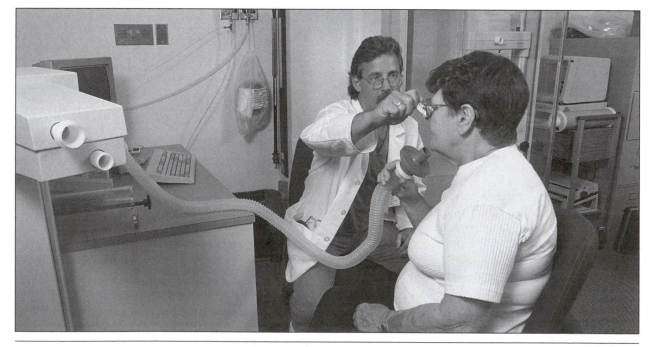

Figure 7.3 Measuring lung capacity. Tests such as this one should be used regularly with patients with ALS. Frequent monitoring for lung problems enables the health professional to provide treatment when it is needed.

Early symptoms of AD are memory loss and subtle personality changes. As the disease progresses, motor dysfunction occurs, and the patient develops weakness and problems with motor planning. With further progression of the disease, the patient loses functional mobility and the ability to perform basic activities of daily living (ADLs).

Alzheimer's disease is the fourth leading cause of death in adults, who have a 7- to 11-year life span after diagnosis. Falling is a frequent problem because of generalized weakness and abnormal movement patterns. Modification of the home environment is needed in order for the patient to maintain independence and safety. Many exercise programs have been developed for patients with AD. Most often patients participate in these programs in small groups and perform activities in a seated position. Emphasis is on automatic reactions, such as tossing or kicking a ball to activate motor unit activity, but the effectiveness of the activities is unknown.

Huntington's Chorea

Huntington's chorea is a hereditary progressive disease characterized by abnormal movement, changes in personality, and dementia. This dis-ease, which often begins in middle age, results from degenerative changes within the basal ganglia, specifically the caudate and putamen. Atrophy of these two nuclei within the basal ganglia contributes to enlargement of the ventricles within the CNS that carry cerebrospinal fluid. The enlargement of the ventricles compresses underlying neuronal tissue and interrupts the normal transmission of impulses. In addition, atrophy of the caudate and putamen alters the balance of neurotransmitters available in the basal ganglia (especially gamma-aminobutyric acid [GABA] and acetylcholine [ACh]) for normal synaptic activity. The depletion of GABA and ACh produces an imbalance with the neurotransmitter dopamine, resulting in excessive excitation of the thalamocortical pathway.

Huntington's chorea is a hereditary progressive disease characterized by abnormal movement, changes in personality, and dementia.

Clinically the patient presents with **chorea**, or abnormal movement superimposed on volitional movement. Coordinated activities become difficult because of the excess movement (figure 7.4). The gait pattern of a patient with Huntington's

Figure 7.4   Huntington's chorea. The standing posture (a) of a patient with Huntingon's chorea is characterized by a wide base of support and choretic movements in the upper extremities. Attempting ambulation (b) without the use of stabilizing devises such as a walker causes the patient to be off-balance due to chorea in the lower and upper extremities.

chorea is characterized as a "prancing-dancing" type of gait. The feet are in a wide base of support, and coordination of stepping movements is impaired, with patients excessively lifting their feet and staggering. In addition to the motor dysfunction and resulting gait deviations, the patient is dysarthric and dysphasic, and may display personality and behavioral changes.

A team approach including medical intervention, social services, and genetic counseling is important in the management of Huntington's chorea. Treatment is symptomatic, with antipsychotic medication administered to block dopamine transmission. Rehabilitation focuses on stability of proximal muscles to diminish choretic movement as well as modification of the patient's functional mobility to maintain independence as long as possible. Huntington's has a slow progression, with death occurring within 15 to 20 years of onset.

Exercise is beneficial in the early stages of the disease to maintain motor unit firing, strength, and muscle flexibility. As the disease progresses the patient becomes more sedentary as a consequence of the inability to control movement, which leads to difficulty with ambulation. In advanced stages of the disease, people must use a wheelchair for mobility.

Parkinson's Disease

Like Huntington's chorea, **Parkinson's disease** is a progressive degenerative disorder of the basal ganglia. Beyond this there is much dissimilarity between the two disease states. Parkinson's disease affects the substantia nigra whereas Huntington's affects the caudate and putamen; Parkinson's disease is a deficiency in dopamine whereas Huntington's is a deficiency in the neurotransmitters GABA and ACh; and Parkinson's disease results in **akinesia** (loss of voluntary movement), rigidity, and tremor whereas Huntington's produces chorea. Huntington's chorea is a hereditary disorder; the etiology for Parkinson's disease is still unknown. Parkinson's most often affects people after 60 years of age, although it has been found in adults at 30 years of age. Some toxic agents or toxicity from medications can cause Parkinson-like syndromes, but these symptoms resolve once the toxicity is cleared.

Parkinson's disease is a progressive degenerative disorder of the basal ganglia.

Parkinson's disease involves the substantia nigra, a nucleus within the basal ganglia, which undergoes degenerative changes resulting in diminished production of dopamine. The deficiency of dopamine causes an imbalance of neurotransmitters, specifically with ACh, which is now in excess in relation to the dopamine available. This imbalance between dopamine and ACh produces excitation of skeletal muscle and the fusimotor system, resulting in rigidity and bradykinesia. Other pathological changes resulting from the dopamine deficiency have been found in the midbrain and within the cerebral cortex.

The classical symptoms of Parkinson's disease are rigidity, tremor, and akinesia or bradykinesia. In addition, patients may develop a masklike face, forward-flexed posture that alters the body's center of gravity, weakness, dementia, and depression. With all of these symptoms, the person with Parkinson's disease can become quickly disabled.

Medical intervention focuses primarily on pharmacologic management of the imbalance in the neurotransmitters. Effective pharmacologic agents include levodopa, Sinemet, and amantadine. In some cases, surgical intervention (pallidotomy or thalamocortical lesion) has also been used with success. In these two procedures, the emphasis is on interrupting the pathways to prevent overexcitation of ACh synapses.

➤ Pallidotomy is surgical destruction of the globus pallidus to reduce involuntary movements.

Parkinson's disease is slow in its progression and does not reduce normal life span. Rehabilitation is effective in increasing and maintaining muscle flexibility and strength, improving postural abnormalities, and promoting safe ambulation and mobility. Patients who have Parkinson's disease are at risk of disuse atrophy and viscoelastic changes within muscle fibers resulting from the rigidity and bradykinesia. Patients with Parkinson's disease should continue in some form of exercise program to prevent these secondary complications.

Multiple Sclerosis

Multiple sclerosis affects people in the prime of their life. Although the etiology of MS is unknown, it is thought to be a combination of a genetic susceptibility and viral infection. The disease can develop in young adults any time after puberty and is two times more prevalent in followed by a period of remission. The period of remission is variable among and within patients and may last months or years. Remyelination can occur if oligodendrocytes (cells that produce myelin) remain in the inflamed area. With each exacerbation and recurrent demyelination of the same axon, scarring occurs, resulting in the "sclerotic plaques" that are the hallmark of MS and that prevent normal transmission along the axon. When this occurs, impairments that had previously been episodic may become permanent. Primary progressive MS is a continual demyelination without periods of remission. Disability in this patient occurs quickly, in contrast to disability in the patient with relapsing remitting MS. Secondary progressive MS occurs when a patient with relapsing remitting MS becomes progressive. Research suggests that this happens in the majority of patients with relapsing remitting MS.

Impairments that accompany MS can be quite variable, depending on the axons that are demyelinated. White matter (myelin) is diffuse throughout the entire nervous system, acting as the insulator on the axon for effective nerve conduction. Anywhere there is myelin there can be inflammation and demyelination. Common impairments seen with MS are **paresthesia** (abnormal sensation), **diplopia** (double vision), **optic neuritis** (inflammation of the optic nerves), **spasticity** (exaggerated muscle tone and reflexes), paresis, and incoordination, all of which can lead to problems with functional mobility and ADLs, as well as to bowel and bladder incontinence.

It was previously thought that patients with MS should not exercise because the increase in activity and generation of body heat would make their symptoms worse, but research has demonstrated just the opposite. People with MS can derive many of the positive benefits of exercise that a nonclinical population experiences. In addition, maintaining joint flexibility, strength, and cardiovascular endurance may enable the person with MS to recover faster after an exacerbation. Muscles can be strengthened in people with MS, and a combination of weight training and aerobic exercise seems to diminish the fatigue associated with the disease. Caution should be taken to maintain a cool environment for exercise, to use cooling collars or cold towels to help control body temperature (figure 7.5), and to have the intensity of exercise progress slowly over several weeks.

Figure 7.5 Exercising with cooling collar. The use of a cooling collar will help people with multiple sclerosis exercise safely, enabling them not only to enjoy the benefits of improved general health, but also to recover from episodes of demyelination more quickly than they would without exercise.

females than in males. Multiple sclerosis is prevalent in the northern latitudes of the world, with the highest incidence in the United States in Seattle (Washington) and Buffalo (New York), suggesting at least some environmental connection.

Multiple sclerosis, an autoimmune reaction, causes **demyelination** (the loss or destruction of myelin) within the CNS. Myelin in the CNS and peripheral nervous system (PNS) acts as an insulator around the axon, enabling quick, uninterrupted transmission of impulses. Damage to myelin impairs the nerve conduction of the axon. New evidence has shown that destruction of the axon also occurs after repeated episodes of demyelination.

> Multiple sclerosis, an autoimmune reaction, causes demyelination within the CNS.

There are different types of MS; the three most common are relapsing remitting, primary progressive, and secondary progressive. In relapsing remitting MS the patient has an exacerbation, or an acute episode of demyelination,

People with MS can derive many of the same positive benefits of exercise that a nonclinical population experiences.

Case Study

Mr. Jones has been active all his life. Though not in a regular exercise routine, he has participated in a variety of recreational sports. He continued to play golf into his 50s, when he began to notice difficulty with the smoothness of his swing and frequent tripping or catching of his feet while walking the fairways. His muscles felt tight, so he started a stretching program that was limited to the hamstrings. His wife observed that he was displaying a tremorlike activity in both hands, more noticeably in the right hand. Mr. Jones continues to work 40 hours per week as an accountant and spends most of his time at his desk. He has seen his primary physician, who referred him to a neurologist. What diagnosis could be the source of Mr. Jones's problems?

Mr. Jones is currently stretching his hamstring muscles. What other muscles should he address in his stretching? What muscles should he strengthen? What types of activities other than stretching and strengthening exercise should he include in his exercise routine?

In the clinical examination you identify rigidity in the biceps, hamstrings, and gastrocnemius. In addition, there is tremor activity in both hands and bradykinesia. How would you explain to Mr. and Mrs. Jones the pathophysiology of Parkinson's disease that produces these symptoms?

The most likely cause of Mr. Jones's symptoms is Parkinson's disease. In addition to the hamstring stretching, which is a good start, Mr. Jones should be given exercises to stretch the gastrocnemius and soleus, the iliopsoas, the quadriceps, the biceps, and all muscles involved in flexion of the trunk. Strengthening exercises should be prescribed for the back and neck extensors, the gluteus maximus, and the anterior tibialis. Because exercise can be important for patients with Parkinson's, Mr. Jones should be encouraged to engage in cardiovascular activities such as exercising on the stationary cycle, walking, exercising on the upper-body ergometer with emphasis on backward motion, throwing/

catching a large gymnasium ball, and even practicing his golf swing.

You should tell the couple that the disease is caused by a deficiency of a substance called dopamine that is involved in transmission of nerve impulses. A deficiency of this substance, which is produced in the brain, results in an imbalance of the neurotransmitters acetylcholine and dopamine.

Vascular Lesions: Stroke

Vascular lesions involving the CNS can strike at any age. Children with congenital aneurysm or arteriovenous malformation (AVM) may develop hemorrhagic stroke. In many instances an aneurysm or AVM may go undetected through adulthood and may not produce any negative consequences. Adolescents or adults using illegal drugs such as crack cocaine can suffer hypertensive stroke. The good news is that the majority of strokes can be prevented. Exercise, diet, management of hypertension or diabetes as prescribed by a physician, avoidance of smoking or use of illegal drugs, and moderate intake of alcohol all contribute to decreased incidence of stroke.

Types and Causes of Strokes

Stroke is the third leading cause of death in the United States and the leading cause of disability. Stroke, or cerebrovascular accident (CVA), occurs when a blood clot (thrombus or embolus) occludes an artery (causing ischemia) or when an arterial wall in the brain ruptures (resulting in hemorrhage) (see table 7.3 and figure 7.6). The most common strokes result from a thrombus that builds up on the intimal wall of the artery. As the clot grows larger, blood supply through the artery becomes diminished and transient ischemic attacks (TIAs) may occur. Transient ischemic attacks are temporary periods of neurologic symptoms resulting from decreased blood supply to the brain. The outcome of a TIA is full resolution of symptoms; however, a TIA serves as a strong warning sign of potential for stroke.

➤ Intimal refers to the innermost part of the vascular wall.

Stroke is the third leading cause of death in the United States and the leading cause of disability.

Table 7.3 Vascular Lesions

Disease	Etiology	Symptoms	Limitations	Rehabilitation
Stroke	Blockage of a cerebral artery or rupture of a vessel	• Dependent on the part of the brain that is affected	• Can range from paralysis, aphasia, sensory loss to loss of very specific motor and cognitive function	• Intensive therapy for early rehabilitation • Deconditioning
Transient ischemic attack	Temporary blockage of cerebral vessel	• Dependent on the part of the brain that is affected	• Can include temporary weakness, numbness, coordination deficits or speech loss	• Generally spontaneous recovery

Embolic strokes occur when a clot formed elsewhere in the body travels to an artery in the brain that it is unable to pass through, resulting in a blockage. A common source of emboli is the heart; these emboli can result from atrial fibrillation. Risk factors associated with stroke are smoking, lack of exercise, high-fat diet, hypertension, and increasing age. Many risk factors such as diet, exercise, and smoking can be modified to decrease the risk. Hemorrhagic stroke results in bleeding into the parenchyma or subarachnoid or subdural space. A hemorrhagic stroke can be due to an anomaly in a vessel (aneurysm or arterial venous malformation), hypertension, side effects of thrombolytic therapy (pharmacologic agents to dissolve clots), or use of cocaine or amphetamines.

The artery most commonly involved in ischemic stroke is the middle cerebral artery (MCA). The MCA, a large artery branching off of the internal carotid, supplies a major region of the cerebral cortex. Occlusion of the MCA can result in paralysis or paresis of the arm more than the leg, **hemianesthesia** (one-sided sensory loss), **homonymous hemianopsia** (loss of one half of the vision in one or both eyes), cognitive limitations, and aphasia (if the left hemisphere is involved). Any combination of impairments can result in disabilities of ADLs and functional mobility.

Results and Treatment of Strokes

Stroke frequently results in loss of cortical activation of alpha motor neurons, producing paralysis or paresis. People with stroke display atrophy in muscle units on the paretic side. Remaining

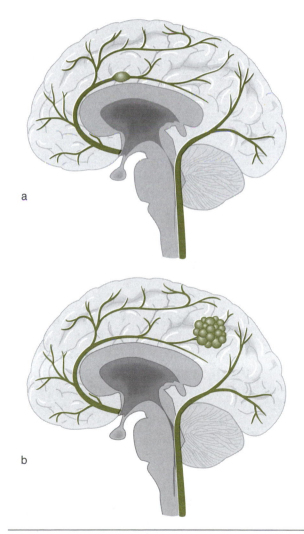

a

b

Figure 7.6 Stroke. Causes of stroke are (*a*) narrowing of the vessel or blockage of a cerebral vessel by a mass of undissolved material (such as a blood clot or fat globules); and (*b*) cerebral hemorrhage.
Reprinted from Wilmore and Costill 1994.

motor units require more time to contract, and they fatigue more rapidly. Weakness in this case may result from altered recruitment and decreased motor unit firing.

Passive and active range of motion (ROM), when appropriate, are critical to maintaining muscle flexibility and activation of motor units during the acute phase of stroke. Patients who experience flaccidity (loss of muscle tone) are at greater risk for atrophy of motor units. Rehabilitation focused on retraining muscles in functional positions for weight bearing or movement is critical. This treatment approach facilitates reactivation of previously existing cortical and subcortical motor programs, as well as generation of action potentials along the neuroaxis for stimulation of motor units and muscle activity.

Exercise should focus on isometric contractions first and then eccentric contractions to increase force generation in the muscle without activating unwanted synergy patterns. Synergies are often seen when a patient with stroke attempts a concentric contraction with muscles that are too weak to complete the task. Overactivation of neighboring alpha motor neurons by the cortex or spinal cord produces unwanted contraction of synergistic muscles. Care should be taken to instruct the patient in exercises that produce normal movement patterns and avoid synergistic activity. Long-term management of the patient with stroke should include participation in exercise for cardiovascular endurance and strength training.

Peripheral Nervous System

The motor component of the PNS consists of the alpha motor neuron, whose cell body is located in the anterior horn of the spinal cord, and its long axon, which leaves the spinal cord and innervates motor units within the muscle. Common disorders in the PNS are due to inflammation or trauma (see table 7.4). Inflammatory conditions can result from exposure to toxic material, metabolic deficiencies, or an antigen response to viral agents. This section deals with one of the most commonly occurring inflammatory lesions, Guillian-Barré.

Guillian–Barré

The counterpart of MS in the PNS is **Guillian-Barré disease (GB)**, or **acute inflammatory demyelinating polyradiculoneuropathy**. Guillian-Barré, like MS, is an autoimmune-mediated disorder; it usually follows the flu, a vaccination, or an infection. In GB, antibodies attack myelin on the peripheral nerves, resulting in demyelination. This quickly progressing disease produces quadriplegia within two weeks of onset. In some cases the demyelination involves the phrenic nerve, requiring that the patient be placed on a ventilator. Patients

Table 7.4 Diseases of the Peripheral Nervous System

Disease	Etiology	Symptoms	Limitations	Rehabilitation
Guillian-Barré	Autoimmune-mediated disorder following the flu, vaccination, or infection	• Progressive paralysis	• Loss of muscle function, including respiratory failure	• Patients usually have full recovery • Rehabilitation consists of restoring muscular function
Myasthenia gravis	Acquired immuno-logical abnormality	• Specific muscle weakness, oculofacial muscles in particular • Progressive weakness	• Double vision • Weakness and easily fatigued as disease progresses	• Patient maintains muscle function but since neuromuscular transmission is disrupted effectiveness limited

must be monitored closely for respiratory failure and may receive plasmapheresis to cleanse the blood of the attacking antibodies.

➤ Plasmapheresis is removal of whole blood, which is then separated so the plasma can be reinfused. Reinfusion prevents loss of blood volume.

The counterpart of MS in the PNS is Guillian-Barré disease, or acute inflammatory demyelinating polyradiculoneuropathy.

Unlike patients with MS, the majority of patients with GB have full resolution of their symptoms (primarily paralysis) and are able to return to normal activities after extensive rehabilitation. The full recovery is due to the presence in the PNS of Schwann cells, which produce myelin and repair inflamed or damaged axons. Remyelination of peripheral nerves enables normal transmission of action potentials and reactivation of motor units.

Rehabilitation during the acute phase requires passive and active ROM to maintain joint and muscle flexibility. As remyelination takes place, exercise should again focus on isometric contractions progressing to eccentric and then concentric. Functional training such as bed mobility, sitting activities, standing, and eventually ambulation enable recovery of mobility and serve as another means of strengthening muscles in functional postures. Like stroke, GB can cause muscles to become flaccid. But whereas flaccidity in stroke is due to cortical damage and loss of corticospinal and corticobulbar activation of alpha motor neurons, the lack of muscle tone in GB is due to loss of alpha motor neuron transmission to the muscle.

Myasthenia Gravis

The motor end-plate is the connection between neuronal activity and muscle fiber activation. This region depends on an action potential generated along the axon, which triggers the release of a neurotransmitter into the synaptic cleft. The motor end-plate is activated with the bonding of the neurotransmitter ACh on its receptors.

Myasthenia gravis (MG), the most common condition of the motor end-plate, is an autoimmune disorder affecting the neuromuscular junction. Risk factors associated with an increase in

MG are hypothyroidism, thymic tumor, and thyrotoxicosis. In MG, the number of ACh receptors at the neuromuscular junction is reduced, and those receptors that remain are flattened so that their efficiency is decreased (see figure 7.7).

Myasthenia gravis, the most common condition of the motor end-plate, is an autoimmune disorder affecting the neuromuscular junction.

A normal muscle contraction occurs when an action potential is propagated along the axon to the terminal bouton, resulting in release of ACh in the synaptic cleft. The ACh binds with the receptors located on the motor end-plate, activating the action potential along the muscle fiber. This triggers the release of Ca^+. A deficiency of receptors in the motor end-plate for ACh limits the ability of the muscle fiber to undergo repeated contractions. Generation of one maximal contraction is usually successful, but attempts at repeated contractions produce fatigue.

Skeletal muscle is most commonly involved in MG, with patients reporting greater energy and strength in the morning and progressive weakness and fatigability as the day progresses. Current

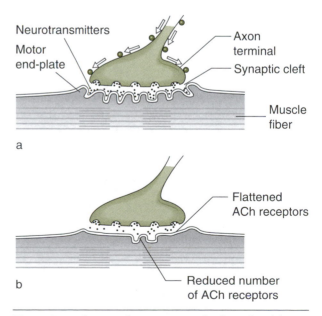

Figure 7.7 A normal neuromuscular junction (*a*) contrasted with a neuromuscular junction that has been affected by myasthenia gravis (*b*). Note the greatly reduced receptor surface available for bonding of ACh.

management of the disease is with medications such as neostigmine, which decreases the presence of acetylcholinesterase in the synaptic junction. This allows more ACh to stay in the synapse for use with repeated contractions. Energy conservation is important in the patient with MG. In addition, maintaining good posture during eating is important because the involvement of the motor endplates of cranial nerves can potentially produce problems with swallowing.

Case Study

Ms. Smith, 32 years old, has had MS for 12 years. She uses her wheelchair for much of her community mobility but can continue to walk short distances in her home with her walker. She feels as if she is continuing to get weaker, although her neurologist states that she is not in an exacerbation. She would like to start an exercise program but is not sure where to begin. She comes to your clinic for short-term rehabilitation and initiation of an exercise program.

What could be the source of Ms. Smith's perceived weakness? In the physical therapy examination of the musculoskeletal system, what would you look for specifically in MS?

You confirm in your examination that Ms. Smith has spasticity in her gastrocnemius and tight iliopsoas bilaterally. In addition, she has weakness in the gluteus maximus, quadriceps, and anterior tibialis muscles bilaterally. What is the pathologic cause or source of the weakness? What type of exercise program would be most beneficial to address the patient's perceived weakness and the spasticity and weakness you identified in the clinical exam, and what considerations should there be regarding the exercise environment?

Deconditioning due to limited mobility and use of the wheelchair is probably the cause of the patient's weakness. Your examination should look for weakness, spasticity, and incoordination. The most likely cause of weakness is demyelination of corticospinal tracts. The exercise program that you prescribe for Ms. Smith must include elements of stretching and of resistance and cardiovascular exercise. The exercise should be supervised and Ms. Smith needs to remain cool; you should consider the use of cooling collars. The exercise should start at a mild to moderate intensity, with short periods consisting of 2 min of aerobic exercise (e.g., stationary bike) followed by 2 min rest, then another 2 min of exercise. Resistance exercise should start at low weight and low repetitions. People with MS can progress to a normal exercise routine, but the exercise sessions should build up slowly over three to four weeks and should include frequent rest periods.

What You Need to Know From Chapter 7

Key Terms

akinesia
Alzheimer's disease (AD)
amyotrophic lateral sclerosis (ALS)
aphasia
ataxia
Brudzinski's sign
cerebrovascular accident (CVA)
chorea
dementia
demyelination

diplopia
dysarthria
dysphasia
encephalitis
fasciculations
Guillian-Barré disease (GB, or acute inflammatory demyelinating polyradiculoneuropathy)
hemianesthesia
homonymous hemianopsia
Huntington's disease

(continued)

Kernig's sign

meningitis

multiple sclerosis

myasthenia gravis (MG)

optic neuritis

paresis

paresthesia

Parkinson's disease

spasticity

stroke

transient ischemic attacks (TIAs)

Key Concepts

1. You should understand the etiology of weakness as it relates to common nervous system disorders.

2. You should understand how common nervous system disorders affect exercise capacity and lead to impairment and disability.

3. On the basis of your understanding of nervous system disorders, you should be able to outline therapeutic approaches to treatment.

Review Questions

1. A 26-year-old professional rugby player missed practice for one week with the flu. Four weeks later he began to develop weakness in his legs, which started distally and progressed proximally. Within a few days he was hospitalized with a diagnosis of acute inflammatory demyelinating polyradiculoneuropathy (GB). Now that he has been medically stabilized, physical therapy treatment has been ordered.

a. What is the focus of physical therapy treatment early on during the patient's acute rehabilitation in the hospital?

b. In some cases of GB, patients must go on a ventilator. Why would ventilatory support be needed?

c. Flaccidity is a common symptom during the acute phase of the disease. Why does flaccidity occur?

d. What cells are present in the PNS that enable remyelination?

e. What is the long-term prognosis for this patient? Will he be able to return to playing professional rugby?

2. Tom is a 16-year-old athlete who was working part time in a butcher shop during the summer. In July he suddenly developed weakness and numbness from his chest down. The weakness prevented him from walking without assistance, and his lower trunk and both legs were completely numb. He was hospitalized and was diagnosed with transverse myelitis, a condition similar to encephalitis but affecting only the spinal cord, not the brain. He is being treated medically with antibiotics administered intravenously and methylprednisolone (steroid to combat the inflammation). He is discharged from the hospital after 10 days, walking with a cane.

a. What region (tract) of the spinal cord is the likely cause for the weakness?

b. Why does Tom experience sensory loss?

c. In this condition would you expect Tom to have loss of deep tendon reflexes?

d. What is Tom's prognosis?

3. Emma is a 52-year-old housewife who works part time at a local retail store. While at work she began to develop problems speaking and was slurring her words. This progressed to include double vision and difficulty swallowing. She denied weakness or numbness on either side of her body and was not experiencing headache. Her coworkers rushed her to the hospital, where she was admitted to the intensive care unit. Emma has a history of diabetes and hypertension.

a. What is a possible cause (diagnosis) for her symptoms?

b. Why would Emma experience problems with speech, swallowing, and vision, but not have weakness or sensory loss?

c. Since weakness is not a primary symptom, would you anticipate any limitations to participation in an exercise program to assist in managing Emma's hypertension once she is discharged from the hospital?

Bibliography

Adams RD and Victor M. *Principles of neurology.* New York: McGraw-Hill, 1985.

Bennett SE and Karnes JL. *Neurological disabilities: Assessment and treatment.* Philadelphia: Lippincott, 1998.

Craik RL. Abnormalities of motor behavior. In *Contemporary management of motor control problems. Proceedings of the II Step Conference.* Alexandria, VA: Foundation for Physical Therapy, 1991.

Fredericks CM and Saladin LK. *Pathophysiology of the motor systems: Principles and clinical presentations.* Philadelphia: Davis, 1996.

Goodman CC and Boissonault WG. *Pathology: Implications for the physical therapist.* Philadelphia: Saunders, 1998.

Kandel ER, Schwartz JH, and Jessell TM. *Principles of neural science.* 4th ed. New York: Elsevier, 2000.

Merritt, HH. *A textbook of neurology.* Philadelphia: Lea & Febiger, 1979.

Miller BL, Chang L, Oropilla G, and Mena I. Alzheimer's disease and fontal lobe dementias. In *Textbook of geriatric neuropsychiatry,* ed. CE Coffey and JL Cummings, pp. 390–400. Washington, D.C.: American Psychiatric Press, 1994.

Rosenfalck A and Andreassen S. Impaired regulation of force and firing pattern of single motor units in patients with spasticity. *J Neurol, Neurosurg, Psychiatry,* 1980, 43:907–916.

Simon RP, Aminoff MJ, and Greenberg DA. *Clinical neurology.* Norwalk, CT: Appleton & Lange, 1989.

Skeletal Muscle Structure and Function

This chapter will discuss

➤ muscle development;

➤ the structure of skeletal muscle and the relationship between structure and function;

➤ the ways in which contractile proteins interact to produce force; and

➤ the mechanical function of muscle fibers and the contractile characteristics of muscles.

Skeletal muscle composes approximately 40% of the total body mass in humans, its diversity of function reflected in a wide variety of shapes, sizes, and architecture. Skeletal muscles are the effectors of movement behavior and are dependent on signals from the somatic nervous system for their precise control. These muscles are prime examples of the intimate relationship between structure and function; they are elegantly designed to produce movement of the body segments, independently or simultaneously, and to generate high forces of brief duration or low-force contractions of long duration. If one examines a longitudinal section of skeletal muscle microscopically, one is struck by the way the smallest functional units, the **sarcomeres**, are organized—each one exactly the same as the next,

and in perfect register. This structural organization allows the small forces produced by each myosin cross-bridge to sum and produce force in the entire muscle as efficiently as possible. Even after repeated contractions, which distort the structural arrangement greatly, the sarcomeres return to their original position and are once again ready to perform.

As we will see in this and subsequent chapters, skeletal muscle is also very plastic; that is, it has a tremendous capability to adapt to a wide range of stimuli. It is common knowledge that skeletal muscle will **hypertrophy** (get larger) in response to resistance training just as it will **atrophy** (get smaller) as a result of disuse or deconditioning. Less well established is whether muscle fibers can be **transformed** from one fiber type to another (e.g., from fast to slow) through training. As pointed out in previous chapters, it is important to understand the basic mechanisms that underlie these qualities of skeletal muscle in order to make informed decisions about fitness assessment and exercise programming, whether for performance enhancement or rehabilitation. Before we examine these adaptive capacities of skeletal muscle, we will briefly explore the origins of skeletal muscle cells and influences on their growth. Then we will look at the macro- and microscopic levels of muscle structure and their relation to function.

Muscle Development

Muscle development is a complex process involving several stages:

• **Blastula stage.** Muscle cells originate from the dorsal **mesoderm** (the middle of the three primary germ layers of the embryo) in the **blastula stage** (an early stage of embryonic development). Cells in the mesoderm form blocks of tissue called **somites** that are adjacent to the neural groove. Under the influence of various growth factors, primitive **myoblasts** in the somites proliferate and join together to form small cylindrical clusters of cells called **myotubes**. The first, or primary, myotubes adhere to one another to form groups of primitive structures (or primordia). A "pioneering" motor nerve axon approaches the cluster of myotubes and innervates each cell in the group. At this time the myotubes separate, and as reported by Kelly and Rubinstein, a secondary generation of myoblasts becomes interposed among the primary cells, using the membranes of the primary cells as a scaffolding to form their own myotubes. The secondary myotubes then separate from the primary cells and eventually become innervated by their own motor axons (figure 8.1). Tertiary myotubes then form on the membranes of secondary myotubes and eventually separate as well. New muscle cells are formed in this manner in the first six to seven months of gestation (Stickland), laying down the framework for motor unit organization—the mosaic pattern of intermingled fiber types typical of adult muscle.

• **Formation of contractile elements.** Early myotubes are limited in their ability to contract. **Contractile elements** (myosin and actin) are laid

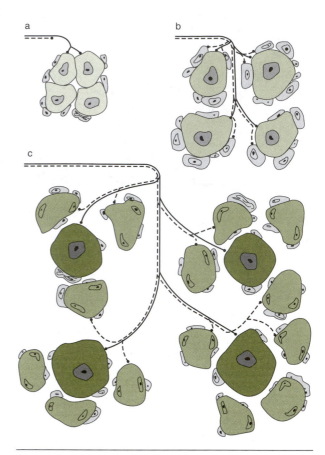

Figure 8.1   Muscle development. Primary myotubes are innervated by a pioneering motor nerve (*a*); they then separate (*b*). Secondary myotubes increase in size and soon become innervated by trailing nerve axons (*c*). Tertiary myotubes form on the membranes of secondary cells and will eventually become innervated and mature. In this manner, motor units form the mosaic appearance characteristic of mature skeletal muscle.
Adapted from Kelly and Rubinstein 1986.

down in the periphery of the myotube first; they then progressively fill in toward the center of the cell. Immature muscle cells can be identified by the central location of their nuclei (each myotube contains a nucleus from each myoblast that has fused). As the muscle cell matures and more sarcomeres form, nuclei migrate to the periphery of the fiber. As cell proliferation slows in the later stages of gestation, most of the increase in cross-sectional area of the muscle results from enlargement of existing fibers by addition of contractile protein. Recall from chapter 6 that motor nerve activity is the key factor in directing the expression of specific phenotypes in these early muscle cells (that is, motor nerve activity determines fiber type). So, fiber differentiation begins early in development—as soon as the correct nerve becomes attached. According to Kelly and Rubinstein, it appears that the primary and early secondary myotubes are predestined to be slow fibers whereas late secondary- and tertiary-generation fibers end up being fast.

> Since motor nerve activity is the key factor in determining fiber type, fiber differentiation begins early in development—as soon as the correct nerve becomes attached.

• **Axial growth.** After birth, muscle cells continue to increase in girth and length. Elongation of muscles (**axial growth**) is accomplished by addition of sarcomeres at the ends of the fibers. Increases in muscle girth are achieved by addition of contractile protein to existing fibers rather than formation of new fibers. That is, increases or decreases in muscle cross-sectional area after birth occur because of increases or decreases in the number of sarcomeres in parallel within each fiber, not because of splitting of fibers or formation of new muscle fibers. This principle becomes important in our upcoming discussions of conditioning and deconditioning of skeletal muscle. It is apparent, then, that fiber-type characterization is predestined to a certain degree and that the fiber-type distribution of a particular muscle is largely established in prenatal and neonatal stages of life.

> Increases or decreases in muscle cross-sectional area after birth occur because of increases or decreases in the number of sarcomeres in parallel within each fiber, not because of splitting of fibers or formation of new muscle fibers.

• **Formation of satellite cells.** Early in the process of **myogenesis** (the formation of new muscle fibers), each myotube becomes surrounded by a **basal lamina** (an amorphous extracellular layer applied to the basal surface of investing muscle cells). As myoblasts continue to divide during maturation of the cell, some get trapped between the basal lamina and the cell membrane. These myoblasts, called **satellite cells**, remain in their primitive, undifferentiated state, even in adult muscle fibers, until they are stimulated to begin to replicate and proliferate. Satellite cells provide the fully differentiated muscle fiber

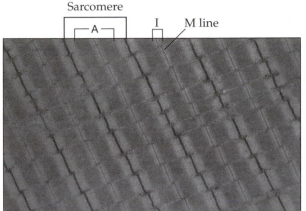

Figure 8.2 Longitudinal section of skeletal muscle illustrating characteristic striated pattern of alternating light and dark bands. A = A band consisting of both myosin and actin filaments. I = I band consisting of just the actin filaments. M = M line which corresponds to the structural proteins that join the myosin filaments together. Notice the perfect transverse register of sarcomeres, which is necessary to add forces of the individual sarcomeres together. The greater the number of sarcomeres in parallel, the greater the force production of the fiber.

with much of its "**plasticity**" because they enable repair of damaged cells (even if the whole cell is destroyed) and play a major role in the hypertrophic process. These processes are examined in chapter 9.

> Satellite cells provide the fully differentiated muscle fiber with much of its "plasticity" because they enable repair of damaged or destroyed cells and play a major role in the hypertrophic process.

Muscle Structure

Humans are capable of a diverse array of movements largely because of wide ranges in muscle size and shape, tendon attachments, and contractile characteristics. Muscle contraction itself is a very simple process involving the energy-dependent interaction of two key proteins, actin and myosin. These two proteins are the major constituents of muscle cells, and it is the very precise arrangement of the thick myosin and thinner actin filaments that produces the alternating dark and light bands from which skeletal muscle gets its characteristic striated appearance (figure 8.2). Let's examine muscle structure in more detail and consider the integral relationship between structure and function.

Macrostructure

Skeletal muscle fibers are cylindrical, ranging in width from 10 to 100 μm and varying widely in length—some are as long as 20 cm (7.8 in.). According to Gans and deVree, fiber architecture is an important determinant of a muscle's force-generating capacity. Many human muscles contain fibers arranged in parallel, or along the longitudinal axis of the muscle. This arrangement enables all or most of the force developed by individual fibers to pull directly on the tendon. Other muscles, especially larger ones, have fibers arranged at an angle in relation to the direction of force generation, an arrangement termed **pennation**. Under these circumstances only a portion of the force developed by individual fibers is transmitted to the tendon, but the advantage is that more fibers can be packed into the same cross-sectional area. Because force production depends on the total cross-sectional area of all fibers, a muscle with the fibers in a pennated arrangement can develop more force than a muscle with fibers arranged in parallel (figure 8.3). A muscle that is bi- or multipennated can generate even more force. Thus, it is the number of fibers packed into a given muscle cross-sectional area that determines overall force-generating capacity.

> It is the number of fibers packed into a given muscle cross-sectional area that determines overall force-generating capacity.

Intramuscular Connective Tissue

As discussed by Borg and Caufield, connective tissue in muscle is divided into three levels of organization providing a framework that binds the muscle together, helping to ensure the proper alignment of fibers. Individual muscle fibers are surrounded by the **endomysium**, and bundles of fibers called fascicles are encircled by **perimysial** connective tissue. The entire muscle is ensheathed by the **epimysium**. The high tensile strength of connective tissue is well suited for transmission of active force during muscle contraction. Slow muscles generally have a higher tensile strength and more stiffness than fast muscle—attributable in part to a greater amount of connective tissue.

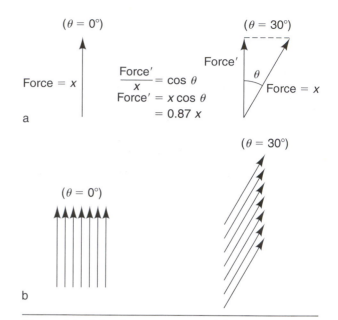

Figure 8.3 Muscle pennation and force production. When muscle fibers are parallel to the long axis of the muscle, all the force is transmitted to the tendon (*a*). When fibers are oriented at an angle, only a portion of the force is transmitted to the tendon (in this example, force is reduced by approximately 13%). The advantage of orientation of fibers at an angle to the long axis is illustrated in (b). More fibers can be accommodated in the same muscle volume, so total force is increased. Adapted from Leiber 1992.

Functionally this is important in facilitating continuous support of the body by slow postural muscles. On the other hand, as reported by Kovanen et al., fast muscles have lower amounts of collagen and greater compliance, which facilitate the efficiency and speed of contraction characteristic of these muscles.

> Because of a greater amount of connective tissue, slow muscles generally have a higher tensile strength and more stiffness than fast muscle. Fast muscles have lower amounts of collagen and greater compliance, which facilitate the efficiency and speed of contraction characteristic of these muscles.

➤ Collagen is the major protein found in connective tissue.

➤ Compliance is a measure of the ease with which a structure can be deformed.

Microstructure

The basic functional unit of skeletal muscle, the sarcomere, is bordered at each end by a rigid, fibrous structure called the **Z line**. **Actin** filaments are anchored in the Z line and provide a solid base from which myosin **cross-bridges** exert their tension. Many of these sarcomeric units are joined together in series to form a cylindrical myofibril (figures 8.4 and 8.5). Each muscle fiber is made up of hundreds to thousands of myofibrils; the greater the number of myofibrils, the greater the cross-sectional area of the fiber and the greater the force-generating capacity. Think of the muscle fiber, with all its myofibrils encircled by the cell membrane (**sarcolemma**), as analogous to a phone cable, made up of numerous thin wires held together by an outer covering. **Myosin** molecules, consisting of long tails and globular heads, are packed together to form the myosin filament. The rod portion of the myosin molecules in each half of the filament points toward the center, with the globular heads protruding. Because of the alignment of the myosin molecules (in an anti-parallel fashion), attachment and swiveling of the myosin heads pull the Z lines toward the middle of the

sarcomere. In this manner, force is generated and the muscle contracts.

The greater the number of myofibrils, the greater the cross-sectional area of the fiber and the greater the force-generating capacity.

Because myosin molecules are aligned in an anti-parallel fashion, attachment and swiveling of the myosin heads pull the Z lines toward the middle of the sarcomere, shortening the muscle.

Cytoskeleton

When you examine a muscle fiber in a longitudinal section (figure 8.2), the perfect alignment of sarcomeres from myofibril to myofibril is striking. Now, imagine the distortion of the sarcomeres that takes place upon a forceful contraction. But when the muscle relaxes, the sarcomeres return to their original position—again in perfect register. This tells us that support structures must be in place to hold the actin and myosin in position. If they were not, the sarcomeres would be in disarray after each contraction, particularly those generating considerable force. Functionally, this alignment of sarcomeres is important because the amount of force generated is proportional to the number of sarcomeres arranged in parallel. If the sarcomeres are not in register, then the force produced by the entire fiber is less than expected given the cross-sectional area of the fiber. For example, muscle fiber damage as a result of forceful contractions (see chapter 9) involves a breakdown of the Z line (called **Z-line "streaming"**). The Z-line damage causes sarcomeres to move out of register. The result is a decrement in force production, even though the contractile elements of the fiber (actin and myosin) may not be damaged. Of course the Z line is not the only structure that provides support and stability for contractile apparatus. In fact, a complex array of **cytoskeletal proteins** not only provides an infrastructure of support within the fiber but also plays a major role in transmission of force.

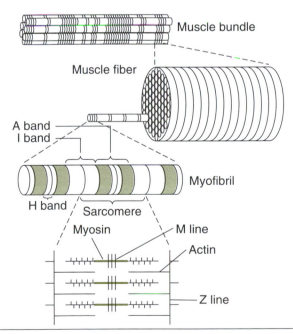

Figure 8.4   Muscle fiber. A muscle fiber is made up of cylindrical myofibrils, which exhibit the characteristic banding pattern attributable to the arrangement of thick myosin and thin actin filaments.
Adapted from Guyton 1991.

Functionally, the alignment of sarcomeres is important because the amount of force generated is proportional to the number of sarcomeres arranged in parallel.

Myosin filaments are linked together in the central region by structural proteins that span the gap between filaments (called the **M line**). Myosin is also stabilized by the protein **titin**, which attaches the end of the filament to the Z line. Patel et al. report that in addition to providing structural stability, titin is important in transmitting force from myosin to the Z line and is also a major determinant of passive tension in skeletal muscle. Other intermediate filaments connect sarcomeres and myofibrils laterally to further stabilize the contractile machinery and also to transmit force radially. The protein **desmin** joins Z lines between adjacent myofibrils, as well as making a connection with a specialized structure in the sarcolemma called the **costamere**. According to Patel et al., costameres form the link between the intracellular and extracellular space, whereby force is transmitted from the sarcomeres through the sarcolemma to the extracellular matrix, then axially to the tendon. Among the many structural proteins that make up costameres is **dystrophin**, a membrane stabilizer that is absent in patients with muscular dystrophy. A very large protein called **integrin** is embedded in the cell membrane and is an integral part of the costamere. The endomysium, described earlier, is also connected to integrin,

providing a link between extracellular connective tissue and the intracellular infrastructure (figure 8.5).

This arrangement of intracellular and extracellular structural proteins provides an intricate meshwork through which force is transmitted radially (laterally) as well as serially (longitudinally). Indeed, as Patel et al. point out on page 343 in their review of force transmission in skeletal muscle, "[I]t is an incorrect oversimplification to assume that muscle fibers generate force . . . only serially to the muscle tendon." This observation—interdependence of intracellular and extracellular structures—gives us insight into the function of skeletal muscle under normal conditions or under circumstances in which generation or absorption of mechanical force is compromised. Muscle function is altered as a result of diseases like muscular dystrophy and under many conditions observed by health professionals, such as immobilization, exercise-related muscle damage, denervation, and other forms of disuse. As we can see from the description of the cellular infrastructure and its intimate contact with extracellular connective tissue, alterations in these structures can have profound effects on muscular activity ranging from athletic performance to activities of daily living.

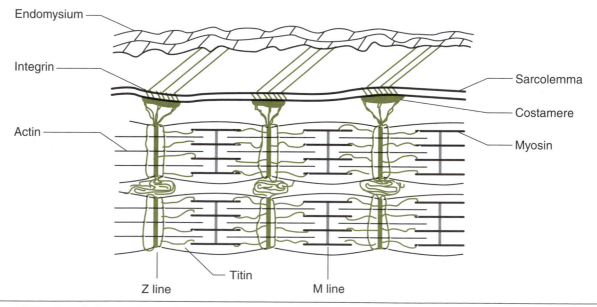

Figure 8.5 Cytoskeleton of muscle fiber. Structural proteins provide an infrastructure to stabilize contractile proteins. Support structures outside the cell (endomysium) are linked to the intracellular infrastructure through integrin.
Adapted from Waterman-Storer 1991.

The interlocking arrangement of intracellular and extracellular structural proteins provides an intricate meshwork through which force is transmitted radially (laterally) as well as serially (longitudinally). Changes in this structure will alter muscle function. Changes will result from diseases like muscular dystrophy as well as from exercise-related muscle damage, denervation, immobilization, and other forms of disuse.

Contractile Function

We have examined the structural components of skeletal muscle and discussed the interdependence between structure and function. Now let's briefly examine how the contractile proteins interact to produce force.

Sliding Filament Mechanism

Swiveling of the globular head of myosin (called the power stroke) moves the actin filament relative to myosin only a very short distance (5-10 nm). So, to shorten a muscle over relatively large distances, myosin heads must attach and detach many times, a process called cross-bridge cycling. The continual cycling of cross-bridges enables actin to slide past myosin—hence the term "sliding filament mechanism." When the muscle fiber is activated, then, each myosin head

1. attaches to actin,
2. swivels to produce movement of the actin relative to the myosin,
3. detaches from actin, and
4. returns to its original position and is ready to reattach (figure 8.6).

Since all cross-bridges cycle independently of one another, it is obvious that not all cross-bridges are attached and are generating force at the same time. In fact, only about 50% of cycling cross-bridges are producing force at any given moment. With myosin heads at different stages of the cross-bridge cycle at different times, a uniform movement of the actin during concentric or shortening contractions is achieved, or uniform force during isometric contractions is produced. Under certain circumstances, a muscle fiber can become potentiated by increasing the number of cross-bridges in the power-stroke phase of the

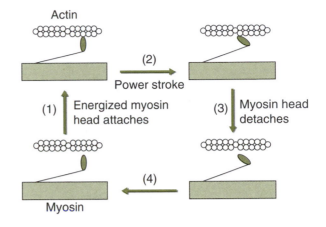

Figure 8.6 Cross-bridge cycling. In the presence of calcium, the active site on actin is uncovered; and the myosin head (1) attaches, (2) swivels, (3) detaches, and (4) returns to its original resting state. If calcium concentration remains high, cycling continues. Many cycles are needed to move actin an appreciable distance.

Adapted from Squire 1990.

cycle. According to Grange and Houston, this mechanism involves phosphorylation of myosin light chains and increases the force that the fiber, or muscle, can generate.

➤ Phosphorylation is the chemical process of adding a phosphate to a compound.

To facilitate cross-bridge cycling, each myosin head contains sites for binding

- adenosine triphosphate (ATP),
- myosin adenosinetriphosphatase (ATPase), and
- actin.

Hydrolysis of one of the three inorganic phosphates (Pi) on ATP (catalyzed by myosin ATPase) provides the energy to tilt the myosin head (ATP $\longrightarrow$ ADP + Pi). Adenosine diphosphate (ADP) and Pi are released from the myosin bridge during the power stroke, and another ATP molecule must attach to myosin to dissociate, or separate, actin and myosin (figure 8.7). Thus, ATP has two crucial roles in the cross-bridge cycle. It provides energy for the power stroke, and it dissociates the myosin head from actin.

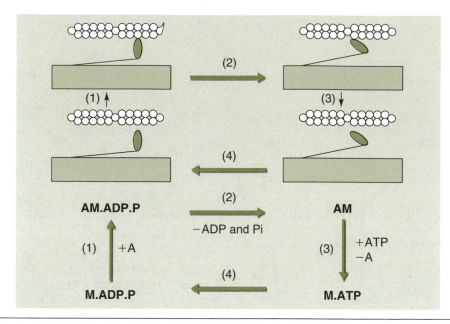

Figure 8.7   Role of adenosine triphosphate in cross-bridge cycle. Adenosine triphosphate is necessary to provide energy for the power stroke (2) and to dissociate the myosin head from actin. Note that energy is transferred to the myosin head before attachment to actin. M = myosin; A = actin; AM = actomyosin complex.
Adapted from Squire 1990.

> ATP has two crucial roles in the cross-bridge cycle: It provides energy for the power stroke, and it dissociates the myosin head from actin.

A common misconception regarding muscle fatigue concerns the role of ATP. People often attribute fatigue to "ATP depletion," but the scenario just described demonstrates that this cannot be true. What would happen to the myosin cross-bridges if ATP were depleted, or even significantly reduced? In the resting state a myosin head has one ATP molecule attached to it, and when the fiber is activated, ATP is hydrolyzed to provide energy for the power stroke. Another ATP molecule must attach to myosin to separate it from actin. If there is little or no ATP available, then the cross-bridge remains attached to actin. Eventually all of the cross-bridges become attached and the muscle enters a state of rigor. You may have guessed that this state exists only once—at the end of our lifetime. Here is a mechanism that can be used to dispel the common myth that ATP is "used up" during exercise, causing "fatigue." In fact, several other mechanisms that do cause fatigue (decrease in force for a given input) are in place to conserve supplies of ATP. The message here is clear: as reported by Green, during any type of exercise, including prolonged exercise to exhaustion, there is no support for the contention that a significant reduction in the muscle concentration of ATP occurs or that ATP plays a major role in muscle fatigue.

> There is no support for the suggestion that muscle concentration of ATP diminishes significantly during any type of exercise or that ATP depletion plays a major role in muscle fatigue.

Excitation–Contraction Coupling

Let's return to the cross-bridge cycle. We have already learned that a single myosin cross-bridge cycles many times during a muscle contraction. When the muscle is relaxed, though, what prevents the myosin from binding to the actin? The state of muscle activation is a function of two major proteins, **troponin** and **tropomyosin**, plus the concentration of calcium ion in the sarcoplasm (cell fluid). Tropomyosin is a rod-shaped molecule that covers the myosin-binding sites on actin, and troponin is a globular-shaped protein that sits on top of tropomyosin (figure 8.8). Calcium binds to troponin and causes it to rotate and pull the tropomyosin off the myosin-binding sites on actin (think of turning the knob [troponin] on a door [tropomyosin] to open it). When the con-

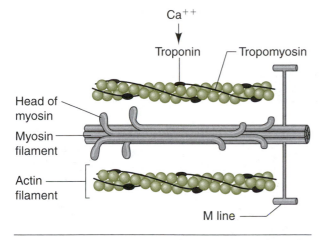

Figure 8.8 Major contractile proteins. When the muscle fiber is relaxed, tropomyosin covers the active sites on actin. Activation of the fiber results in calcium binding to troponin, which rotates and pulls tropomyosin off the active sites and allows the myosin head to bind to actin.

Adapted from McCardle et al. 1990.

centration of calcium in the sarcoplasm is low, tropomyosin prevents myosin from binding to actin, and the muscle fiber is relaxed. Conversely, when Ca^{++} concentration is high, the calcium binds to troponin, and tropomyosin is moved from actin, allowing the myosin head to attach. As long as the active sites are uncovered, myosin heads continue to cycle and attach to any site they can reach. The key to this whole process, then, is Ca^{++} concentration within the sarcoplasm. Calcium, acting as a second messenger here, couples the activation signal (action potential) to the mechanical event (cross-bridge cycling)—a process called excitation-contraction coupling. Where does the Ca^{++} come from? It is released from stores that lie within the muscle cell, in the sarcoplasmic reticulum (SR). To relax the fiber, Ca^{++} is pumped back into the SR via an energy-dependent mechanism at a rate that is largely determined by the specific activity of SR Ca^{++}-activated ATPase (figure 8.9).

▶ The sarcoplasmic reticulum is a network of tubules within the muscle fiber that store and release calcium.

The last thing to consider in this process is the simultaneous delivery of the action potential to the contractile machinery that lies in the center of the fiber, as well as that in the periphery. Recall that the contractile elements are contained within cylindrical myofibrils. To activate a myofibril at the center of the fiber at the same time as one that lies near the outside, we must have a delivery system that spans the entire cross section of the fiber. This simultaneous activation is accomplished by transverse tubules (T tubules). The membrane of these tubular structures is continuous with the sarcolemma; and because T tubules transect the entire fiber, action potentials are speedily transmitted to the interior. The lateral sacs of the SR (containing Ca^{++}) are in close proximity to the T tubules to facilitate Ca^{++} release upon the arrival of an action potential (figure 8.9).

Action potentials are generated at the surface of a muscle fiber in a one-to-one ratio to action potentials delivered from the motor axon. Muscle fiber action potentials travel along the surface of the fiber, but are also delivered to the interior of the fiber by T tubules to cause an almost simultaneous release of Ca^{++} from the SR in all areas of the fiber. Calcium binds to troponin, causing a change in conformation and a subsequent movement of tropomyosin off the myosin-binding sites on actin. Myosin heads, now allowed to interact with actin, begin to cycle (attach and detach). Adenosine triphosphate provides energy to power the movement of the myosin head; then ADP is released and another ATP binds to the myosin head, triggering the separation of myosin and actin. As Ca^{++} concentration drops, fewer myosin heads are able to cycle, and force production declines until a critical level is reached and the muscle relaxes. Calcium is actively pumped back into the SR at a rate dependent on the specific activity of Ca^{++}-activated ATPase. The level of muscle activation, then, is dependent on cytosolic Ca^{++} concentration, which is a function of (a) the number of action potentials arriving per unit time and (b) the rate of Ca^{++} uptake back into the SR.

Action potentials are generated at the surface of a muscle fiber in a one-to-one ratio to action potentials delivered from the motor axon.

Contractile Properties

We now have the fundamentals in place for a better understanding of the mechanical function of muscle fibers (how muscle force is generated)

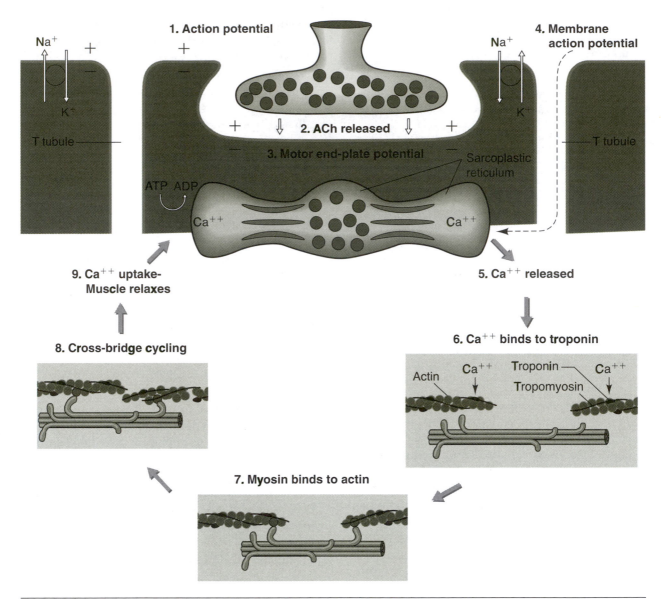

Figure 8.9 Excitation-contraction coupling. Myosin cross-bridge cycling is coupled to the electrical signal through the release of calcium from the sarcoplasmic reticulum (SR). Calcium is removed from the sarcoplasm by active transport (requiring adenosine triphosphate) back into the SR, and the fiber relaxes. Follow this process from the delivery of action potentials by the motor nerve (1), to calcium release from the SR (5) and binding to troponin (6), and finally to the reuptake of calcium by the SR (9) and cessation of cross-bridge cycling.

and the reasons fast muscles are different from slow muscles in their contractile characteristics.

Differences Between Fast and Slow Muscles

The two key functional differences between fast and slow muscles are rate of force development (or contraction time) and rate of relaxation. Remember that during muscle contraction, each myosin head cycles many times to move the actin filament an appreciable distance or to produce a measurable force. The rate of force production, then, is determined by the rate of cross-bridge cycling, which in turn depends largely on the activity of myosin ATPase. That is, the hydrolysis of ATP, which is catalyzed by myosin ATPase, is the major rate-limiting step in cross-bridge cycling. Thus, a fast muscle fiber has rapid-acting myosin ATPase, so the cycling rate is fast, as is the rate of force development. The fastest muscles in humans (i.e., ocular muscles) have a time-to-peak tension of around 10 ms. Slow muscles have a much slower cross-bridge cycling rate, and the slowest muscles (e.g., the soleus) have contraction times of approximately 100 ms. Fast and slow muscles also have different relaxation times. The information presented in the previous section makes it clear that relaxation rate is related to the rate at which Ca^{++} is taken back into the SR. The rate of SR uptake is regulated by the enzyme Ca^{++}-activated ATPase. The long relaxation time of a slow muscle is illustrated in figure 8.10 and is related to a slower rate of pumping Ca^{++} into the SR, as compared to the shorter relaxation time of fast muscles with their faster rate of Ca^{++} reuptake. What does this mean in terms of recruitment or function of fast and slow fibers? Before considering this question we need to examine the basic differences between a muscle twitch and tetanus.

The hydrolysis of ATP, which is catalyzed by myosin ATPase, is the major rate-limiting step in cross-bridge cycling.

Active State

The state of muscle activation has to do with cytosolic (fluid portion of cell) Ca^{++} concentration. A single action potential releases enough Ca^{++} from the SR to saturate all, or most, of the troponin in a fiber and to trigger cycling of all, or most, myosin heads. But, as we learned previously, Ca^{++} is pumped back into the SR after being released, so the amount of calcium available to trigger cross-bridge cycling decreases quickly.

The mechanical response to a single action potential is called a **muscle twitch**. If another action potential is delivered before the muscle fiber relaxes, more Ca^{++} is released and is added to the Ca^{++} already in the sarcoplasm. Consequently the forces from the first and second twitch are **summated**, and a higher force results (figure 8.11). Recall from chapter 6 that action potentials are delivered to muscle fibers in trains, or groups, and at a specific frequency, depending on various properties of the motor neuron. Action potentials that are far apart produce a ripple effect during contraction because the fiber partially relaxes before the next action potential arrives—a state

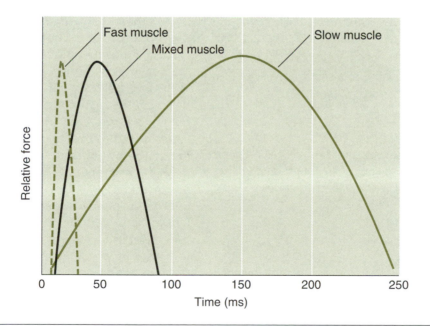

Figure 8.10   Contractile properties. The rates of force development and relaxation during a muscle twitch relate to the rate of cross-bridge cycling and rate of calcium reuptake into the sarcoplasmic reticulum (SR), respectively. Myosin adenosinetriphosphatase (ATPase) and SR ATPase of fast muscle have high activity rates resulting in rapid force development and relaxation. Myosin and SR ATPase in slow muscle have much lower rates of activity, resulting in a slower rate of force development and relaxation.
Adapted from Guyton 1991.

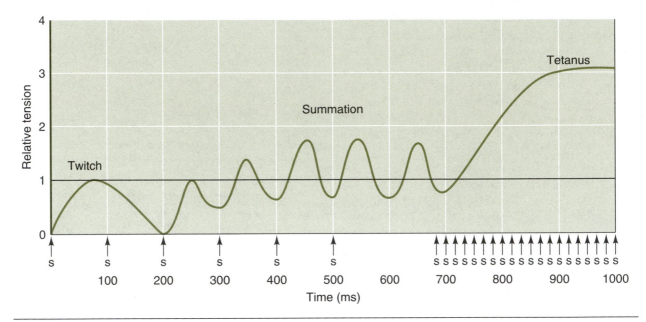

Figure 8.11 Summation of forces. The mechanical response to a single action potential is a muscle twitch. When an action potential is delivered to a muscle fiber before it completely relaxes, forces summate. If the stimuli are close enough together, the maximum force potential of the fiber is reached—muscle tetanus. s = stimulus. Adapted from Vander et al. 1994.

called unfused tetanus. As the action potential frequency increases, force production also increases to a point where any further increase in action potential frequency does not produce an increase in force. This maximum force level is called **tetanus**.

The mechanical response to a single action potential is called a muscle twitch.

As the action potential frequency increases, force production also increases until further increases in action potential frequency do not produce an increase in force. This maximum force level is called tetanus.

The peak force produced by an isometric twitch contraction is about three to four times less than the maximum force produced during an isometric tetanic contraction. But, if enough Ca^{++} is released from a single action potential to trigger cycling of all myosin cross-bridges, why isn't the peak force in a twitch equal to the maximum tetanic force? The answer lies in the elastic components of the contractile and support structures within the muscle. When a muscle is first activated, the force applied to the Z lines of an individual sarcomere (internal force) reaches a maximum almost immediately. However, time is needed to stretch the elastic components in the muscle before all of the force generated at the level of the sarcomere can be transmitted through the tendon to the bony lever (external force). During a muscle twitch, the elastic components are not stretched quickly enough to transmit all of the internal force externally to the bony lever. But if another action potential arrives before the fiber has completely relaxed, more Ca^{++} is released and the cross-bridges continue to cycle. If the action potential frequency is high enough, the elastic components are stretched to their maximum, and the external force now equals the internal force. Thus, the arrival of an action potential activates the muscle fiber, but several action potentials in quick succession are needed to maintain the active state and produce maximum external force. A tetanic contraction, then, is essentially the maintenance of the active state.

A tetanic contraction is essentially the maintenance of the active state.

As an example to help you understand this principle, use a rubber band to move a paperweight across a table. Loop the rubber band around the paperweight and slowly increase the tension on the band. The elastic begins to stretch, but the weight does not begin to move until the slack has been taken up and enough force is transmitted through the elastic band to move it. In a similar sense, the elastic components in muscle must be stretched to their limit before force generated at the tendon equals the force generated at the level of the sarcomeres.

Frequency–Force Relationship

We learned in chapter 6 that fast motor nerves have higher discharge rates than slow motor nerves. The functional significance of this difference becomes apparent when we examine the mechanical response of fast versus slow muscle fibers to various stimulation frequencies. In figure 8.9 you can see that fast muscle fibers, which have a short **time-to-peak tension (TPT)** and relax quickly, have a different response to a single electrical impulse than mixed or slow fibers, which have slower TPTs and relaxation times.

If we stimulated each of these fibers at a low frequency, say 10 Hz (10 pulses/s), what would be the mechanical response of each? It is helpful to first determine the time it takes for each fiber to contract and relax during a single twitch. Since the impulses arrive 100 ms apart, the fast fiber would contract and then have time to relax in response to the first impulse long before the next stimulus arrived. The intermediate fiber would do the same, and a series of twitch contractions would be the result. But the mechanical response for the slow fiber is different. Why? The slow rate of Ca^{++} uptake into the SR fiber means that some cross-bridges are still cycling when the second stimulus arrives (remember that the impulses arrive 100 ms apart), so the forces in the slow fiber would summate. If we increase the stimulus frequency to 20 Hz (impulses are 50 ms apart), what is the response now? The forces in the intermediate fiber begin to summate and the slow fiber is at, or near, fused tetanus. The fast fiber is still producing a series of twitches because the stimuli are still far enough apart to allow Ca^{++} reuptake between stimuli. When stimulus frequency is increased to 50 Hz (stimuli are now 20 ms apart), the fast fiber begins to summate, and both the slow and intermediate fibers are in fused tetanus.

It is important to remember that once a fiber has reached its maximum force potential, any

further increase in stimulation frequency will not produce additional increases in force. Thus, the manner in which fibers of different motor units respond to different rates of stimulation depends on their basic contractile characteristics, most importantly the rates of cross-bridge cycling and of Ca^{++} uptake into the SR. This information is vital to an understanding of muscle function, not only in healthy people, but also under conditions of disuse, deconditioning, disease, and injury. Indeed, in their review of electrical stimulation for denervated muscle, Eberstein and Eberstein point out on page 1463 that electrical stimulation is most effective for the preservation or restoration of normal contractile properties of denervated muscle "when the stimulation pattern resembles the firing pattern of the normal motoneuron." Effective use of electrical stimulation therefore requires not only a knowledge of the fiber-type makeup of the muscle but also an understanding of the different responses of motor units to different frequencies of stimulation.

> Once a fiber has reached its maximum force potential, any further increase in stimulation frequency will not produce additional increases in force.

Specificity of Motor Unit Recruitment

Now, let's put all of this information together and examine how motor units are recruited specifically for different tasks and how this impacts muscle adaptation to training. Recall that one of the factors determining motor unit recruitment is the force requirement of the task as shown in figure 6.16. This figure illustrates a hypothetical muscle made up of only four motor units (remember that most human muscles contain hundreds to thousands of motor units). If small amounts of force are needed for a task, then motor unit 1 is recruited and none of the others are. If more force is needed, motor unit 2 is recruited in addition to motor unit 1. Motor units 3 and 4 are not activated until large amounts of force are needed. Notice that each motor unit has a threshold stimulation frequency for recruitment. Once threshold stimulation has been exceeded, each motor unit will summate forces as stimulation frequency is increased. Thus, each motor unit has a range of stimulation frequencies over which it will respond. Motor unit 4 has the highest threshold for recruitment and the highest maximum force potential. On the basis of this information, we can surmise that motor unit 4 has the most fibers per unit and that it contains fast fibers, which have the largest diameter. Motor unit 1 would consist of fewer fibers per unit; the fibers would be slow and would have a small diameter. We could also guess that motor unit 4 would be recruited only occasionally (for high-force or "power" movements) whereas motor unit 1 would be recruited frequently—probably for most activities of daily living.

These basic principles of motor unit recruitment help us understand specificity of training. If a specific task requires a given force or power output from a muscle or group of muscles, the motor units that make the largest contribution must be recruited during the period of training or conditioning in order for meaningful adaptation to occur. For example, motor unit 4, which is the major contributor to a task requiring large force output, would not be recruited during training at low levels of force output (at least not until motor units 1, 2, and 3 had fatigued) and thus would adapt very little, or not at all. On the other hand, motor unit 1 would show considerable adaptation to low-intensity, long-duration training because it is designed to produce low levels of force over long periods of time. Conversely, activities that require high forces or high power output are accomplished largely by the faster motor units (those that have large cross-sectional area), and these motor units would show the most adaptation to high-intensity training. The slower motor units would be recruited but would contribute little to the overall movement and would not adapt much. This is a very simplistic approach to specificity of training. Muscle recruitment and subsequent adaptation are affected by myriad factors including speed, power output, and type of exercise, but the principle itself is very basic: in order for optimum adaptation to occur, muscles should be trained in a manner similar to the way they are expected to perform. We will consider other factors that affect specificity of training in chapter 9. Here we will examine other properties of skeletal muscle that impact on their function.

If a task requires a given force or power output from a muscle or group of muscles, the motor units that make the largest contribution must be recruited during training or conditioning for

meaningful adaptation to occur. Thus, muscles should be trained in a manner similar to the way they are expected to perform.

Case Study

On one of your regular visits to your fitness facility, you overhear two U.S. football players discussing in loud voices how much they can bench press and how much this will help them next week during the game. You ask them what positions they play—they are both offensive linemen. You ask what part of their performance they think increasing their bench press will help. They respond by showing you the "up" position as if they were pushing against an opponent. What would you tell them that relates to the idea of training specificity?

You should tell them that lying down on a bench and pushing straight up likely recruits very different muscle fibers than the motion they have indicated. Further, training a specific set of motor units in the muscles using a bench press really is not optimizing the training that needs to be done for the other motor units that are recruited during the actual event. In avoiding training with the more position-specific movements they will not be able to optimize performance. Instead, simulating this exact movement using a blocking sled or on one another would be more specific.

Length–Force Relation

A muscle's resting length affects the amount of force it can generate (figure 8.12). Recall that actin overlaps myosin and that force is produced when myosin heads attach to and pull on actin. The greatest potential for cross-bridge attachment exists at optimum resting length (L_o) when there is maximum overlap of myosin and actin. If a muscle is stretched, the actin is pulled away from the myosin and fewer cross-bridges can be formed, resulting in lower force generation. At fiber lengths less than optimum, actin filaments from each end of the sarcomere begin to overlap and cover each other's active sites. As length continues to decrease, fewer and fewer cross-bridges can form, and force decreases. Notice that the muscle fiber can generate maximum force over only a very short range of resting muscle lengths; force drops rapidly if the muscle is shorter or longer than the

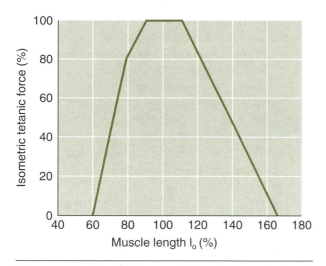

Figure 8.12 Length-force relation. A muscle develops maximum isometric force only over a short range of resting lengths. As resting length is decreased or increased from optimum (L_o), force decreases because fewer myosin heads are able to attach to actin.

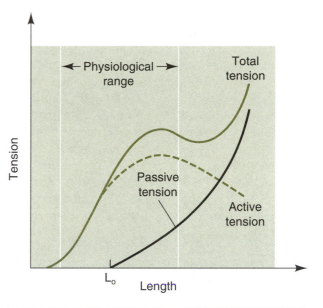

Figure 8.13 Active and passive force. At short muscle lengths, muscle tension is attributed entirely to active tension; but as the muscle is stretched past L_o, total force is a combination of active and passive force. As the muscle length increases, actin and myosin overlap decreases and active tension begins to decrease as passive tension rises exponentially.
Adapted from Kandel et al. 1991.

optimum resting length. Human muscles are, of course, limited in the amount they can be shortened or lengthened by their attachment to bones. The range of motion (ROM) of a bony lever thus limits the change from optimum resting length to approximately 30%.

Muscle fibers can generate maximum force over only a very short range of resting muscle lengths; force drops rapidly if the muscle is shorter or longer than the optimum resting length.

Total force exerted by a muscle is not solely a function of active force. The elastic components of the muscle, in particular the cytoskeletal protein titin (see page 126) and the extracellular connective tissue, exert a passive force when stretched (figure 8.13). At shorter muscle lengths, all of the force exerted by the muscle is due to cross-bridge attachment; but as the muscle is stretched past optimum length, more and more of the total force is related to the stretch of the elastic components.

Total force exerted by a muscle is not solely a function of active force. The elastic components of the muscle, in particular the cytoskeletal protein titin and the extracellular connective tissue, exert a passive force when stretched.

It is important to use caution when applying the force-length relationship to dynamic contractions. Note that figure 8.12 refers only to isometric contractions. That is so because the force-length relationship can be applied without modification only to isometric contractions. During concentric contractions, for example, it is a mistake to believe that force will increase as the muscle shortens from a lengthened position. When the muscle shortens, the speed with which actin slides past myosin affects force, or conversely, the load on the muscle determines the maximum velocity at which it can be moved.

Force-Velocity Relationship

When muscle force equals the load, or resistance, no movement occurs (**isometric contraction**). As the load is reduced and muscle force overcomes the resistance, the muscle will shorten and move the load at a specific velocity (**concentric contraction**). The more the resistance decreases, the greater the velocity of movement until maximum velocity (Vmax) is reached at zero load. This relationship was first described in 1938 by A.V. Hill, who generated an equation to represent changes in velocity with alterations in load (figure 8.14). To understand why force decreases as velocity increases, think again of the cross-bridge cycle. The force generated by a fiber is related to the number of myosin heads pulling on the actin filaments, and its force-generating potential is determined ultimately by the probability that a cross-bridge will, in fact, form at an actin-binding site. When actin is moving past myosin, the probability that a myosin head will attach to a binding site decreases as the velocity at which actin moves increases. Fewer cross-bridge attachments result in less force.

The force generated by a fiber is related to the number of myosin heads pulling on the actin filaments, and its force-generating potential is determined by the probability that a cross-bridge will form at an actin-binding site.

The force-velocity relationship just described does not apply to contractions in which the load imposed on a muscle exceeds the force generated, causing the muscle to be forcibly lengthened (**eccentric contraction**). You can see from figure 8.14 that, in contrast to the force-velocity relationship of a concentric contraction, force is largely independent of velocity for eccentric contractions. Only at low "negative" velocities is there a relationship with force. Maximum eccentric force is reached at relatively low velocities. Notice also that maximum eccentric force is about 50% greater than maximum isometric force. The higher force during eccentric versus isometric or concentric contractions has been attributed by Colomo et al. to a greater number of cross-bridge attachments. Recall that only about 50% of the cycling cross-bridges are generating force at any given instant during an isometric contraction. When an activated muscle is lengthened, all attached cross-bridges are pulled in the direction opposite the direction to which they are exerting force. This may cause a greater number of cross-bridges to form, thus increasing total force output.

When an activated muscle is lengthened, all attached cross-bridges are pulled in the direction opposite the direction to which they are exerting force.

The lack of a relation between velocity and force during an eccentric contraction has been attributed by some, for example, Morgan, to inhomogeneities in individual sarcomere lengths.

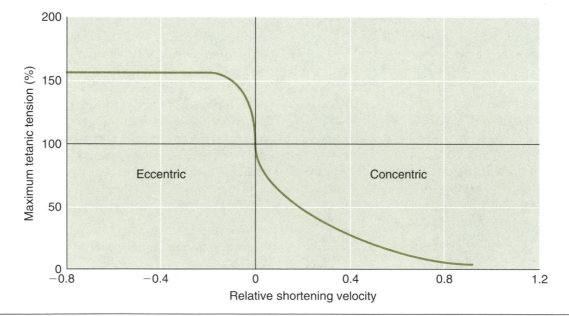

Figure 8.14   Force-velocity relationship. During a concentric contraction, force is inversely related to velocity. During eccentric contractions, force is dependent on velocity only at low velocities.
Adapted from Leiber 1992.

When a fiber is lengthened, longer—and therefore "weaker"—sarcomeres reach their yield point faster than shorter, "stronger" sarcomeres and are on the passive rather than the active part of the length-tension curve (see page 135). If this rapid lengthening of sarcomeres (termed "popping") occurs in order from weakest to strongest, then the total of the active and passive forces in the lengthening fiber always equals the force at the initial length. The velocity with which the muscle is lengthened does not affect passive force or the relative contributions of passive and active force to total force. Thus, during an eccentric contraction, velocity of lengthening has little bearing on the force generated.

It is important to remember the differences in the force-velocity relationship and in force-generating capabilities among the different types of contractions when choosing strengthening exercises, not only in conditioning for athletes, but also (and especially) for rehabilitation purposes. Exercise program prescriptions for people with different requirements should be based on sound physiological principles to maximize the benefits while minimizing musculoskeletal injuries. Many people will not continue an exercise program if they experience persistent muscle soreness or if they incur an injury that could have been prevented.

It is important to remember the differences in the force-velocity relationship and in force-generating capabilities among the different types of contractions when choosing strengthening exercises for athletes and especially for rehabilitation purposes.

Case Study

You are using an **isokinetic** exercise machine to strengthen the quadriceps of a patient recovering from a knee injury. An isokinetic machine allows you to regulate the velocity with which the patient moves a lever arm through the entire ROM. You are able to choose either concentric or eccentric contractions in your program. Which would you choose, and why?

An isokinetic dynamometer allows you to regulate velocity of the movement. Thus, you would choose a concentric movement early in the exercise program because in concentric movements, force is dependent on velocity (high velocity = low force). You would not choose an eccentric movement because in these movements, force is largely independent of velocity (i.e., at almost all velocities, force is very high). As muscle strength increases, eccentric movements may be appropriate, but caution is always warranted.

Moment Arm

As we have discussed, a muscle exerts a force through the tendon to the bone and thus causes movement. The muscle exerts force in a linear fashion, but movement is normally achieved by rotation about a joint axis. Therefore, the muscle produces a rotational force or torque that is a product of the muscle force and the perpendicular distance between the muscle's line of action (or pull) and the axis of rotation. The latter is referred to as the moment arm. To clarify this concept, think of a person performing knee extensions on a Cybex dynamometer. The lever that the subject pushes against rotates about an axis, and the knee is in line with this axis. The quadriceps supply the force to move the lever; but to measure this force, the transducer would have to be positioned directly in line with the action of the force. Since it is not, we are actually measuring torque and not force. Torque is the product of the two vector quantities: force × moment arm (figure 8.15). Basically, as shown in figure 8.16, torque can be affected in three ways:

1. By altering force production by the muscle
2. By changing the moment arm
3. By changing the angle of force application

➤ Torque is the rotational force generated on a moment arm.

Lever System

The muscles, bones, and joints of the body are organized to produce various lever systems, which can advantage or disadvantage a muscle. Figure 8.17 illustrates the mechanical disadvantage placed on skeletal muscle in order to operate a lever. In this example, a 10 kg (22 lb) weight is held in the hand, but the force needed to support the weight is seven times greater than the downward force of the weight itself. Why? The length

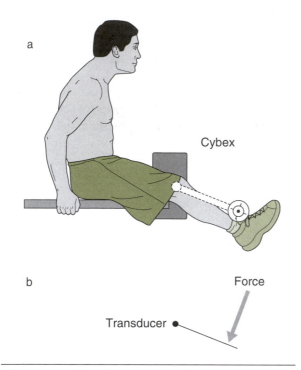

Figure 8.15   Muscle torque. A Cybex dynamometer is used to measure quadriceps strength during a knee extension (*a*). The quadriceps supply force to move the lever; but because the line of action of the force generation is perpendicular to the axis of rotation (*b*), it is actually torque that is measured.
Adapted from Enoka 1992.

of the lever, in this case, dictates how much force is required to support the weight ($5 \times x = 35 \times 10$). Now, think of the force that must be generated to support the weight if the arm is completely extended: much greater, again, than if the arm is flexed at 90°. This principle tells us that it is much more efficient and less fatiguing to carry or hold objects close to the body than extended away from the body. The mechanical disadvantage increases as the length of the lever increases. However, the same lever system allows us to move objects at the end of the lever much faster than the muscle is actually shortening. This mechanical advantage is illustrated in figure 8.18. If the muscle shortens 2 cm (about 0.8 in.), then the object at the end of the lever moves 15 times that distance in the same amount of time. This principle allows us to impart tremendous velocities to objects in the hand. For example, the velocity at which a baseball is thrown or a tennis ball is served is much

higher than the velocity at which the muscles that control the actions are shortening.

> It is much more efficient and less fatiguing to carry or hold objects close to the body than extended away from the body.

Putting all of this basic information together, we can see that the force required to move an object through a specific range is affected by many factors:

1. Resting length
2. Force-velocity relationship
3. Angle of pull on the bony lever
4. Moment arm

A movement task is further complicated by fiber-type distribution and architecture, fiber cross-sectional area, stimulation frequency, physiologic state of the muscle (fatigue, etc.), coactivation of antagonists, and other factors. One begins to understand that production of movement depends on a complex interaction of the specific characteristics of the muscle itself, its attachments to the bony lever, and its relation to other muscles in its proximity. The CNS acts to coordinate or direct the activity, but the outcome is ultimately determined by the physiologic and mechanical factors we have discussed to this point.

Factors Affecting Exercise Performance

A variety of factors, ranging from prior physiological state to muscle fatigue, affect an individual's performance—whether it involves simple activities of daily living or athletic activities. An older patient may describe the symptoms of "fatigue" as a sense of increased effort to accomplish daily tasks, and the problem may be exacerbated by decreased motivation, depression, or pain. These factors act centrally to reduce motor nerve activation. Peripheral factors that

> The ability of skeletal muscle to generate force is impacted by changes in the entire neuromuscular system.

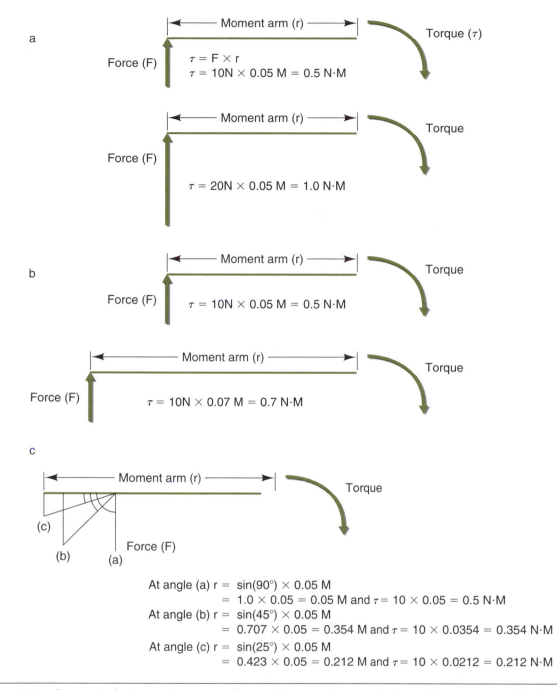

a

$\tau = F \times r$
$\tau = 10N \times 0.05\ M = 0.5\ N \cdot M$

$\tau = 20N \times 0.05\ M = 1.0\ N \cdot M$

b

$\tau = 10N \times 0.05\ M = 0.5\ N \cdot M$

$\tau = 10N \times 0.07\ M = 0.7\ N \cdot M$

c

At angle (a) r = sin(90°) × 0.05 M
      = 1.0 × 0.05 = 0.05 M and τ = 10 × 0.05 = 0.5 N·M
At angle (b) r = sin(45°) × 0.05 M
      = 0.707 × 0.05 = 0.354 M and τ = 10 × 0.0354 = 0.354 N·M
At angle (c) r = sin(25°) × 0.05 M
      = 0.423 × 0.05 = 0.212 M and τ = 10 × 0.0212 = 0.212 N·M

Figure 8.16   Strategies for increasing torque. Torque is increased as muscle force (*a*), moment arm (r) (*b*), and angle between the force application and axis of rotation (*c*) are increased. Moment arm is affected by the angle of force application (r = D × sin[θ]).
Adapted from Leiber 1992.

act at the neuromuscular junction or on the contractile apparatus of the muscle fiber also affect performance. Thus, the ability of skeletal muscle to generate force is impacted by changes in the entire neuromuscular system. **Central fatigue** is related to decreased excitation of motor neurons, and **peripheral fatigue** occurs at the level of the muscle fiber.

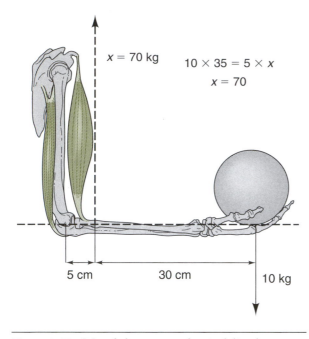

$x = 70$ kg

$$10 \times 35 = 5 \times x$$
$$x = 70$$

5 cm   30 cm   10 kg

Figure 8.17 Muscle lever—mechanical disadvantage. A weight held in the hand (10 kg) must be balanced by a much greater force in the muscle (70 kg) because of the lever principle.
Adapted from Vander et al. 1994.

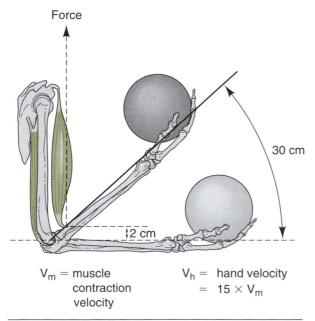

Force

30 cm

2 cm

V_m = muscle contraction velocity

V_h = hand velocity = $15 \times V_m$

Figure 8.18 Muscle lever—mechanical advantage. When the muscle shortens 2 cm, an object held in the hand will move a greater distance (30 cm) in the same amount of time, thereby having 15 times greater velocity than the muscle.
Adapted from Vander et al. 1994.

Central Motor Drive

As discussed in chapter 6, the quality of a movement is affected by the individual's desire, or level of motivation. The limbic system has a direct impact on the level of activation of neurons in the motor cortex, thereby affecting movement behavior. Pain, a subjective feeling of discomfort, is often difficult to assess because people vary greatly in their tolerance. Musculoskeletal disorders that are painful, particularly those affecting the muscle itself or joints, cause a central inhibition of motor neurons and adversely affect performance.

Prior Physiologic State

The amount of work an individual can perform is influenced by the prior state of body systems, that is, their condition before the work task begins. Among important elements of prior physiological state are glycogen stores, state of hydration, and muscle temperature.

Perhaps the most important factor determining performance of long-duration activities is muscle glycogen stores. If stores have been depleted by a prior activity and have not been sufficiently replenished, long-term performance is adversely affected. Conversely, the practice of glycogen loading or "supercompensation"—use of a combination of exhaustive exercise and dietary manipulation—will prolong muscle endurance. This phenomenon is covered in more detail in chapter 4 (pages 54-55).

Dehydration and a loss of electrolytes, common in hot environments especially when combined with physical activity, cause a decrease in performance (figure 8.19). Consumption of alcohol or caffeine, both diuretics, can also cause the body tissues to become dehydrated. When body tissues are dehydrated before the individual begins a work task or before an athletic event, cardiovascular and muscle function are both diminished. If electrolyte imbalances accompany dehydration, nerve and muscle excitation will decline.

➤ A diuretic is any agent that increases the amount of urine production.

A third element of prior physiologic state is muscle temperature. Optimal rates for chemical reactions in muscle fibers occur in a relatively narrow range of temperatures that spans 37° C.

Figure 8.19   Hydration before exercise. Many of your clients will not realize how vital good hydration is to their ability to perform physical activity. Emphasize its importance when discussing their exercise program.

Higher muscle temperatures increase ATP utilization and may cause an uncoupling of oxidative phosphorylation, which decreases production of ATP. Lower muscle temperatures result in impaired excitation and decreased enzyme reaction rates, leading to decreased force production. Manual dexterity is especially adversely affected by cold. Thus, both high and low muscle temperatures prior to the start of the work task will impact on performance.

Muscle Fatigue

The decline in force-producing capabilities over time (muscle fatigue) is basically a protective mechanism that prevents us from exhausting metabolic reserves within muscle and limits the buildup of harmful metabolic products. Muscle fatigue also reduces the likelihood that continual generation of high forces will cause damage to the contractile elements. There are many causes of muscle fatigue, and many are not completely understood. We will briefly examine only a few: mechanisms that relate to stimulation of muscle at high and at low intensities; those that relate to type of contraction; those that relate to the work-rest cycle; lactic acid production; and glycogen depletion.

> Muscle fatigue protects against the exhaustion of metabolic reserves and against damage to the contractile elements.

High- and Low-Frequency Fatigue

Mechanisms that underlie muscle fatigue from prolonged, low-intensity activity are different from those that cause force reduction as a result of high-intensity exercise. **Low-frequency fatigue** is the selective loss of force in slow motor units as a result of stimulation of the muscle at low frequencies; fast motor units are largely unaffected. Recovery is very slow—it can take as long as 24 hr for the slow fibers to regain their force-producing capabilities. Conversely, **high-frequency fatigue** specifically targets the fast motor units, and recovery is very rapid. Figure 8.20 illustrates these two phenomena. Although the experiments represented were conducted on isolated skeletal muscles, the information is useful in application to muscle fatigue in humans. The graphs show that a muscle fatigued at low-frequency stimulation (e.g., 10 Hz) has almost no force production at the low frequencies at 1 and 60 min following the fatigue protocol, but the force production at high frequencies is largely unaffected (figure 8.20*a*). In humans, as just mentioned, fatigue of the slow motor units may last up to 24 hr. It is obvious that when the same muscle is fatigued at a stimulation frequency of 80 Hz, the slow motor units are less affected than the fast (notice the low force 1 min after the fatiguing protocol when the muscle is stimulated at 80 Hz in figure 8.20*b*). Recovery from high-frequency fatigue is rapid as evidenced by achievement of maximum force within 1 hr of administration of the fatiguing protocol.

Recovery is slow from low-frequency fatigue; conversely, recovery is very rapid from high-frequency fatigue.

This information can be valuable when you are assessing productivity in a workplace where the work is repetitive and prolonged. If the workers seem to be suffering undue fatigue from day to day, it may be wise to have them perform higher-intensity and lower-intensity jobs on alternate days. You should also keep in mind the 24 hr duration of low-frequency fatigue when planning workouts for athletes. Interspersing high-intensity with low-intensity workouts may produce better performance and training results.

Type of Contraction

Even though we are capable of producing more force with eccentric than with isometric or concentric contractions, eccentric contractions are less fatiguing at submaximal levels of force generation. How can we explain this apparent paradox? Let's begin with the simplest principle. Isometric contractions actually produce the fastest time to fatigue because they reduce or completely

shut off muscle blood flow. Think of the compression on the blood vessels within the muscle during performance of an isometric contraction. Blood flow is completely shut off at surprisingly low percentages of maximum force (between 40% and 60% of maximum voluntary contraction). The difference among muscles may be related to variance in fiber orientation, fiber-type distribution, and capillary density. The ischemia caused by an isometric contraction interrupts O_2 delivery, leading to hypoxia, and results in a buildup of metabolites that cause decrements in force. Dynamic contractions are less fatiguing because blood flow is not shut off or is interrupted for only a brief time. Submaximal eccentric contractions are less fatiguing than concentric contractions because more force is produced per muscle fiber, so fewer muscle fibers are needed to produce the same amount of absolute force. With eccentric versus concentric contractions, less energy is used and more motor units are held in reserve, available for use as contracting fibers fatigue.

Work-Rest Cycle

More total work can be performed when the work task is intermittent rather than continuous.

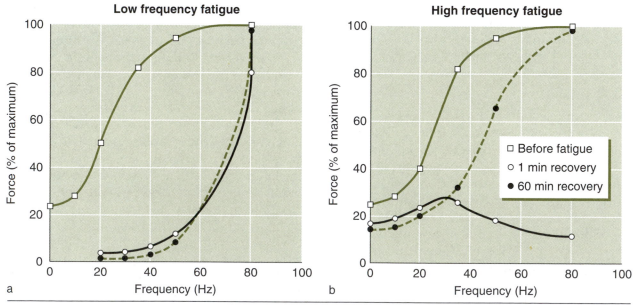

Figure 8.20 Low- and high-frequency fatigue. Force/frequency curve of an isolated muscle before fatigue and after 1 and 60 min of recovery following stimulation to exhaustion at 10 Hz (*a*) and 80 Hz (*b*). (*a*) Force at the low frequencies has not recovered even after 1 hr; but force at the highest frequency (80 Hz) is high even at 1 min of recovery and is back to normal after 1 hr, indicating that low-frequency fatigue is selective for slower motor units and recovery is slow. (*b*) Force after 1 min at the higher frequencies is very low but has recovered by 60 min, indicating that high-frequency fatigue affects fast motor units more and that recovery is rapid.
Adapted from Faulkner 1983.

Productivity can also be improved through alterations in the length of work versus rest periods or the ratio of work time to rest time. These principles are illustrated in figure 8.21. In this example, the subject was exercising at an intensity of 412 W—work that he could continuously perform for only 3 min. The object was to have the subject perform 247 kilojoules (kJ) of work in 30 min with the least amount of physical stress. If the work was interspersed with rest periods at a ratio of 1:2, he could work for longer periods before becoming exhausted. Blood lactic acid levels (on the y axis) are a good indicator of the subject's fatigue level. If the work period was 60 s with 120 s of rest, the subject was exhausted in 24 min and could not complete the work task. If the ratio remained the same but the work was reduced to 30 s, the subject completed the work and was somewhat fatigued at the end. When the work was reduced to 10 s and rest to 20 s, the worker not only completed the work, but was fatigued only slightly (note the low levels of blood lactic acid after 30 min) and likely could have continued. When the work:rest ratio was changed from 1:2 to 1:4 (i.e., 60 s work and 240 s rest), there was hardly any benefit in terms of fatigability. Of course the worker then had to spend twice the time to complete the same amount of work. Although it is unrealistic to work for 10 s and rest for 20 in real-life situations, the principle here is important. The critical factor in completion of work with the least amount of stress is the length of the work period. The duration of the rest periods and the total time spent resting are of secondary importance.

The length of the work period is the critical factor in completing work with the least amount of stress.

Lactic Acid Production

Significant lactic acid accumulation occurs during high-intensity exercise in which non-oxidative metabolism is the primary source of energy production. Lactic acid is labile and breaks down into lactate and hydrogen ion (H+). The H+ is what is thought to interfere at many levels of excitation-contraction coupling to reduce force production. The major means by which H+ affects the contractile process are (1) inhibition of phosphofructokinase, a rate-limiting enzyme for glycolysis; (2) inhibition of the transformation of phosphorylase b to a, which decreases glycogen breakdown to glucose; (3) reduced force generation by inhibition of myosin ATPase; and (4) decreasing muscle fiber membrane excitability. Hydrogen ion also inhibits mobilization of fatty acids from adipose tissue, necessitating increased use of glycogen stores.

Glycogen Depletion

Ultimately, muscle stores of glycogen limit the length of time a prolonged exercise can be performed. If glycogen stores are low, endurance performance is hampered; if they are high, endurance time is increased (see chapter 4).

Two other mechanisms are also worth noting. Inorganic phosphate (Pi), a by-product of

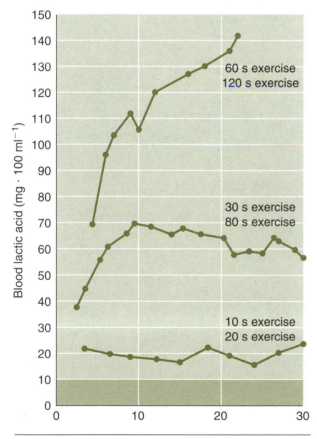

Figure 8.21 Intermittent work. A subject is trying to complete 30 min of work at an intensity of 412 W. If he performs continuously, he is exhausted in 3 min. If the work is intermittent, at a work:rest ratio of 1:2, the least physical stress is seen when the subject works for 10 s and rests for 20. Blood lactic acid accumulation (on the y axis) is a good indicator of fatigue.
Adapted from Astrand and Rodahl 1986.

ATP hydrolysis, is associated with a decline in force production. As Pi concentration increases within the cell, the contractile apparatus is affected directly to reduce force output. During intense exercise, buildup of H^+ interacts with Pi to augment the reduction in force. During high-frequency stimulation, loss of potassium from inside the cell and accumulation in the extracel-lular space are thought to produce decreased membrane excitability. Accumulation of ammonia (from the reaction in which adenosine monophosphate is combined with water to yield inosine monophosphate and ammonia [AMP + $H_2O \rightarrow IMP + NH_3$]) during prolonged exercise is thought to inhibit the Krebs cycle, thereby reducing energy production.

What You Need to Know From Chapter 8

Key Terms

actin
atrophy
axial growth
basal lamina
blastula stage
central fatigue
concentric contraction
contractile elements
costamere
cross-bridges
cytoskeletal proteins
desmin
dystrophin
eccentric contraction
endomysium
epimysium
high-frequency fatigue
hypertrophy
integrin
isokinetic
isometric contraction
low-frequency fatigue
mesoderm
M line
muscle twitch
myoblasts
myogenesis
myosin
myotubes
pennation

perimysial
peripheral fatigue
plasticity
sarcolemma
sarcomeres
satellite cells
somites
summation
tetanus
time-to-peak tension (TPT)
titin
transformed
tropomyosin
troponin
Z line
Z-line "streaming"

Key Concepts

1. You should be able to describe muscle development, including the formation of satellite cells.

2. You should be able to describe basic micro- and macrostructure of skeletal muscle, including the role of the cytoskeleton and that of connective tissue.

3. You should be able to describe the processes of muscle fiber activation, cross-bridge cycling, and excitation-contraction coupling.

4. You should be able to describe the differences between fast and slow muscle fibers in relation to contractile properties, as well as the mechanisms that underlie these differences.

5. You should be able to describe the mechanical responses of different types of muscle

fibers to a single electrical stimulus and a series of stimuli delivered at different frequencies.

6. You should be able to list the factors that determine patterns of motor unit recruitment.

7. You should be able to describe various factors that affect muscle force production, including length-tension relation, force-velocity relation, moment arm, and the lever system.

8. You should be able to distinguish between central and peripheral fatigue.

9. You should be able to describe how the person's prior physiological state affects work or exercise performance.

10. You should be able to distinguish between low- and high-frequency fatigue.

11. You should be able to describe various factors that cause muscle fatigue.

Review Questions

1. How is the mosaic pattern of intermingled fiber types that is observed in adult humans established during skeletal muscle development?

2. How are satellite cells formed, and what is their function?

3. Describe how fiber orientation (angle of pennation) affects muscle force production.

4. What is the difference in composition of intramuscular connective tissue between slow and fast muscle, and how does this relate to the function of those two fiber types?

5. After each muscle contraction, sarcomeres return to their original position when relaxed—in exact register from myofibril to myofibril. Describe the arrangement of stabilizing proteins within the fiber that keeps sarcomeres in exact register.

6. If it is true that a single cross-bridge cycle moves an actin filament relative to the myosin filament only a very short distance, how is it that a whole muscle can shorten a great distance (to move a limb through an entire ROM), and how is uniform movement or uniform force produced?

7. Describe the two roles of ATP in cross-bridge cycling, and explain why ATP "deple-

tion" cannot be a major cause of muscular fatigue.

8. Describe mechanisms that determine differences in rate of force development and rate of relaxation between slow and fast muscle fibers.

9. What are the mechanical responses (force production) of a muscle fiber to a single action potential? To a series of action potentials at a low frequency? To a series of acting potentials at high frequency? If all, or most, troponin is saturated with Ca^{++} with a single action potential, why is twitch force considerably less than tetanic force?

10. Describe how prior levels of muscle glycogen, body water, and muscle temperature can affect work output or exercise performance.

11. Explain which of the three types of contractions (isometric, eccentric, and concentric) is the most fatiguing and which is the least fatiguing during submaximal contractions.

12. Graph the mechanical responses of fast and slow muscle fibers to stimulation frequencies of 10, 20, and 50 Hz. For a muscle twitch (the mechanical response to a single action potential), the fast fiber has a TPT of 10 ms and is completely relaxed in 20 ms; the slow fiber has a TPT of 100 ms and is relaxed in 200 ms.

13. In your job at a work-hardening clinic in a large factory, your responsibilities include rehabilitation and conditioning of injured workers. You know that fatigue is a major cause of job-related injuries, so part of your job is injury prevention. At one of the work stations you notice the following scenario. A worker has to reach across a table and lift a 10 kg (22 lb) object with the elbow angle at 130°, then turn and carry the object 9 m (30 ft) (with the arms in the same position), return to the table, and repeat the process. The worker has trouble lifting the object off the table and also complains of undue muscle fatigue after only a few trips. In terms of muscle mechanics, explain why the object is difficult to lift, and explain the probable cause of muscle fatigue. Outline what you think could be done to address the problem.

(continued)

14. You have a sedentary patient who has fractured her tibia. The leg will be casted for four weeks, effectively immobilizing the limb and causing muscle atrophy. Which fiber type will be most affected in the immobilized muscles and why? What would be the difference in response between this patient and one who is a highly trained sprinter?

15. A worker is expected to complete 300 kJ of work in 1 hr. The work is of fairly high intensity, so his supervisor tells him to work for 3 min and rest for 6. The worker becomes exhausted in about half an hour and can't continue. The supervisor suggests increasing the rest period to 7 min. As a consultant, do you agree with the supervisor? Why or why not? In your answer, include the cause of fatigue.

16. Company X has hired you to improve productivity. Workers are complaining of undue muscle fatigue at the end of a shift, which carries over to the next day. The work is repetitive, low intensity, and long duration. What is the major problem affecting these workers, and what is your recommendation for improving productivity?

Bibliography

Astrand I, Astrand P-O, Christensen EH, and Hedman R. Intermittent muscular work. *Acta Physiol* 1960, 48:443–449.

Bigland-Ritchie B, Cafarelli E, and Vollestad NK. Fatigue of submaximal static contractions. *Acta Physiol Scand* 1986, 128 (suppl. 556):137–148.

Borg TK and Caulfield JB. Morphology of connective tissue in skeletal muscle. *Tissue Cell* 1980, 12:197–207.

Colomo F, Lombardi V, and Piazzesi G. The mechanisms of force enhancement during constant velocity lengthening in tetanized single fibres of frog muscle. *Advances Exper Med Biol* 1998, 226:489–502.

Ellenbecker TS, Davies GJ, and Rowinski MJ. Concentric versus eccentric isokinetic strengthening of the rotator cuff. *Am J Sports Med* 1988, 16:64–69.

Eberstein A and Eberstein S. Electrical stimulation of denervated muscle: Is it worthwhile? *Med Sci Sports Exerc* 1996, 28:1463-1469.

Faulkner JA. Fatigue of skeletal muscle fibers. *Proceedings of the 3rd Banff International Hypoxia Symposium*, pp. 243–255. New York: Liss.

Gans C and de Vree F. Functional bases of fiber length and angulation in muscle. *J Morphol* 1987, 192:63–85.

Grange RW and Houston ME. Simultaneous potentiation and fatigue in quadriceps after a 60–second maximal voluntary isometric contraction. *J Appl Physiol* 1991, 70:726–731.

Green HJ. How important is endogenous muscle glycogen to fatigue in prolonged exercise? *Can J Physiol Pharmacol* 1991, 69:290–297.

Harry JD, Ward AW, Heglund NC, Morgan DL, and McMahon TA. Cross-bridge cycling theories cannot explain high-speed lengthening behavior in frog muscle. *Biophysical J* 1990, 57:201–8.

Irving M and Piazzesi G. Motions of myosin heads that drive muscle contraction. *News Physiol Sci* 1997, 12:249–254.

Kelly AM and Rubinstein AM. Development of neuromuscular specialization. *Med Sci Sports Exerc* 1986, 292–298.

Kovanen V, Suominen H, and Heikkinen E. Mechanical properties of fast and slow skeletal muscle with special reference to collagen and endurance training. *J Biomech* 1984, 17:725-735.

Maclaren DP, Gibson H, Parry-Billings M, and Edwards RHT. A review of metabolic and physiological factors in fatigue. *Exerc and Sport Sci Rev* 1989, 17:29–61.

Morgan DL. New insights into the behavior of muscle during active lengthening. *Biophysical J* 1990, 57:209–221.

Patel TJ, Richard MS, and Lieber RL. Force transmission in skeletal muscle: From actomyosin to external tendons. *Exerc and Sport Sci Rev* 1997, 25:321–359.

Rubinstein NA and Kelly AM. Development of muscle fiber specialization in the rat hindlimb. *J Cell Biol* 1981, 90:128–144.

Stickland NC. Muscle development in the human fetus as exemplified by m. sartorius: A quantitative study. *J Anat* 1981, 132:557–579.

Muscle Plasticity

This chapter will discuss

➤ changes in muscle structure and function as a result of conditioning,

➤ changes in muscle structure and function as a result of deconditioning, and

➤ changes in muscle structure and function as a result of injury and repair.

It is crucial, when one is prescribing an exercise program, to understand the adaptations the subject can expect in response to a particular mode, frequency, duration, and intensity of exercise. Programs should always be tailored to the person. The same exercise conditioning may produce widely different adaptations in people with different capabilities and limitations, so expectations should be realistic for each. Muscle adaptation in response to different stimuli is called muscle plasticity. Before exploring plasticity, let's examine the various types of muscle contractions and how they impact on exercise training.

Basic Principles of Muscle Training

The term "contraction" is a bit of a misnomer when used to refer to all of the conditions under which the muscle is activated and generating force. The muscle is really contracting or shortening under only one of three possible conditions. As mentioned in the previous chapter, the muscle torque relative to the external resistance, or load, determines the type of muscle contrac-

tion. When the torque equals the load, the muscle length does not change; this is an **isometric contraction**. When the torque exceeds the load, the muscle shortens, producing a **concentric contraction**; and when the load is greater than the torque, the muscle is forced to lengthen—an **eccentric contraction** (figure 9.1). If we were to replace the term "contraction" with the term "muscle activation," we would be conveying more clearly what is really happening. Because of tradition, however, we will continue to use the term contraction, but with the understanding that it refers to muscle activation.

Muscle contractions can further be classified as **static** (isometric) or **dynamic** (concentric and eccentric). An easy way to distinguish between concentric and eccentric is to refer to their literal meanings—toward the center (concentric) and away from the center (eccentric). A term that is commonly used to describe dynamic contractions is "**isotonic**." This term means "same force" and, as we learned previously, there are no dynamic contractions in which muscle force does not change throughout a range of motion (ROM). It is a mistake, then, to refer to muscle contractions in "real life" as isotonic.

Static Versus Dynamic Contractions

Although large gains in muscle strength can be achieved with isometric training, the applicability to movement in real life is quite limited. Strength gains are largely specific to the joint angle at which the isometric training is conducted, and most physical activities are dynamic in nature. It is more practical to choose dynamic contractions as a basis for increasing strength, although weight lifters use isometrics to overcome "sticking points" in a particular lift. These sticking points relate to the joint angle in a particular ROM at which the muscle is weakest. That is, because of the various mechanical limitations of the bony levers, the torque generated by the muscle is less at a particular point in the ROM than at any other point. Thus, the weakest joint angle, or the sticking point, limits the load that the lifter can move through the entire ROM. By overloading the muscle isometrically at that specific joint angle, the strength trainer will be stronger at that angle and able to overcome this limitation to a certain extent.

Torque production during concentric contractions and eccentric contractions is subject to many of the same constraints imposed by joint angle, initial muscle length, and contractile characteristics. But one should consider two basic differences between torque production in these two types of contractions when choosing a particular strengthening exercise. Recall that eccentric contractions have the potential to produce much more force than concentric or isometric contractions—and that whereas muscle force is directly related to contraction velocity with concentric contractions, it is largely independent of velocity for eccentric contractions. Therefore, the high forces achieved with eccentric contractions may not be appropriate for a deconditioned or sedentary person. Even at submaximal levels of force generation, more stress (force/cross-sectional area) is imposed on muscle fibers with an eccentric versus a concentric contraction. Each muscle fiber can generate more force eccentrically, so fewer fibers (or motor units) are required to generate the same absolute force with an eccentric move-

Figure 9.1 Types of muscle contraction. During a concentric muscle contraction (*a*), the muscle shortens. During an eccentric muscle contraction (*b*), the muscle lengthens. During an isometric muscle contraction (*c*), the muscle length remains unchanged.

ment as compared to the same movement performed concentrically.

Eccentric contractions have the potential to produce much more force than concentric or isometric contractions.

Isokinetics

Differences in force-velocity relationships between concentric and eccentric contractions are important with use of an isokinetic dynamometer, an exercise machine that regulates the velocity of a movement and that is often used to rehabilitate weakened muscles. During concentric movements, force output of the muscle can be controlled by altering angular velocity. The faster the limb is moved through the ROM, the lower the torque production. However, as already discussed, eccentric contractions produce very high torque at all but the slowest velocities. The high stresses put all the components of the musculoskeletal system at risk for injury. In addition to high stresses on bone, tendons, and ligaments, damage to the muscle fibers themselves is linked almost exclusively to eccentric, rather than isometric or concentric, contractions. We should also note that the magnitude of the change in length (strain) during eccentric contractions is a major determinant of the extent of muscle injury. Although the high torque production associated with eccentric contractions is appealing for maximizing strength gains, it is wise to use caution when prescribing training exercises biased largely toward eccentric activities. One must also remember that an increase in "strength" does not relate solely to an increase in the muscle mass, but also includes neural adaptations that enable the strength trainer to apply forces effectively in a particular movement pattern. This involves intricate timing, maximal recruitment and synchronization of motor units, inhibition of antagonists, and disinhibition of the Golgi tendon organ (GTO) (remember that the GTO prevents generation of large forces by directly inhibiting the contracting muscle). Much more is involved in the effort to optimize strength gains than merely increasing the size of the muscle.

➤ Disinhibition is the removal of an inhibitory influence.

As mentioned earlier, isokinetic exercise allows one to regulate the concentric torque pro-

duction through an entire ROM by regulating the velocity of the movement. As the subject applies force to move the lever arm through the ROM, the machine "accommodates" or resists with the same force. This type of resistance exercise is attractive because, as already discussed, normally the muscle is constrained mechanically at particular joint angles, and the torque produced changes throughout any given ROM.

The theoretical advantage of the isokinetic dynamometer over other training methods is the ability to overload the muscle through the entire movement without being limited by the sticking point, or weakest joint angle. By setting the isokinetic velocity relatively high (e.g., 540°/s), the health professional can guarantee that the muscle is overloaded through the entire ROM, but at a low level of torque production. This ensures a good training response and minimizes chances of musculoskeletal injury, making the technique appealing for rehabilitation purposes.

Isokinetic exercise training does have some limitations, however. The velocities that the machine is capable of producing are relatively slow as compared to the velocities needed for movements in most real-life activities. Therefore, optimum training results for physical tasks requiring high power outputs may be more achievable using other techniques. Another limitation of isokinetic training, especially at the lower speeds, is that strength gains acquired at slow speeds do not translate to equivalent gains in strength for activities performed at fast speeds. Training at fast speeds, as reported by Coyle et al., produces more generalized strength gains at both slow and fast speeds.

By setting the isokinetic velocity relatively high (e.g., 540°/s), the therapist can guarantee that the muscle is overloaded through the entire ROM, but at a low level of torque production.

Plyometrics

Plyometric training (also called explosive-jump training) is used to improve "instantaneous" power, or the ability to generate high levels of muscle force over a short period (figure 9.2). Plyometrics involves overloading the muscle rapidly with an eccentric contraction, which is followed immediately by a concentric contraction. For example, a person jumps down from a bench (eccentric contraction), absorbs the impact, then jumps back up (concentric contraction). The concentric contraction is more powerful than normal (i.e., than if the person performed the vertical jump from a standing position) because the preloaded concentric contraction utilizes elastic recoil and the stretch reflex (see chapter 6) from the rapid stretch of the muscle to enhance force production. Komi has shown that people with plyometric training have higher levels of muscle activation when performing a drop jump than untrained people. This finding indicates that gains in power are, at least in part, related to neural adaptations, probably an enhancement of the stretch reflex. Plyometric training for the upper extremities uses heavy objects such as a medicine ball.

Plyometrics involves overloading the muscle rapidly with an eccentric contraction, which is followed immediately by a concentric contraction.

Figure 9.2 Plyometric training. Plyometric exercises such as this high leap provide optimal gains in strength and power. However, they also present a much higher potential for soft tissue injury than conventional strength training, and therefore must be carefully supervised.

As with all exercises involving high muscle forces, caution is in order when one is incorporating plyometrics into a training program. Plyometrics should not be used for increases in general health or for improvement of performance in recreational activities. A person should be able to leg press two times body weight before being introduced to this type of training, and increases in the height of the jump should be gradual. Jumping from heights of 0.45 to over 0.9 m (1.5 to over 3 ft) was shown by Lundin and Berg to be effective in increasing jumping strength. It is recommended that novices begin at the lower end of the scale, 8 to 10 repetitions, two to three times per week. Gradually increasing the jump height increases training effects, but increasing the frequency or volume is not recommended. The dangers of plyometric training are obvious. Tremendous forces are generated when a person jumps down from a considerable height, putting ligaments, tendons, and muscle fibers under immense stress. The knee is especially vulnerable to injury as a result of misuse of the technique. The temptation is always there—if jumping down from 0.5 m produces small benefits, why not increase the height to 1 m, then 2 m? Younger athletes are especially at risk for injury through overzealous use of plyometrics. Used correctly, plyometric training can yield tremendous gains in muscle power. These gains in instantaneous power are of course very specific to the type of activity performed.

Plyometrics should not be used for increases in general health or for improvement of performance in recreational activities.

Used correctly, plyometric training can yield tremendous gains in muscle power.

Case Study

An adolescent athlete has just been taken to the hospital for a ruptured anterior cruciate ligament. He had returned from a soccer camp and was excited about some new training techniques. The coaches at the camp had introduced the players to the concept of plyometric conditioning and given out a list of exercises to try at home. This young athlete had doubled the height of the drop jump to take advantage of the potential for greater gains in power. You have been asked by a local television station to comment on this situation. How will you explain the injury, and what will your advice be?

You need to explain that although the potential for strength and power gains is greater with plyometric exercises than with more traditional forms of training, so too is the potential for injury. The increased forces associated with plyometrics, particularly if performed from a great height, put tremendous stress on the musculoskeletal system. The ligaments that stabilize the knee are especially vulnerable because of the high forces generated by eccentric contraction of the quadriceps muscles during the drop jump. This is especially true for younger athletes. You should warn the general public, and athletes in particular, to avoid plyometric exercises except under proper supervision.

Training Specificity

Throughout much of our discussion of muscle function we have touched on the principle of training specificity. The principle itself is very simple: skeletal muscles should be conditioned in a manner similar to that in which they are to perform. We have already discussed the fact that different motor units are designed to perform different tasks; and we have seen that in order for an increase in function to occur, the relevant motor units need to be recruited during training. The extent of muscle adaptation relates directly to the frequency (number of times per week) with which particular motor units are recruited and the nature of the activity (intensity and duration). For example, slow oxidative motor units are recruited primarily for low-intensity, long-duration activities, and this type of training improves oxidative capacity. Little or no adaptation will be observed in fast glycolytic motor units. Why? The reason is that they are not recruited, or are recruited only briefly, when slower motor units fatigue. Conversely, high-intensity, short-duration activities require predominately fast glycolytic motor units, and the primary adaptation to this type of training is improvement in glycolytic flux and energy stores and/or fiber hypertrophy. With this type of training, even though slow oxidative fibers are recruited, they contribute little to the overall force production and subsequently will not adapt as much as the faster, larger-diameter fibers—the

primary contributors to this activity. Of course, most physical activities or athletic events require a combination of endurance and strength or power, and some require that different muscle groups be trained in different ways (e.g., triathlon, which involves swimming, biking, and running in succession). Specificity, though, applies not only to athletic performance but is important also in rehabilitation—whether in strengthening specific muscles in an older patient to enhance activities of daily living or in conditioning a worker to enable timely return to the workplace. Training programs, regardless of the purpose, need to be structured to meet the specific requirements of different movement patterns within a particular physical task. This necessitates knowledge of the requirements of the task and an understanding of the particular muscle adaptations that can be anticipated. (See chapter 1.)

The extent of muscle adaptation relates directly to the frequency with which particular motor units are recruited and the nature of the activity.

Training programs, regardless of the purpose, need to be structured to meet the specific requirements of different movement patterns within a particular physical task.

Clearly, specificity of training requires that the conditioning stimulus should mimic the exact pattern of movement or movements in which improvement is desired. Let's use an example to illustrate the importance of understanding and applying the principle of specificity.

This example comes from an experiment conducted by Thorstensson et al. The object was to increase quadriceps strength through eight weeks of squat training. As expected, the subjects increased quadriceps strength as measured by the maximum amount they could lift in one squat (called a 1 repetition maximum [RM]) (figure 9.3). But strength increases over the eight weeks in a leg press and knee extension (two exercises that utilize the same muscle group as the squat) were considerably less. In fact, the increase in knee-extension strength was negligible. These findings underline the importance of training the muscle in the same movement pattern in which improvement is desired. For specific strength or performance gains it is not enough to just increase the size of the muscle.

Figure 9.3   Training specificity. Subjects were tested for squat, leg-press, and knee-extension strength, then trained for eight weeks using squats only. Even though the same muscle group is used for all three movements, increases in strength with squat were much greater than with nonspecific leg press and knee extension.

Reprinted from Thorstensson et al. 1976.

Case Study

The rotator cuff muscles of the shoulder are often injured by repetitive overhead motion such as the tennis serve. It is important, then, to strengthen these muscles specifically for the movement pattern they are to perform. Your patient is a tennis player with pain in the long head of the biceps. After initially treating the tendinitis you prescribe an exercise program, using free weights, designed to strengthen the rotator cuff muscles. After a few weeks the athlete has returned to tennis and is pain free. She comes to you for advice on how to increase the power of her serve. You know the following facts:

- The power of the serve is supplied by a concentric contraction of the internal rotators.

- Eccentric contractions of the infraspinatus and teres minor provide the deceleration of the arm.

- Increases in strength are greater with the use of eccentric contractions during training than with concentric contractions.

You have a Cybex isokinetic dynamometer that enables the athlete to strengthen the rotator cuff muscles either concentrically or eccentrically. What is your approach to increase the power of her serve? Justify your choice. Design a simple experiment whereby you can test the hypothesis that either concentric or eccentric contractions increase the power of a tennis serve.

Compared with concentric training, training only with eccentric contractions will result in greater strength increases of the internal rotators. Recall that more force can be generated with eccentrics which results in a greater stimulus for strength gains. But, because greater increases in performance occur when training mimics the exact pattern of movement, concentric training will provide the greatest increase in the power of the serve. A very simple test of this is to have two groups train with either concentrics or eccentrics only and compare the speed of the tennis ball during the serve before and after the period of conditioning. (See Ellenbecker et al.)

The importance of training specificity cannot be overstated. To achieve optimum gains in muscle strength or performance, it is essential that the subject be trained in a manner that simulates the movement patterns of the activity. This principle is especially applicable to reconditioning after a sports injury. After diagnosis and treatment of the injury, rehabilitation should begin with general strength training followed by very specific conditioning to ensure optimum gains in performance.

> To achieve optimum gains in muscle strength or performance, it is essential that the subject be trained in a manner that simulates the movement patterns of the activity.

Training Overload

The second general training principle that is a key for optimizing performance gains is **overload**. Again, the concept is very simple: the muscle must be taxed beyond some critical level to achieve increases in structure and/or enzymatic activity. Increased synthesis of specific cellular components—in particular, proteins (enzymes, contractile, structural)—relates to the nature of the conditioning stimulus. For example, you would expect an increase in oxidative enzymes in response to endurance training, whereas you would expect an enhanced synthesis of contractile and structural protein in response to resistance training. Regardless of the type of training, the conditioning stimulus must increase gradually over time to produce consistent increases in muscle function. If the conditioning stimulus is fixed, the rate of change in the muscle will plateau and no further adaptation will occur. Altering training frequency, intensity, or duration, or more than one of these, can serve to increase the conditioning stimulus.

A common mistake in training programs is not allowing the body to recuperate sufficiently between exercise bouts. The conditioning stimulus should be discontinuous: periods of rest are needed for optimum increases in structure and function of the muscle cells. Overtraining actually hampers adaptive responses and often leads to overuse injuries. In addition, people who train when injured often alter the mechanics of their movement patterns. This may lead to an altered dispersion of impact forces (such as in running) and cause secondary injury in tissues unaccustomed to the newly imposed stresses.

In practice, the overload principle is achieved through use of **progressive resistance exercise**. For beginning strength trainers, it is important to start at a low intensity and volume to avoid undue fatigue and muscle soreness, then to overload the muscle gradually with sufficient rest periods.

The overload principle is achieved through use of progressive resistance exercise.

Conditioning and Deconditioning

To this point we have examined the major principles underlying skeletal muscle function and conditioning responses. In this section we will consider some of the mechanisms related to strength and power, as well as deconditioning responses.

Strength

A widespread misconception is that increases in strength are related solely to an increase in muscle size. As we are about to see, this is untrue.

Adaptations Resulting in Increased Strength

By examining figure 9.4 closely, it becomes clear that increases in strength are related to a blend of neuromuscular adaptations, not just to an increase in the size of the muscle fiber. Pay particular attention to the time course of the adaptations. It is apparent that early increases in strength (3-4 weeks) can be attributed almost exclusively to neural adaptations, whereas fiber hypertrophy is the reason for later changes. Neural adaptations are not well understood. In general, however, we know that unconditioned people are unable to recruit the fastest motor units (recall that motor units are recruited in order, from slowest to fastest), but that in the first few weeks of training they "learn" to recruit all the units. Remember also that motor units are usually recruited asynchronously; but, again, in the early stages of training, recruitment during maximum lifts becomes more synchronous. The last neurally related mechanism, inhibition of the GTO, is very intriguing. It has to do with a basic mechanism of the nervous system called disinhibition or gating—a principle of neuronal function that you will come across frequently.

➤ Hypertrophy is an increase in the size (cross-sectional area) of the muscle.

Early increases in strength can be attributed almost exclusively to neural adaptations, whereas fiber hypertrophy is the reason for later changes.

Case Study

A client in your fitness area makes the observation that his girlfriend increased her bench press, when expressed as a percentage change, more than he increased his. Her increase was made with very little apparent increase in muscle mass. How can this happen?

Several mechanisms at work here enable an increase in strength without an increase in the size of the muscle. One of these involves changes in the role of the GTO. Review the normal function of the GTO (page 89) before you determine how the GTO is most likely involved in disinhibition. What are some of the other potential adaptations?

The GTO is located in the myotendinous junction and is in series with the muscle fibers; therefore it is responsive to changes in force. Its normal function is to inhibit the muscle to some extent and to excite antagonists to prevent large forces from causing damage. But, as a person begins to train regularly, the input of the GTO is inhibited by input from higher brain centers (gating) at the level of the spinal cord. Inhibition of the inhibitory input of the GTO is called disinhibition. This and other neural adaptations (increased synchronization of motor units, better activation of the fastest motor units) to resistance training enable the person to get stronger without an increase in the size of the muscle.

Neural adaptations that enable increases in strength begin to peak at the same time the muscle fibers begin to increase in size (hypertrophy). Our knowledge of long-term increases in strength is scant because few long-term longitudinal studies have been conducted (most training studies last 8-10 weeks).

Studies using animal models have shown that fiber splitting and/or the generation of new fibers occurs when skeletal muscle is overloaded by one of a variety of methods. However, it is unclear whether the addition of new fibers (hyperplasia) in response to resistance training occurs in humans. Even if it does, the hyperplasia is probably limited and likely contributes little to the increase in size of the muscle. It is apparent, that the increase in size of human muscle in response to resistance training results from hypertrophy of individual muscle fibers, not fiber splitting or formation of new fibers (see page 123).

Resistance training also induces increases in tensile strength of the connective tissue and stimulates changes in the cytoskeleton (infrastructure that keeps the contractile elements aligned). Without these changes, increases in force-generating capabilities of the muscle would easily damage the support structures.

Training for Strength Versus Endurance

Strength gains are optimum when the resistance, or load, is high (85-90% of 1 RM), while muscle endurance is enhanced when loads of 20 RM or more are used (figure 9.5). Two to three sets of each exercise appear to produce the greatest benefit for strength gains; three to six sets are recommended for increases in muscle endurance. The length of rest periods between sets is important for optimum adaptations. Greatest strength gains occur when time for recovery of energy stores is sufficient (2-3 min between sets). For increases in muscle endurance, adaptations are

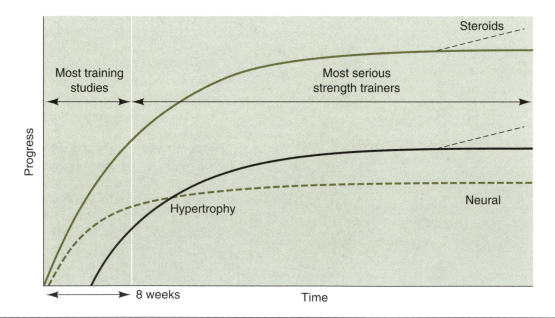

Figure 9.4 Strength training. Initially, strength gains are mediated by neural mechanisms, which peak at three to four weeks. At that time muscle size begins to increase (hypertrophy), soon predominating as the major contributor to strength gains. Administering ergogenic aids such as steroids can accelerate gains in strength.
Adapted from Sale 1988.

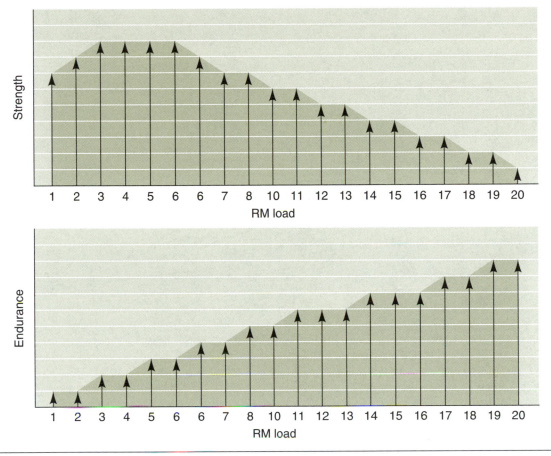

Figure 9.5 Muscle endurance versus strength. Greater increases in strength are achieved at low-repetition-maximum loads (high intensity). Muscle endurance is enhanced when intensity is lower and repetitions are higher.

better when rest periods are 30 s to 1 min. Note that adaptations resulting from muscle endurance training are different from benefits produced by cardiovascular training. Even with performance of several sets of resistance training with 30 s rest periods (circuit training), adaptations are largely local (confined to the muscle). Getman reports that increases in $\dot{V}O_2$max associated with circuit training are limited (approximately 5%). Much larger increases in cardiovascular endurance and $\dot{V}O_2$max are observed with aerobic-type training, and the adaptations are largely central (in the heart).

Strength gains are optimum when the resistance, or load, is high (85-90% of 1 RM), while muscle endurance is enhanced when loads of 20 RM or more are used.

Power

Power is a function of force and velocity ($P = F \times D/T$), making fast motor units the main contributors to powerful movements.

Peak Power

Fast muscle fibers are especially suited for explosive power generation. Examine figure 9.6 and you can see the advantage that fast fibers (largely a function of the high contraction velocity) have over slow fibers in generating instantaneous or peak power. The figure contains force and velocity curves for fast and slow fibers, with power curves superimposed. Note that peak power is reached at about 33% of both maximum velocity and force. There is such a large discrepancy between power capabilities of slow and fast fibers that the slow fibers contribute little if anything to

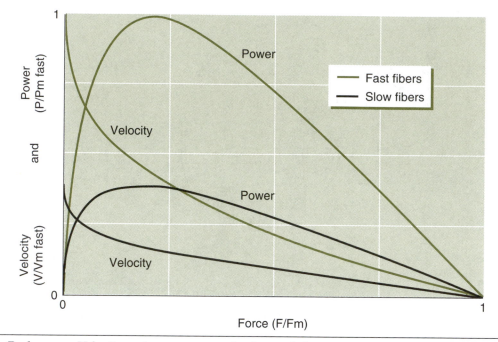

Figure 9.6   Peak power. Velocity and power curves for fast and slow muscle fibers. Fast muscle fibers are better suited to instantaneous power because of their higher speed of contraction. Peak power of slow fibers is about 30% that of fast. Note that peak power is achieved at about 33% of maximum force output. Velocity, power, and force are expressed relative to maximum (m) values for the fast fiber.
Reprinted from Faulkner 1984.

a quick, powerful movement. This relates to the basic contractile characteristics discussed earlier.

Sustained Power

Maintaining power over an extended period requires a different strategy. Figure 9.7 demonstrates that the fast glycolytic motor units fatigue very quickly and cannot sustain high levels of power longer than a few seconds. The fast oxidative motor units, however, have an initial drop in power, but then level off at slightly below 50% of maximum. Although the time scale ends at 5 min, an extrapolation of the curve indicates that the fast oxidative fibers may be able to sustain power for an extended period (Faulkner et al.). The key point is that fast oxidative fibers are the primary contributors to sustained activity that requires a moderate power output. This may even include running a marathon (at the higher speeds, of course). It is a common misconception that slow fibers are the primary contributors to all aerobic activities. This is clearly not true, since the power capabilities of the slow fibers are much lower than those of fast fibers. Refer to the figure and notice that the power of the slow motor units is very well maintained over time, but begins at a low level.

The concepts discussed here apply to the principle of specificity, and you should be able to use this information in training or rehabilitation.

Fast oxidative fibers are the primary contributors to sustained activity that requires a moderate power output.

Muscle Hypertrophy

As discussed previously, an increase in the size of individual muscle fibers (hypertrophy), rather than an increase in the number of fibers (hyperplasia), is responsible for the increase in muscle mass seen with resistance training. Even though animal studies have demonstrated an increase in fiber number with training, the contribution to increases in muscle mass in humans is minimal at best. Muscle hypertrophy involves the response of the cell to a mechanical stress. According to Clarke and Feeback, it appears that the initial event is microscopic tears in the sarcolemma. This **sarcolemmal "wounding"** initiates a cascade of events that includes activation of satellite cells and release of hormones and eventually results in the addition of contractile and cytoskeletal pro-

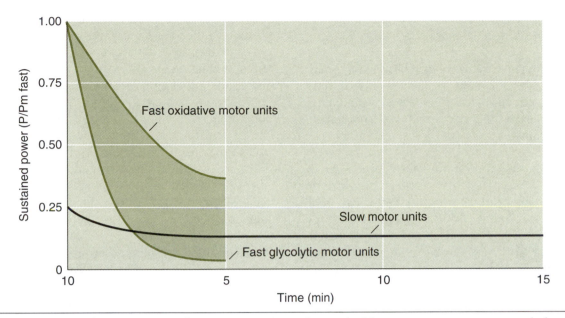

Figure 9.7 Sustained power. Fast oxidative motor units are best suited for power output over extended periods. Fast glycolytic fibers fatigue quickly, and slow fibers have low power outputs initially. Reprinted from Faulkner 1984.

teins. Figure 9.8 illustrates the sequence of events that occur in muscle hypertrophy.

An increase in the size of individual muscle fibers (hypertrophy), rather than an increase in the number of fibers (hyperplasia), is responsible for the increase in muscle mass seen with resistance training.

It is important to distinguish here between **anabolism** and **catabolism**. As just mentioned, the hypertrophic response is initiated by microscopic tears in the muscle fiber cell wall, but the cell itself does not break down—the cell membrane maintains its integrity. When muscle cells are damaged, the sarcolemma is often torn; this allows cellular contents to leak out and substances outside the cell, like calcium, to leak in. Under these conditions the cell will begin to break down (catabolism) and may even be completely destroyed. In conditions that stimulate hypertrophy, the cell adds protein (anabolism) and is not broken down. It is important to distinguish between tearing or injuring the contractile elements of the fiber (catabolism) and sarcolemmal wounding, which leaves the cell intact and stimulates hypertrophy.

The two growth factors involved in this process are **fibroblast growth factor (FGF)** and **insulin-like growth factor I (IGF-I)**. Fibroblast growth

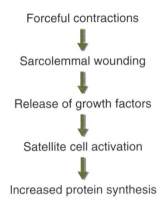

Figure 9.8 Muscle hypertrophy. The signal triggering muscle hypertrophy is sarcolemmal wounding. Small microtears in the cell membrane initiate release of fibroblast growth factor, which activates satellite cells.

factor is located in the cell membrane, is released in proportion to the amount of wounding, and activates satellite cells. Recall from chapter 8 that satellite cells are primitive muscle cells that are trapped between the basal lamina and cell membrane during development. When activated, these cells divide, proliferate, and add protein to that already existing in the cell. Insulin-like growth factor I, on the other hand, is released from inside the cell, and its major function is to increase amino acid transport into the cell. Amino acids are the building blocks of proteins.

It is important to distinguish between tearing or injuring the contractile elements of the fiber (catabolism) and sarcolemmal wounding, which leaves the cell intact and stimulates hypertrophy.

Now that we understand these processes, the question arises: which is the best stimulus for muscle hypertrophy? You may have guessed already that the most important factor is the amplitude of the mechanical stress or the intensity. Frequency and duration of the activity are also important, but less so. As a matter of fact, most people tend to train too frequently, which promotes the catabolic processes and may lead to overuse injuries. The nature of the training stimulus is also important. That is, eccentric contractions produce a greater hypertrophic response than concentric contractions at the same relative stimulus, and have a lower threshold for inducing a response. What are the differences between eccentric and concentric contractions that may account for the differences in inducing the hypertrophic response? Keep in mind the nature of each type of contraction (more stress and strain with eccentric contractions).

➤ Stress is force per cross-sectional area.

➤ Strain is change in length relative to resting length.

Many people train too frequently, which promotes the catabolic processes and may lead to overuse injuries.

Muscle Atrophy

Muscle mass is maintained under normal conditions by a balance between protein synthesis and protein degradation. In the hypertrophic response, it is obvious that protein synthesis is accelerated, resulting in a net gain in protein. Muscle **atrophy** (reduction in size), on the other hand, is a response to reduced use or disuse of the muscle. Immobilization or unloading a limb, denervation, and prolonged bed rest are typical conditions that produce muscle atrophy. Of course, a decrease in or cessation of training (deconditioning) also results in a reduction in muscle mass. Typically, no matter what the stimulus is, atrophy begins with an almost immediate reduction in protein synthesis. Protein degradation begins to increase several days after the initial stimulus, and the two together produce a very quick loss of muscle mass. For conditions of immobilization or unloading, the greatest loss of muscle mass occurs in the first three weeks. This is a key point for therapists, because it takes several weeks or months to produce the hypertrophic, or conditioning, response. In addition, recovery of muscle mass with reconditioning, or remobilization, typically takes two to three times longer to occur than the atrophic response does. Confounding the situation is an increase in intramuscular connective tissue concentration that accompanies atrophy. This increases muscle passive stiffness, which predisposes the person to connective tissue tearing if the reconditioning program is too vigorous or if muscle stretching is not emphasized as an important part of the reconditioning process.

Typically, no matter what the stimulus is, atrophy begins with an almost immediate reduction in protein synthesis.

Unloading and Bone Mass

Prolonged bed rest or other forms of unloading cause loss of bone mass as well as muscle atrophy. Bloomfield indicates that because bone turns over more slowly than muscle, the decrease in bone mass with unloading lags behind decreases in muscle mass (figure 9.9). The highest rate of bone loss occurs in the first 6 to 7 weeks after unloading; the rate of loss begins to level after 18 weeks. If limb unloading is prolonged, recovery of bone mass may take months (figure 9.10). As a matter of interest, the earlier astronauts were observed to have decreased bone density up to five years after the completion of their space flight.

Prolonged bed rest or other forms of unloading cause loss of bone mass as well as muscle atrophy.

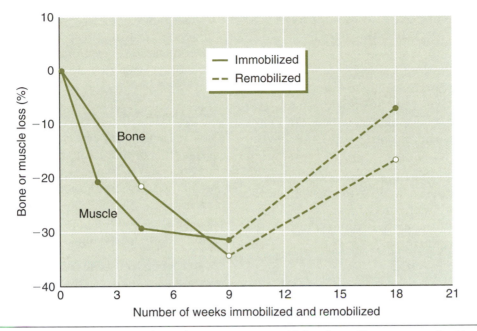

Figure 9.9 Bone and muscle loss. Loss of muscle mass is more rapid than loss of bone as a result of immobilization, but bone mass is regained at a slower rate after remobilization.

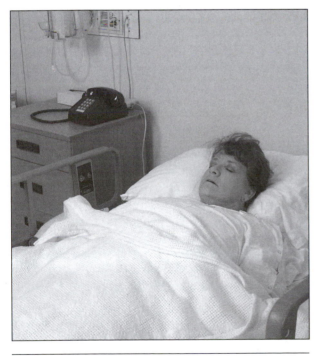

Figure 9.10 Enforced bed rest. The "use it or lose it" principle is well illustrated by prolonged bed rest. Significant periods during which bone and muscle are unchallenged will result in major losses of both muscle and bone mass. Regaining lost strength and bone mass will take considerably longer than losing it did.

General Principles of Rehabilitation

The previous discussion underlines the importance of intervention therapy in conditions under which muscle atrophy occurs. Electrical stimulation or very specific resistance exercises may be used to delay or prevent the atrophic response. Remember, because the atrophy response is so rapid, the longer the time is between the initial insult and the first treatment, the less effective any treatment becomes.

It is essential that you always keep the following key points in mind as you design any rehabilitation program:

1. Recall from our previous discussion that prior level of conditioning is a major determinant of which muscle types will be most affected by reduced use. Knowing this will help the health professional selectively train the muscle groups or motor units most affected by the deconditioning stimulus.

2. An increase in intramuscular connective tissue that accompanies muscle atrophy produces an increase in passive stiffness. Carefully

monitored stretching exercises need to be incorporated into the reconditioning program.

3. After prolonged unloading or immobilization, muscle mass and strength will return to normal before bone mass and tensile strength. The danger here is an imbalance between muscle strength and bone mass that may result in bone fracture. Increases in muscle strength need to be gradual.

4. All these problems are magnified in persons who are elderly; people in this category, according to Bloomfield, may have balance, kinesthetic, and endurance deficits in addition to increased passive stiffness.

Case Study

You have an active older patient (72 years) who is a runner (24 to 32 km/week, or 15-20 miles/week); she also does squats and knee-extension exercises three times per week. Suddenly she is confined to bed rest for five to six weeks. Outline the deconditioning responses in the lower extremities that you would expect and also how you would approach a strengthening and endurance program. What precautions do you need to take when reconditioning this patient?

In response to disuse, rate of muscle protein synthesis begins to decrease within 24 hr, and rate of protein degradation increases within two to three days. Together, these two processes produce a dramatic decline in muscle mass in the first three weeks. Thereafter, muscle mass continues to decline but at a much slower rate. The muscles of the lower extremity are more affected (both fast and slow) than upper-extremity muscles because of the patient's prior training. You can intervene here with electric stimulation or, if possible, low-intensity dynamic resistance exercise (especially in lower extremity) to delay atrophic response. The rate of bone loss lags behind that of muscle loss but is still significant after five to six weeks. Intramuscular connective tissue concentration will increase, so when rehabilitation begins, the patient should start with ROM exercises to increase flexibility. Ambulation will reintroduce weight bearing and arrest loss of bone, but be cognizant of balance problems. Introduce low-resistance exercises, and gradually increase load over time. Keep in mind the imbalance between muscle strength and bone density—muscle strength returns more quickly.

Exercise-Induced Muscle Injury

Most muscle injury associated with overexertion or forceful contractions produces minor discomfort and does not limit performance to any great extent. So why should we bother examining the processes involved? Interest in this area has increased tremendously in the past few years and relates to the following:

• **Oxygen free radical damage.** More and more evidence is emerging that oxygen free radicals (OFR) produced during exercise may damage muscle tissue but, more importantly, promote cellular breakdown in other organs including the heart, liver, and kidney. Oxygen free radicals, which are slightly altered forms of oxygen, are very reactive with, and break down, lipids, proteins, and nucleic acids. The body has natural defense mechanisms that neutralize OFR, but these are easily overwhelmed in conditions under which OFR production is increased (exercise, inflammation, cigarette smoke, air pollutants, hyperoxia, and many diseases). At-risk populations for oxidative damage are people deficient in vitamins and zinc, smokers, people with chronic inflammatory diseases and/or lowered immune function, and those with high levels of blood low-density lipoproteins (LDL). Regular aerobic exercise heightens the body's natural defense against oxidative stress, and supplementation with antioxidant substances like vitamins A, C, and E also appears to prevent some of this damage.

➤ Low-density lipoprotein (LDL) is a protein lipid complex that transports the lipid in the blood.

Regular aerobic exercise heightens the body's natural defense against oxidative stress, and supplementation with antioxidant substances like vitamins A, C, and E also may prevent some of this damage.

• **Long-term muscle damage.** A single bout of damage-producing exercise may not be cause for concern, but more and more evidence shows that repetitive exercise-induced muscle injury eventually results in a decreased ability to repair the tissue. This may relate to a declining population of satellite cells. Over time, as muscle is injured again and again, it is replaced with fibrous connective tissue.

• **Kidney and liver damage.** Muscle injury associated with unaccustomed forceful eccentric contractions, which result in a large efflux of protein into the blood, has caused kidney and liver failure. In rare cases the condition has been fatal.

• **Altered running mechanics.** Repetitive muscle damage that leads to muscle weakness in the lower extremities is linked to an alteration in running mechanics that predisposes the person to other more serious injuries such as stress fractures, as well as ligament or tendon damage.

• **Age-related damage.** Recent evidence has shown that elderly people are more susceptible to exercise-induced muscle injury and have a diminished capacity to repair the tissue. Damaged muscle fibers have been observed in older exercisers up to six months after a single bout of exercise. The problem is heightened as a result of increased awareness of the benefits of regular exercise programs for persons who are elderly and subsequent increased participation.

> Elderly people are more susceptible to exercise-induced muscle injury and have a diminished capacity to repair the tissue.

Two Phases of Injury

Figure 9.11 presents the typical time course of an eccentric contraction-induced injury. The index used to measure injury is maximum muscle force generation, the most accurate measure available. Note the initial decline in force, followed by an increase (the initial decline is related to muscle fatigue plus cell damage) at 3 to 4 hr. From this point, force decreases again over the next two to three days before the repair process begins to restore contractile function. The recovery of maximum force production usually takes two to three weeks and is dependent on the severity of the initial injury.

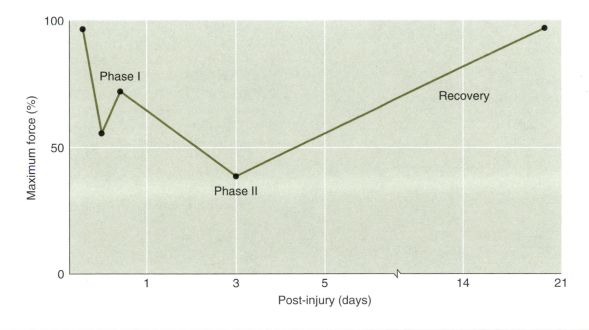

Figure 9.11   Typical course of eccentric contraction-induced injury. Muscle is mechanically injured initially; this disrupts contractile elements. The initial decline in force is related to fatigue plus cell breakdown. Within two to three days, further damage is observed and is associated with inflammatory processes. Repair process takes days or weeks depending on severity.
Adapted from Faulkner 1989.

The injury process, then has two phases:

• **Phase I.** The initial injury is mechanical and is related to structural damage, mostly to the Z line area of the sarcomeres (figure 9.12). In more severe cases the sarcolemma is also damaged, and this may lead to destruction of the entire fiber. The initial damage relates more to "active strain," that is, the change in fiber length relative to resting length, than to the absolute force. The more a fiber (or muscle) is stretched when activated, the more damage occurs. Also, more damage occurs when the eccentric contraction starts from a longer resting length as compared to a shorter resting length. For example, the biceps incurs more damage when a person lowers a weight from 90° of elbow flexion to 180° than from full elbow flexion to 90°. This point is important to remember when you are prescribing strengthening exercises that have an eccentric component.

The more a fiber (or muscle) is stretched when activated, the more damage occurs.

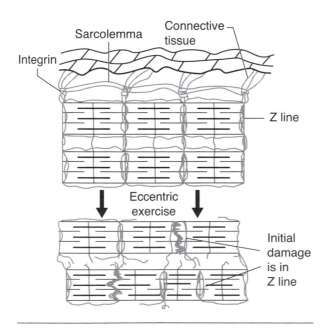

Figure 9.12 Eccentric contraction-induced injury. Initial damage to muscle fiber is located at the Z line, resulting in Z-line "streaming" or splitting of the structure. This results in a loss of transverse register of sarcomeres and a decline in force production capability.

• **Phase II.** The second phase of the injury relates to the inflammatory process and the further degradation of injured cells by invading macrophages. It is this phase that is associated with the clinical manifestation of muscle injury—**delayed-onset muscle soreness** or **DOMS**. Inflammatory mediators such as bradykinin, histamine, and serotonin all stimulate pain receptors and thus produce the pain associated with this phase. In addition, edema formation may cause an increase in tissue pressure and add to the soreness.

Repair Process

Most exercise-induced muscle damage is focal and is easily repaired without widespread remodeling. Occasionally, though, severe muscle injury causes widespread destruction of entire muscle fibers. A muscle fiber that literally has been destroyed has the ability to regenerate completely, forming a new fiber that is indistinguishable from the original. In cases of both minor and extensive damage to the muscle fiber, activation of satellite cells begins. These primitive muscle cells that are trapped between the basal lamina and cell membrane during development (see chapter 8), start to divide and fuse, eventually replacing the cell constituents that were destroyed. If the cell is severely injured, the entire cell will break down and will be replaced in its entirety by a new cell. This process is dependent on the arrival of macrophages, which phagocytize the debris from the damaged cell (figure 9.13). As already mentioned, though, repeatedly injuring a cell may result in a loss of the ability of satellite cells to complete the repair process.

➤ Macrophages are a class of mononuclear phagocytic cell involved in the inflammatory process.

➤ To phagocytize means to ingest bacteria, other cells, and foreign particles.

Prevention—Phase I

The initial injury from eccentric contractions relates to breakdown of structural components within the muscle fiber as the activated muscle is forcibly lengthened. The mechanism, according to Morgan, appears to be linked to inhomogeneity of sarcomere lengths, which leads to "stronger"

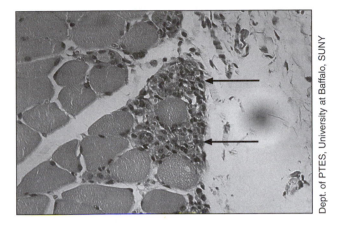

Dept. of PTES, University at Buffalo, SUNY

Figure 9.13 Cross-section of skeletal muscle that has been injured by repeated eccentric contractions. Damaged fibers have been invaded by macrophages (see arrows), which phagocytize cellular debris to enable fiber regeneration to occur.

sarcomeres pulling apart "weaker" ones, mostly at the Z lines. It is almost impossible to prevent this type of damage, but it is possible to minimize injury by

- introducing eccentric activity slowly into the exercise regimen and
- avoiding stretching the muscle from long resting lengths.

Two other factors can reduce the amount of damage. One is the prophylactic effect of a prior bout of exercise. Apparently the muscle makes an adjustment when it is forcibly stretched repeatedly. Sarcomeres are added to the muscle in series, and as Morgan indicates, this has the effect of making the sarcomeres more homogeneous in length. With the removal of the length discrepancies, sarcomeres are more evenly matched in force production, and less tearing at the connections (Z lines) occurs. Other evidence has emerged that changes in intramuscular connective tissue impact the severity of eccentric contraction-related damage. Recall that the endomysium (surrounding each fiber) is connected to the Z line by way of the cell membrane protein integrin. It makes sense, then, that any strengthening or weakening in the connective tissue will have a direct impact on contraction-induced injury. Changes in connective tissue concentration and tensile strength have been observed with immobilization, unloading, denervation, aging, and training.

Any strengthening or weakening in the connective tissue will have a direct impact on contraction-induced injury.

Prevention—Phase II

Recall that the second phase of the injury process is mediated by inflammatory substances and that it includes the production of large amounts of OFR. It seems reasonable, then, that treatment with anti-inflammatories and/or antioxidants would attenuate this secondary injury. Surprisingly, the research in this area is equivocal. That is, some anti-inflammatories like bromelain (a natural substance contained in pineapples) and ibuprofen seem to have potential, but some studies have shown negative results. There has been much research in the last few years on the potential for antioxidants (vitamins A, C, and E) to attenuate exercise-induced injury. These vitamins seem to be effective in attenuating cell damage associated with prolonged exercise, but not damage caused by short-term exercise and associated with forceful eccentric contractions. Thus, the mechanism of injury and the potential for protection by antioxidants and/or anti-inflammatories may be related to the nature of the activity. During prolonged exercise, OFR are produced as a result of the prolonged elevation in O_2 consumption. Oxygen radicals, produced in the mitochondria, overwhelm the natural defenses and break down the lipid bilayer in the cell membrane (a process called lipid peroxidation). The reactive oxygen species also attack and break down proteins. Vitamin supplementation may help prevent this type of exercise-related cell damage by neutralizing the oxygen radicals. Vitamin E is especially effective as an antioxidant because it is fat soluble and becomes embedded in the cell membrane.

Vitamin supplementation may help prevent exercise-related cell damage by neutralizing the oxygen radicals.

Case Study

Your friend Brianne is a vegetarian athlete who is vitamin and zinc deficient. Zinc is a component of an enzyme called superoxide dismutase that is part of the body's natural defense against oxidative stress. Runners are exposed to repetitive eccentric contractions, especially upon heel strike, and prolonged O_2 consumption. Explain the differences in the types of muscle damage that Brianne may be exposed to during a marathon, and outline what she can do before running the marathon that may help to prevent the injury.

The major source of zinc in our diet is red meat, so Brianne needs to supplement her diet with zinc. As a runner she should also consider taking vitamin C in combination with vitamin E to prevent damage from OFR associated with prolonged exercise. Vitamin C, which is water soluble, acts as the first line of defense against OFR in the blood. It also regenerates vitamin E. Vitamin E is lipid soluble and is contained in cell membranes. This will help combat free radical damage to cell membranes. During the marathon,

Brianne is exposed to a repetitive eccentric contraction injury that occurs during the heel strike and that most likely affects the tibialis anterior muscle the most. She can help prevent these injuries by performing a downhill run about two weeks before the marathon. This will have a prophylactic effect, which will help prevent the eccentric injury.

We have learned in this chapter that skeletal muscle is very plastic and that it will readily adapt to a variety of stimuli in very characteristic ways. To be able to effectively apply basic principles of exercise conditioning for either enhanced performance, prevention of disease or injury, or rehabilitation, it is vital to understand the relationship between structure and function of skeletal muscle and the underlying cellular responses to increased or decreased use.

What You Need to Know From Chapter 9

Key Terms

isometric contraction

anabolism

atrophy

catabolism

concentric contraction

delayed-onset muscle soreness (DOMS)

dynamic contraction

eccentric contraction

fibroblast growth factor (FGF)

insulin-like growth factor I (IGF-I)

isokinetic contraction

isotonic contraction

overload

oxygen free radicals (OFR)

progressive resistance exercise

repetition maximum (RM)

sarcolemmal "wounding"

specificity

static contraction

Key Concepts

1. You should be able to list the benefits and limitations of using different types of contractions for resistance training.

2. You should be able to explain the importance of the principles of specificity and overload when designing a training program.

3. You should be able to describe the mechanisms that underlie neural and hypertrophic adaptations to resistance training.

4. You should be able to explain the rationale for specific training programs for increases in muscle strength, endurance, and power.

5. You should be able to describe the basic processes of muscle hypertrophy and atrophy.

6. You should be able to explain the general principles of rehabilitation for deconditioned skeletal muscle.

7. You should be able to describe the processes of exercise-induced muscle injury and repair.

8. You should be able to specify the measures to undertake for preventing exercise-induced muscle injury.

Review Questions

1. What are the limitations of the use of isometric versus dynamic resistance training for strength gains?

2. What are some advantages and disadvantages of isokinetic resistance training?

3. What two mechanisms facilitate an increase in vertical jump performance using plyometric training? Over time, what is the increase in muscle power attributed to?

4. Describe the importance of specificity and overload when one is planning a conditioning program aimed at rehabilitating an athlete for optimum performance.

5. Support or refute the following statement: All aerobic activities are performed using primarily slow oxidative fibers.

6. Describe the differences between eccentric and concentric or isometric contractions that result in greater hypertrophy with eccentric training.

7. What is the stimulus for muscle hypertrophy? What are the roles of the growth factors IGF-I and FGF in regulating hypertrophy?

8. Which fiber type is especially suited for movements requiring explosive power? Why?

9. Describe the two phases of contraction-induced muscle injury.

10. From the following choices, identify the movement (involving moving a 9 kg [20 lb] weight held in the hand) that would cause the most injury to the biceps muscle, and explain why: (a) lifting the weight from 90° at the elbow to full elbow flexion, (b) lifting it from full elbow extension to 90°; (c) lowering the weight from full elbow flexion to 90°, or (d) lowering it from 90° to full elbow extension.

11. A local high school athlete has just been taken to the hospital for near kidney failure. He had returned from a gridiron football camp and was all excited about some new training techniques that saved time and gave maximum strength gains. The coaches at the camp had introduced the players to the concept of eccentric resistance training. This young athlete was unaccustomed to performing eccentric-only exercise training, but doubled up on these exercises anyway to take advantage of the potential for greater strength gains. You have been contacted by a local TV station to comment on this situation. How will you explain the hospitalization, and what will your advice be?

Bibliography

Bloomfield SA. Changes in musculoskeletal structure and function with prolonged bed rest. *Med Sci Sports Exerc* 1997, 29:197–206.

Clarke MSF and Feeback DL. Mechanical load induces sarcoplasmic wounding and FGF release in differentiated human skeletal muscle cultures. *FASEB J* 1996, 10:502–509.

Coyle EF, Feiring DC, Rotkis TC, Cote RW, Roby FB, Lee W, and Wilmore JH. Specificity of power improvements through slow and fast isokinetic training. *J Appl Physiol* 1981, 51:1437–1442.

Eberstein A and Eberstein S. Electrical stimulation of denervated muscle: Is it worthwhile? *Med Sci Sports Exerc* 1996, 1463–1469.

Ellenbecker TS, Davies GJ, and Rowinski MJ. Concentric versus eccentric isokinetic strengthening of the rotator cuff. *Am J Sports Med* 1988, 16:64–69.

Faulkner JA, Claflin DR, and McCully KK. Power output of fast and slow fibers from human skeletal muscles. In *Human muscle power*, ed. NL Jones, N McCartney, and AJ McComas, pp. 81–94. Champaign, IL: Human Kinetics, 1986.

Fleck SJ and Kraemer WJ. Resistance training: Basic principles (part 1). *Phys Sportsmed* 1988, 16 (3):160–171.

Fleck SJ and Kraemer WJ. Resistance training: Physiological responses and adaptations (part 2). *Phys Sportsmed* 1988, 16 (4):108–123.

Fleck SJ and Kraemer WJ. Resistance training: Physiological responses and adaptations (part 3). *Phys Sportsmed* 1988, 16 (5):63–74.

Getman LR. Physiological effects on adult men of circuit training and jogging. *Arch Phys Med Rehab* 1979, 60:115–121.

Komi PV. Training of muscle strength and power: Interaction of neuromotoric hypertrophic and mechanical factors. *Int J Sports Med* 1986, 7 (suppl.):10–15.

Kramer WJ and Fleck SJ. Resistance training: Exercise prescription (part 4). *Phys Sportsmed* 1988, 16 (6):69–81.

Lundin P and Berg W. A review of plyometric training. *Natl Strength Cond Assoc J* 1991, 13:22–30.

Lynn R and Morgan DL. Decline running produces more sarcomeres in rat vastus intermedius muscle fibers than does incline running. *J Appl Physiol* 1994, 77:1439–1444.

Morgan DL. New insights into the behavior of muscle during active lengthening. *Biophys J* 1990, 57:209–221.

Sale DG. Neural adaptation to resistance training. *Med Sci Sports Exerc* 1988, 20 (suppl.):s135–s145.

Thorstensson A, Hulten B, Von Doblem W, and Karlsson J. Effect of strength training on enzyme activities and fibre characteristics in human skeletal muscle. *Acta Physiol Scand* 1976, 96:392–398.

Waterman-Storer CM. The cytoskeleton of skeletal muscle: Is it affected by exercise? A brief review. *Med Sci Sports Exerc* 1991, 11:1240–1249.

Muscular Diseases

This chapter will discuss

➤ the pathophysiology of several common muscular diseases with particular emphasis on how these diseases affect exercise and

➤ how exercise might affect these diseases.

You should now understand how muscular activity is controlled and affected by the neuromuscular systems. This chapter presents information on the pathophysiology of muscle diseases and how these pathologies are reflected in exercise responses in these patients. You should pay particular attention to the information on the specific muscular abnormalities associated with each pathology and to the progression of the disease. By understanding the specific effects of the disease you will be better able to evaluate and devise appropriate exercise interventions.

Muscular Dystrophy

The **muscular dystrophies (MDs)** include two common and several less common forms of a genetic disease linked with the X chromosome. Because of this linkage, they are more common in males than in females. The two most common forms of MD are Duchenne's and Becker's.

Duchenne's is the most common and most severe form, occurring in approximately 1 in 3500 males. The progression of Duchenne's MD is faster, with death occurring when patients are in their 20s or 30s; people with **Becker's** may live a near-

normal life span. Muscular dystrophy is characterized by a progressive weakness and degeneration of striated muscles in particular, but cardiac and smooth muscles also are affected. The disease is usually diagnosed in infancy or early childhood, but in some cases not until adulthood.

Diagnosis is made through examination of a muscle biopsy, electromyography, nerve conduction velocity tests, or blood enzyme tests. Examination of the muscle biopsy can show early signs of structural disturbance and, aside from genetic testing, is the test most likely to be used for diagnosis. As muscles degenerate, muscular enzymes can leak from inside the cells into the blood, making blood analysis for the muscle creatine kinase enzyme a valuable diagnostic tool.

Pathophysiology

The defective gene that results in MD encodes a protein called **dystrophin**. This protein appears to be involved in stabilizing the muscle membrane during contraction and relaxation, as part of the linking of the intracellular cytoskeleton and the extracellular matrix. The dystrophin protein is also involved in the differentiation of muscle fibers into the fast glycolytic type. Examination of the muscle biopsy (figure 10.1b) shows a great variety of muscle fiber sizes, with many very large and very small fibers. In addition, there is evidence of fiber splitting and fiber degeneration and regeneration.

> MD is caused by a lack of the muscle protein dystrophin, resulting in a disruption of normal muscle structure.

➤ Phagocytosis is the process by which cellular matter is surrounded by a specialized cell that then ingests and digests that matter. This is one mechanism by which cellular debris is removed from the body.

In the later stages of the disease, contractile tissue is almost completely replaced by connective tissue and fat. Since the diaphragm is a striated muscle, this dystrophin defect results in a progressive compromise in diaphragm function and eventually failure. Cardiac muscle also contains a form of dystrophin that is affected by the mutation. The abnormality causes tissue degeneration and can ultimately lead to heart failure.

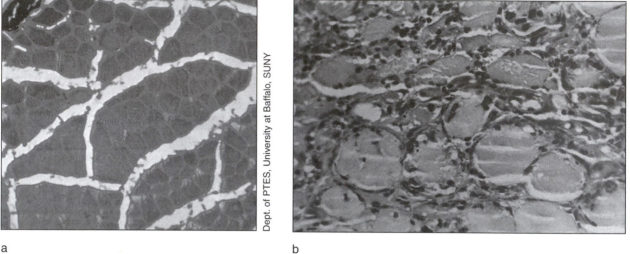

a

b

Dept. of PTES, University at Baffalo, SUNY

Dept. of Pathology, University at Baffalo, SUNY

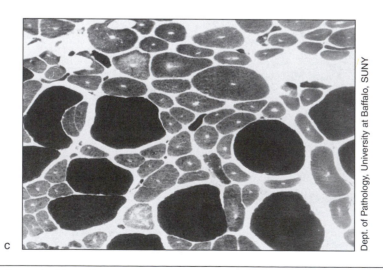

c

Dept. of Pathology, University at Baffalo, SUNY

Figure 10.1 (a) Normal muscle fiber. The connective tissue has been stained to show how completely it surrounds each fiber as well as how clearly differentiated it is from the muscle fibers. (b) Muscle fiber affected by muscular dystrophy. Note the invasion of the fibers by connective tissue, and phagocytosis. (c) Limb girdle dystrophy.

Clinical Manifestations

Weakness and poor development of physical abilities are common signs of MD during the first several years of life. Children with MD walk on their own at a later age and may have difficulty navigating. The weakness is first manifested in proximal muscles and then progresses to more distal muscles. In addition, these children assume a characteristic gait that is described as waddling. To compensate for the loss of lower-limb strength, young children with MD use their arms to assist in standing from sitting on the floor. Finally, an early manifestation of the disease is a **pseudohypertrophy** that is expressed clearly in enlargement of the calf muscle. This muscle hypertrophy is caused initially by enlargement of the muscle fibers, then by replacement of the fibers with fat and connective tissue.

MD is a progressive disease starting with weakness in the proximal muscles and spreading to the distal muscles.

By the early teens the progressive muscle weakness results in the patient's need to remain in a wheelchair. As the weakness progresses through the spinal muscles, severe scoliosis develops; this interferes with normal respiratory movements and, in combination with diaphragmatic and respiratory muscle weakness, can result in respiratory failure.

Treatment

The only medications shown to be of some benefit in MD are the **corticosteroids**. These steroids slow the progression of the disease in some patients. In some studies, corticosteroid treatment of MD has been associated with an extension of the age at which the patient can walk without assistance. Corticosteroids, however, have many downsides, including a diabetes-like condition, weight gain, and ultimately a loss of muscle tissue. Other drug treatments, including anabolic steroids, are currently under evaluation. The transfer of healthy, dystrophin-producing **myoblasts**, or immune muscle cells, into the muscles of MD patients has not been effective.

Exercise has been promoted as a means to stimulate muscle development and possibly delay the effects of MD on patients. Exercise, particularly in patients with slowly progressing MD

(Becker's), seems to improve function for varying periods of time. Exercises that emphasize mobility should be the focus, as they are more likely to allow the patient to sustain independence longer. Since failure of the respiratory muscles is a particular problem with MD, several studies have attempted to use specific respiratory muscle training to slow the progressive decline in respiratory function. Breathing against a resistance for six months results in improvements in respiratory muscle strength, but the gains are lost once the conditioning stops. The effects of prolonged training are unknown. Figure 10.2, based on the results of several studies, illustrates a typical improvement/deterioration rate for an MD patient when she exercises versus when she does not exercise. Moderate exercise appears to have no deleterious effect on the muscles of most patients; and since a lack of activity leads to a deterioration of muscle function, many who treat MD encourage the patient to be active. Some, however, express concern that a possible local exercise-induced inflammatory response might exacerbate the deterioration of MD. Swimming is promoted by many in the field because it is weight supported and would appear to carry less risk of exercise-induced muscle damage.

Exercise has been promoted as a means to stimulate muscle development and possibly delay the effects of MD on patients.

Muscle contractures are common in MD. Contractures can occur when muscle agonists and antagonists deteriorate at different rates. Appropriate rehabilitative exercise may help prevent these contractures. Aggressive flexibility exercises help to minimize the impact of contractures in susceptible patients. Daily full range-of-motion stretching with moderate exercise helps to prevent and slow the progression of contractures.

Congenital Myopathies

Congenital myopathies are distinguished from the MDs in that the dystrophin mutation is not involved. The more common congenital myopathies include

- central core disease,
- minicore/multicore disease,

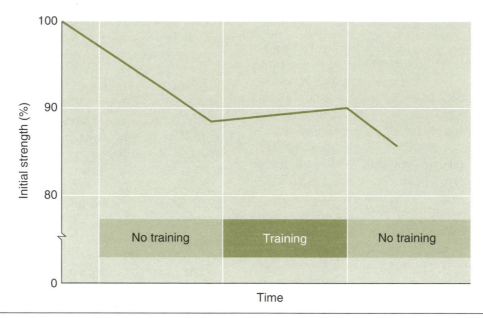

Figure 10.2 Idealized drawing showing the changes in strength during a period with no exercise therapy compared to a period of therapy in patients with muscular dystrophy. Training periods ranged from months to one year. The results are similar for muscles of both the upper and lower body.

- nemaline (rod) myopathies, and
- myotubular myopathy.

An initial diagnosis can be made on the basis of early muscle weakness, but a final diagnosis must be made on examination of a muscle biopsy.

Pathophysiology

All of the pathologies just listed are inheritable. These myopathies are similar in that they all result from alterations in muscle structure. The myopathies are identified by the characteristic morphological changes observed with each. For instance, metabolically inactive cores within muscle fibers characterize **central core disease (CCD)**. **Minicore/Multicore disease** is differentiated from central core disease by the oval shape of the cores. **Nemaline (rod) myopathies** show thin rodlike structures within the muscle fiber, and **myotubular myopathy** is identified by the appearance of fetal-like myotubes on examination of the biopsy (figure 10.3*b*). Congenital fiber-type disproportion (figure 10.3*c*) is less common, but is characterized by exceedingly large, interspersed with very small, groups of fibers. Limb-girdle muscular dystrophies are characterized by involvement of the upper or proximal parts of the limbs. The fiber types affected by the congenital myopathies are variable; type I fiber dominates

in CCD and nemaline rod disease, but type II fibers dominate in multicore disease. In spite of the type I dominance in CCD, the existing fibers are smaller than normal. Congenital fiber-type disproportion (figure 10.3*c*) is less common.

Clinical Manifestations

All the congenital myopathies are associated with muscle weakness. The degree of weakness is widely variable within each disease, and symptoms can appear during infancy or not until later in childhood. Central core disease is associated with hypotonia and characterized by proximal muscle weakness, with the legs affected more than the arms. Minicore/multicore disease generally manifests within the first year of life as weakness in both the upper and lower body. Nemaline myopathies include at least four variations, some becoming evident in infancy and others only in the 5- to 15-year-old age group. They generally present as a distal weakness, particularly in the legs, with the ankle dorsiflexors greatly affected. Myotubular myopathy has its onset in infancy, and the mean age of survival is less than six months. Although all these myopathies either do not progress or progress slowly, some forms are severe at onset; these can interfere with cardiorespiratory function and may be fatal.

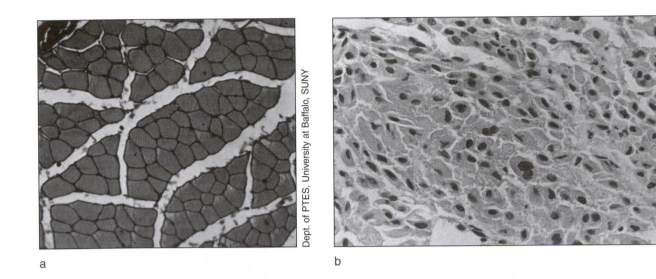

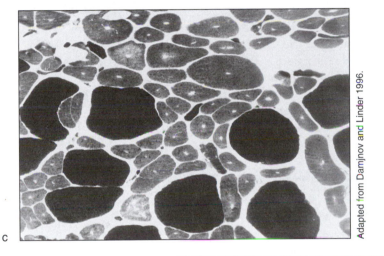

Figure 10.3 Comparison of normal muscle fibers with those affected by various congenital myopathies. (*a*) Normal muscle fiber. (*b*) Muscle fibers affected by myotubular myopathy. (*c*) Muscle fibers affected by centronuclear myopathy; the type 2 fiber hypertrophy and type 1 atrophy also are typical for congenital fiber-type proportion.

Treatment

Treatment of the congenital myopathies is oriented to maintaining muscle function. Whether exercise targeted to stimulate type I fiber development might be an effective treatment is unknown.

Corticosteroid–Induced Alterations in Muscle

Excessive endogenous production or exogenous administration of corticosteroids results in a muscle myopathy. Corticosteroids, also called glucocorticoids, are produced in the adrenal gland. Cortisol, accounting for about 95% of the glucocorticoids produced, is involved in the mobiliza-tion of glucose and in the suppression of inflammatory responses. Because of this latter action, artificially produced glucocorticoids are used to treat a variety of inflammatory diseases, including asthma. Unfortunately, long-term use and particularly high doses of exogenous corticosteroids also have negative effects on the muscle. Because of the metabolic effects, excess glucocorticoids induce a glucose intolerance, diabetes, and hyperlipidemia. By increasing catabolic activity in skeletal muscle this hormone leads to muscle atrophy.

➤ Endogenous means within the body or tissue; exogenous means from outside the body or tissue.

Treatment of muscular diseases with corticosteroids may improve function in some patients with a chronic inflammatory process, but also may result in muscle atrophy.

Pathophysiology

Excessive levels of circulating corticosteroids enhance muscle protein catabolism, resulting in a loss of contractile protein. The resulting muscle atrophy is selective to type II fibers, particularly in muscles with fibers high in this type. In addition, type IIb fibers are more susceptible to this catabolic effect than type IIa fibers. Early changes include enlarged mitochondria and accumulation of glycogen. As degeneration progresses, fibrillar and mitochondrial loss is evident; and the muscle morphology becomes disorganized with fragmentation of contractile elements. With the exception of the liver and smooth muscle, protein stores of almost all skeletal muscle tissues are decreased with chronic or high-dose corticosteroid therapy. Active muscles show fewer changes than inactive muscles. In spite of the continual nature of diaphragmatic contraction, the diaphragm also shows atrophy and disruption of the fiber structure.

Clinical Manifestations

The most obvious sign of excessive levels of corticosteroids is cushingoid signs, that is, increased adiposity and striations. Muscle weakness is characterized by proximal involvement. The pelvic girdle muscles are affected earlier and more severely than the pectoral girdle and distal muscles. Diaphragm and myocardial disruption may eventually result in cardiorespiratory dysfunction.

Treatment

The treatment of corticosteroid-induced myopathy is to reduce or, if possible, eliminate the use of the drug. Attempts to counteract the effect of corticosteroids with growth hormone have not been effective. Exercise, however, is very effective in partially reversing and preventing the myopathy. Both generalized aerobic exercise training and specific muscle training, including that for the diaphragm, increase fat-free mass and improve muscle strength. It also appears possible that targeted exercise may selectively affect the susceptible type II fibers.

Case Study

Your rehabilitation manager has asked you to develop an exercise program for patients with muscle diseases, including MDs. She would like this to be a combination of an outpatient and home exercise program. Your facility has standard aerobic and resistance equipment. You also have a small gymlike area. Patients will range in age from preschool to young adult. On the basis of your understanding of the disease processes, describe the important elements of your programs.

Your program should address several important characteristics:

1. The disease generally progresses from proximal to distal.
2. Type II muscles may be more affected more than type I.
3. Work capacity may be reduced.
4. Your population includes a wide age range.

On the basis of these characteristics, you should consider age group-specific programs. Each program should include a combination of endurance (aerobic) activities and resistance exercises. The resistance exercises should include some exercises done at lower resistances but at higher power to stimulate type II fibers, if possible. For example, resistive exercises with 12 to 15 repetitions performed in one to two sets would be appropriate. Programs for younger age groups should emphasize fun, gamelike activities; those for the older age groups should emphasize activities of daily living and activities to maintain the patient's occupation. You might consider once-a-week group activities with an incentive program to encourage daily activities in the home setting.

What You Need to Know From Chapter 10

Key Terms

congenital myopathies

- central core disease

- minicore/multicore disease

- myotubular myopathy

- nemaline (rod) myopathies

corticosteroids

Duchenne's, Becker's

dystrophin

muscular dystrophies (MDs)

myoblasts

pseudohypertrophy

Key Concepts

1. You should understand the etiology and progression of some of the more common muscular diseases.

2. You should be able to use information regarding muscular diseases to develop exercise/rehabilitation programs for patients with these diseases.

Review Questions

1. The diseases presented in this chapter vary in terms of their progression and in terms of the specific muscle fibers involved. Why is it important to understand these differences when you are developing a therapeutic regimen for one of these diseases?

2. Progression of many of the muscular diseases is not affected by exercise. Why is it still important to consider exercise therapy for these patients?

3. Why should regular exercise be part of the treatment for muscular diseases?

4. Why should stretching be part of the treatment for muscular diseases?

5. Why would you consider using leg extensions with lower resistance, but at higher velocities, for a patient with the muscle diseases described in this chapter?

6. Patients with chronic inflammation, including asthma, are sometimes put on corticosteroid therapy. What are the implications of this for making therapeutic exercise recommendations?

Bibliography

Aitkens SG, McCrory MM, Kilmer DD, and Bernauer EM. Moderate resistance exercise program: Its effect in slowly progressive neuromuscular disease. *Arch Phys Med Rehab* 1993, 74:711–715.

Braith RW, Welsch MA, Mills RM, Keller JW, and Pollock ML. Resistance exercise prevents glucocorticoids-induced myopathy in heart transplant recipients. *Med Sci Sports Exerc* 1998, 30:483–489.

DeLateur BJ and Giaconi RM. Effect on maximal strength of submaximal exercise in Duchenne muscular dystrophy. *Am J Phys Med* 1979, 58:26–36.

DiMarco AF, Kelling JS, DiMarco MS, Jacobs I, Shields R, and Altose MD. The effects of inspiratory resistive training on respiratory muscle function in patients with muscular dystrophy. *Muscle Nerve* 1985, 8:284–290.

Falduto MT, Czerwinski SM, and Hickson, RC. Glucocorticoid-induced muscle atrophy prevention by exercise in fast-twitch fibers. *J Appl Physiol* 1990, 69:1058–1062.

Fowler WM and Gardner GW. Quantitative strength measurements in muscular dystrophy. *Arch Phys Med Rehab* 1967, 48:629–644.

Gozal D and Thiriet P. Respiratory muscle training in neuromuscular disease: Long-term effects on strength and load perception. *Med Sci Sports Exerc* 1999, 31:1522–1527.

Hayes A and Williams DA. Beneficial effects of voluntary wheel running on the properties of dystrophic mouse muscle. *J Appl Physiol* 1996, 80:670–679.

Hayes A and Williams DA. Contractile properties of clenbuterol-treated *mdx* muscle are enhanced by low-intensity swimming. *J Appl Physiol* 1997, 82:435–439.

McCartney N, Moroz D, Garner SH, and McComas AJ. The effects of strength training in patients with selected neuromuscular disorders. *Med Sci Sports Exerc* 1988, 20:362–368.

Smith PEM, Calverly PMA, Edwards RHT, Evans GA, and Campbell EJM. Practical problem in the respiratory care of patients with muscular dystrophy. *N Eng J Med* 1987, 316:1197–1205.

Vignos PJ and Watkins MP. The effect of exercise in muscular dystrophy. *JAMA* 1966, 197:843–848.

Wineinger MA, Abresch RT, Walsh SA, Carter GT. Effects of aging and voluntary exercise on the function of dystrophic muscle from mdx mice. *Am J Phys Med Rehab* 1998, 77:20–27.

Pulmonary System

This chapter will discuss

➤ changes in pulmonary gas exchange that occur to ensure adequate O_2 and prevent the buildup of CO_2 during exercise;

➤ the achievement of pulmonary gas exchange at a minimal energy cost through optimization of breathing mechanics;

➤ various neural and other factors that control ventilation during exercise; and

➤ the transport of O_2 and CO_2 in the blood.

Abbreviations Used in Pulmonary Physiology

Simple abbreviations

A	=	alveolar
a	=	arterial
c	=	capillary
d	=	dead space
E	=	expiration
I	=	inspiration
P	=	pressure
S	=	saturation (of hemoglobin)
v	=	venous
V	=	ventilation
$\dot{V}$	=	ventilation per unit time (minutes or seconds)

More complex abbreviations

Fb	=	breathing frequency
PAO_2	=	partial pressure of oxygen in alveolus
PaO_2	=	partial pressure of oxygen in the arterial blood
$PaCO_2$	=	partial pressure of CO_2 in the arterial blood
Pb	=	barometric pressure
PcO_2	=	partial pressure of oxygen in capillary blood
PIO_2	=	partial pressure of inspired oxygen
PO_2	=	partial pressure of oxygen
PvO_2	=	partial pressure of oxygen in venous blood
$\dot{Q}c$	=	lung perfusion per unit time
SaO_2	=	oxygen saturation of arterial blood
Vd	=	volume of dead space
Vt	=	tidal volume
$\dot{V}_A$	=	alveolar ventilation per minute
$\dot{V}_E$	=	volume of air exhaled per minute
$\dot{V}_I$	=	volume of air inhaled per minute

The respiratory system consists of a series of branching conducting airways that terminate in the gas-exchange units, the alveoli. The alveoli are invaginated and are surrounded by capillaries to enhance gas exchange from the lungs to the blood and back again (figure 11.1). The surface area of the alveoli in an adult male is on the order of 70 m² (84 square yards). To appreciate the potential area for gas exchange in this lung, note that 70 m² is the area covered by a tennis court! This tremendous gas-exchange area is one reason the lungs of a healthy person do not limit the availability of O_2 during exercise. The first section of this chapter demonstrates how the body uses this wonderful gas-exchange system to ensure adequate delivery of O_2 and removal of CO_2 at rest and exercise.

➤ The alveolus is the terminal sac-like portion of the branching structure at the end of the branchings of the pulmonary airways. This is the gas-exchange unit of the lung where the alveolar air comes into contact with the pulmonary capillary blood.

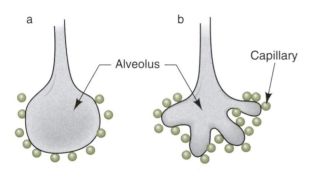

Figure 11.1 (*a*) An idealized alveolus without invaginations; (*b*) an alveolus with typical invaginations. Note how many more capillaries are associated with the alveolus in (*b*), an anatomical feature that gives the lung a large surface area for gas exchange.

Gas Exchange

The sum of the partial pressure of each gas in an environment is equal to the total pressure. In ambient air, with a barometric pressure of about 750 mm Hg, O_2 contributes 20.9% of the pressure, or about 157 mm Hg (750 × .209 = 157). You can easily grasp the concept of gas exchange during exercise if you understand the factors that affect the O_2 pressure. These changes in O_2 pressure are depicted in the O_2 cascade (figure 11.2). Note that there is a drop in the pressure of O_2 as it is transported along the cascade. The factors affecting the drop are also indicated in the figure: the decrease in O_2 from ambient air to the alveoli is affected by the level of ventilation; that from the alveoli to the capillary is affected by diffusion; and that from the capillary to the arterial blood is affected by **ventilation-to-perfusion matching** ($\dot{V}_A/Qc$). Each of these transitions provides a

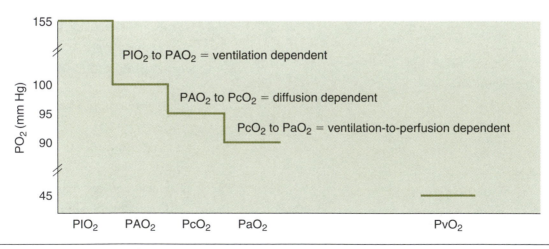

Figure 11.2 The O_2 cascade during resting conditions. This schematic shows the change in partial pressure of O_2 as it is inspired from ambient air into the alveolus, then transferred into the capillary and collected into the arterial blood to the body tissues. The venous O_2 is shown for reference.

potential source for a loss of O_2 pressure. At rest, the loss of pressure is on the order of 155 to 145 mm Hg from ambient air to the alveolus and another 10 mm Hg from the alveolus to the arterial blood. When you consider the small loss in O_2 pressure from the alveolus to arterial blood, you can appreciate the efficiency of gas exchange in the lung.

The "job" that the lung must perform is to add O_2 to and remove CO_2 from venous blood. Table 11.1 shows typical values for arterial and venous O_2 and CO_2 at rest and during moderate and heavy exercise. Note first that the level of O_2 in the venous blood coming back to the lungs becomes progressively lower as the exercise intensity increases. Likewise, the need to remove CO_2 becomes greater. Again, the task for the lungs is to ensure that the O_2 level remains at or above resting arterial levels and that CO_2 remains at or below resting arterial levels. It is readily apparent that arterial blood O_2 and CO_2 levels are in fact maintained or improved upon during heavy exercise. These values indicate that gas exchange is very effective at all levels of activity.

Ventilation

The level of O_2 in the alveolus is dependent on how ventilation occurs and how much occurs. Imagine what would happen to the level of O_2 in your lungs if you held your breath. The drop in O_2 and increase in CO_2 indicate **hypoventilation**. On the other hand, if you were to remain at rest but consciously increase your ventilation, the level of CO_2 in the lungs would decrease, indicating **hyperventilation**, and the level of O_2 would increase. Note that hypo- or hyperventilation is defined only by the change in arterial CO_2; an increase in ventilation with no change in $PaCO_2$, as during moderate exercise, is defined simply as **hyperpnea**

Minute ventilation ($\dot{V}_E$), the total volume of air moved in and out of the lung per minute, increases during exercise as a result of several stimuli such as movement and arterial O_2 and CO_2. In healthy people, the increase in $\dot{V}_E$ at low to moderate levels of exercise is appropriate for the increase in $\dot{V}O_2$ and $\dot{V}CO_2$ so that alveolar gases change little. At these exercise levels, then, the maintenance of blood gases is achieved by an appropriate increase of ventilation to keep alveolar gases steady. During heavier exercise, however, the level of ventilation increases out of proportion to $\dot{V}O_2$ and $\dot{V}CO_2$ (see page 190), resulting in alveolar hyperventilation and thus increases in PAO_2 from resting levels of near 100 to above 125 mm Hg. This increase in PAO_2 during heavy exercise helps to ensure sufficient alveolar O_2 pressure to drive the gas into the capillary blood.

It is clear that how much we ventilate assures that the first step in the O_2 cascade will not limit O_2 availability during exercise. How we ventilate, in terms of the volume of each breath (**tidal volume**, Vt) and **breathing frequency** (Fb), also plays an important role in determining how much air is moved into and out of the gas-exchange units, the alveoli. Air moving into and out of the lungs must pass through the conducting airways before it enters the alveoli. There is no gas exchange in the conducting airways; the lack of exchange creates what is called **dead space** (Vd) or "wasted ventilation." The Vd in the conducting airways in a healthy adult is approximately two times the

Table 11.1 Values for Venous (PvO_2) and Arterial (PaO_2) O_2 Pressure and Arterial ($PaCO_2$) CO_2 Pressure at Rest and During Moderate and Heavy Exercise

	PvO_2 (mm Hg)	PaO_2 (mm Hg)	$PaCO_2$ (mm Hg)	$PvCO_2$
Rest	40	92	40	45
Moderate exercise	32	92	40	45
Heavy exercise	<20	105	<20	<35

person's body weight in kilograms (if weight = 70 kg, Vd = 140 ml or 0.14 L/breath). If the Vt were equal to the Vd, no fresh air would enter the alveoli since we would only be moving air into and out of the conducting airways. For instance, if normal Vt = 500 ml and Fb = 10, normal $\dot{V}_E$ would be 5 L/min. The amount of air actually reaching the alveolar exchange units, called **alveolar ventilation** ($\dot{V}_A$), is the total ventilation ($\dot{V}_E$) minus the Vd; that is, $\dot{V}_A = \dot{V}_E - (Fb \times Vd)$. For a 70 kg person, $\dot{V}_A = 5\,L - (10\,breaths/min \times 0.14\,L)$, or 3.6 L. If we decreased Vt to 0.14 L but increased Fb to 35.8, $\dot{V}_E$ would remain the same at 5 L, but $\dot{V}_A$ would be 0.0 L. Thus, in spite of a continued $\dot{V}_E$ level of 5 L/min, the lack of air moving into and out of the alveolar volume would mean that no gas ex-

change would take place, PAO_2 would decrease, and $PACO_2$ would increase. The normal increase in Vt observed with low to moderate exercise (figure 11.3) helps to ensure that Vd becomes a smaller percentage of each breath. In fact, if we measure the amount of each breath that is "wasted" in the Vd by calculating the Vd/Vt ratio, we can see that it decreases from approximately 30% of each breath at rest to less than 10% during heavy exercise. Figure 11.4 illustrates the efficiency of the first step in the oxygen exchange cascade.

The levels of O_2 and CO_2 in the alveoli are determined by how much and how we ventilate. During exercise, the absolute level of ventilation increases to ensure adequate levels of O_2 in the alveoli and CO_2 elimination. The normal increases

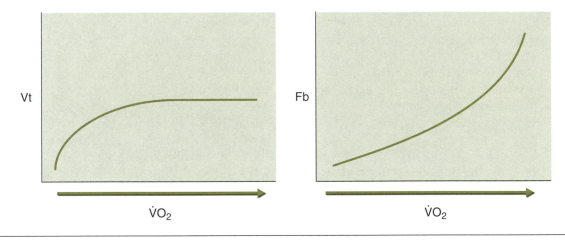

Figure 11.3   Changes in tidal volume (Vt) and breathing frequency (Fb) with increasing energy expenditure ($\dot{V}O_2$).

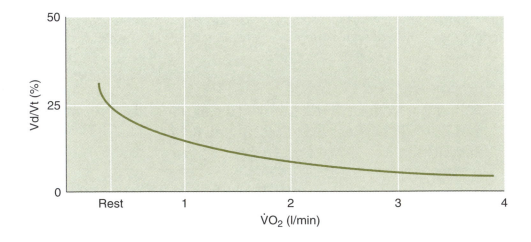

Figure 11.4   The change in the amount of each breath (Vt) that only moves into and out of the conducting airways (Vd), expressed as a percentage, at rest and at increasing levels of $\dot{V}O_2$. Note that the Vd, as a percentage of each breath, decreases as exercise intensity increases.

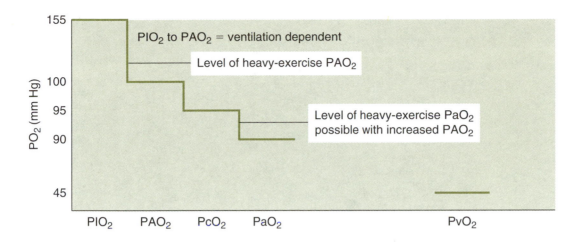

Figure 11.5 The O_2 cascade during heavy exercise. The black line indicates the effects of appropriate ventilatory adjustments on PAO_2 and the consequent theoretical effect this increase in PAO_2 has on PaO_2.

in Vt with exercise also ensure that the effect of Vd is minimized. The first step in the O_2 cascade ensures that sufficient O_2 is available in the alveoli to diffuse into the capillary blood passing through the alveoli (figure 11.5).

The levels of O_2 and CO_2 in the alveoli are determined by how much and how we ventilate.

Diffusion

Diffusion of O_2 through the alveolar wall into the capillaries and red blood cells is affected by driving pressure, surface area, and transit time.

• **Driving pressure.** The diffusion of O_2 from the alveoli into the capillaries is dependent on the difference in pressure of the gas on either side of the membrane, as well as on the surface area where the alveolar and capillary wall interface (figure 11.6). This is called the "driving pressure." We have already seen that the PAO_2 (page 179) for O_2 is increased during exercise. The increased driving pressure is indicated by the greater number and longer length of the arrows on the right side of figure 11.6. Once the O_2 is transferred to the blood, it becomes dissolved in the plasma. This dissolved O_2 is measured as the pressure of O_2 in the blood, or PO_2. Most of the O_2 is carried on the **hemoglobin** (Hb) molecule (see page 192).

• **Surface area.** At rest, the distribution of blood in the lung capillaries is uneven such that many capillaries may not be functional and those that are perfused may have a limited diameter. During exercise, the increased cardiac output flowing through the pulmonary capillaries opens previously unopened capillaries and increases the diameter of all capillaries that are perfused. This expansion of the pulmonary capillary bed is indicated by an increased number and size of capillaries (figure 11.6). Note that the surface area of the interface between the alveolar and capillary walls is increased. It is this increased surface area that greatly enhances the ability of the lungs to diffuse gases during exercise.

• **Transit time.** A third factor affecting the diffusion of gas from the alveoli into the capillary blood is the amount of time the blood, or the red blood cell, spends in the capillary—the **transit time**. At rest, the red blood cell spends an average of 0.75 s in the capillary (figure 11.7). The diffusion of O_2 into the red blood cell takes only 0.3 s, so transit time could be decreased by about 0.5 s before the transfer of O_2 would be compromised. During exercise, if the pulmonary capillary bed did not expand, it is likely that transit time would be compromised. However, the increased size of the capillary bed during exercise ensures that the increased cardiac output can be accommodated with a minimal increase in rate of flow, which would decrease transit time. The situation is analogous to that of water flowing through a hose. If the diameter of the hose is increased, the pressure in the hose and the flow rate will be decreased.

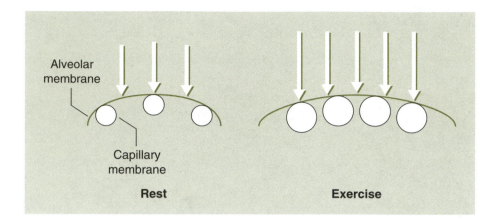

Figure 11.6   Schematic drawing of the factors that affect diffusion of O_2 from the alveolus into the capillary. The driving pressure is represented by the number and length of the arrows. The surface area for exchange is indicated by the interface between the alveolar and the capillary membranes.

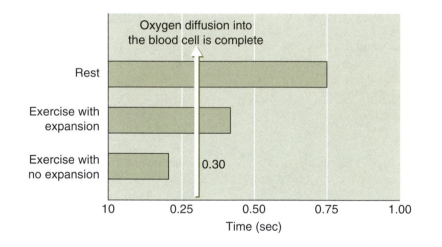

Figure 11.7   The time it takes to equilibrate O_2 into the red blood cell as it passes through the pulmonary capillary bed. The bars show the length of time a red blood cell spends in the pulmonary capillary. The arrow shows the time necessary for equilibration. Expansion of the pulmonary capillary bed with exercise ensures that the time a red blood cell spends in the capillary is sufficient for gas exchange.

The diffusion of O_2 from the alveolus to the capillary is ensured by

- increase in the driving pressure of O_2 in the alveolus,

- increase in the surface area for exchange through increase in the number and size of the perfused pulmonary capillaries, and

- maintenance of the transit time of the red blood cell within the time frame needed for complete transfer, or equilibration, of the alveolar and capillary gas.

Figure 11.8 shows the combined effects on PaO_2 of appropriate ventilatory adjustments to exercise in alveolar gases and of diffusion enhancement.

Ventilation-to-Perfusion Matching

The distribution of alveolar ventilation ($\dot{V}_A$) and lung perfusion ($\dot{Q}c$) is not uniform throughout the lung. If the uneven distribution of $\dot{V}_A$ is not matched to the uneven distribution of $\dot{Q}c$, there will be a poor transfer of gases. To understand the effect of ventilation-to-perfusion matching ($\dot{V}_A/\dot{Q}c$) on gas exchange during exercise, it is important to understand the mechanisms for the uneven

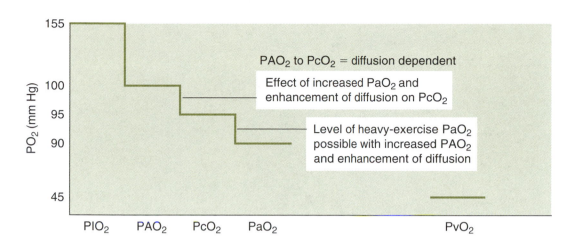

Figure 11.8   The O_2 cascade during heavy exercise. The effects of appropriate ventilatory adjustments on PAO_2 and diffusion, and the consequent theoretical effect on PaO_2, are shown by the black line.

distribution of both $\dot{V}_A$ and $\dot{Q}c$. The primary reason for the variation in $\dot{V}_A$ and $\dot{Q}c$ distribution is gravity. During exercise, this $\dot{V}_A$ /$\dot{Q}c$ variation is largely overcome to effect a more uniform distribution of both $\dot{V}_A$ and $\dot{Q}c$, and therefore better matching.

The Effect of Gravity

A good way to understand the distribution of inhaled air in the lung is to think of the lung as a loose spring or a child's "Slinky" toy (figure 11.9). At the end of a normal resting exhalation, the spring is stretched more at the top than at the bottom. As air is inhaled, and in this case the spring is pulled from the bottom, the space between rings increases more at the bottom than at the top. Part of the reason is that the rings at the top are already stretched and are therefore more difficult to stretch than the ones at the bottom. At rest, the lung is similarly stretched: at the end of a normal expiration there is more air in the top (cephalad), but when air is inspired, more of the air goes to the base of the lung. This uneven, gravity-dependent, distribution of air into the lung occurs in all body positions; the part of the lung closest to the ground is always the gravity-dependent part, and receives the most air on inspiration. In the supine position there is a better overall cephalad-to-caudad distribution of $\dot{V}_A$ in the lungs (figure 11.10) but still a gravity-dependent gradient in the anterior-to-posterior axis.

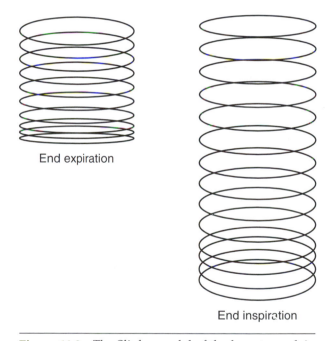

Figure 11.9   The Slinky model of the lung to explain the uneven distribution of air breathed into the lung. The relatively unstretched bottom part of the lung at the end of a normal expiration can expand more during inspiration so that more air goes to the bottom than to the top of the lung.

The increased top-to-bottom expansion of the lung during exercise results in a more even distribution of the ventilation (figure 11.10). The exercise-induced improvement in $\dot{V}_A$ distribution, along with the previously discussed overall

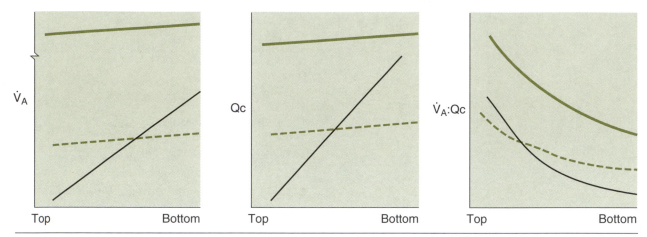

Figure 11.10   Schematic of the distribution of ventilation ($\dot{V}_A$) and perfusion ($\dot{Q}c$) and the distribution of the matching of $\dot{V}_A$ and $\dot{Q}c$ upright at rest (black line), during exercise (green line), and in the supine position at rest (dashed green line). See text for further explanation.

increase in $\dot{V}_A$, ensures that virtually all alveoli receive an exchange of air with each breath. In addition, the better distribution of $\dot{V}_A$ and increase in $\dot{V}_A$ ensure that blood perfusing the lungs will be exposed to the lung O_2.

The distribution of blood perfusing the lung also is gravity dependent (figure 11.11). Blood is pumped from the right ventricle into the pulmonary artery and subsequently distributed throughout the vessels surrounding the alveoli. In the upright position, this blood distribution is subjected to the force of gravity such that more blood enters the vessels in the basal (dependent) part of the lung than in the upper part. In fact, during diastole, pressure in the pulmonary arterial system may be insufficient to force blood to the upper parts of the lung, and flow there is cyclical (i.e., it ebbs and flows with the cardiac cycle). As with ventilation, when body position changes from upright to supine there is a better overall cephalad to caudad distribution of blood in the lungs, but the dependent part of the lung also changes so that increased flow occurs in the anterior-to-posterior direction (figure 11.10).

> The exercise-induced improvement in $\dot{V}_A$ distribution, along with overall increase in $\dot{V}_A$, ensures that virtually all alveoli receive an exchange of air with each breath.

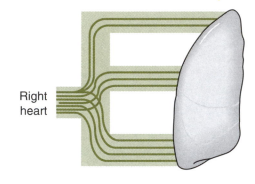

Figure 11.11   The blood flowing into the lungs is distributed according to gravity. More flow enters the bottom of the lung than the top.

Effects of Poor and Even Matching

Figure 11.12 shows schematically the significance of poor matching of $\dot{V}_A$ and $\dot{Q}c$. Assume that blood returning to the heart has a PO_2 of 40 mm Hg and that the volume of blood past lobes A, B, and C is the same. Blood flowing past the part of the lung represented by lobe A will exit the system with a PO_2 of 95 to 100 mm Hg. Lobe B represents a part of the lung that is completely obstructed. Blood flowing past this part of the lung will exchange little or no gas so that it leaves the lung with a PO_2 of 40 mm Hg. If this blood were to mix with the blood from lobe A, the final, arterial blood would have a PO_2 of about 70 mm Hg (since the final PO_2

would be half of the difference between 100 and 40: 100 − 40 = 60; 60 ÷ 2 = 30; 100 − 30 = 70). The effect of poor matching of $\dot{V}_A$ and $\dot{Q}c$ can be readily seen. The blood flowing past lobe C, which is only partially obstructed, will exchange gases; but since blood flow is greater than ventilation, less O_2 will be present in that lobe and the blood leaving it also will have a low PO_2. When this blood mixes with blood coming from areas of the lung having other $\dot{V}_A/\dot{Q}c$ ratios, the arterial PO_2 will have a value that reflects all the PO_2 values mixed together. For optimal gas exchange, a perfect match between $\dot{V}_A$ and $\dot{Q}c$ ($\dot{V}_A/\dot{Q}c = 1$) would

be necessary. Values of $\dot{V}_A/\dot{Q}c >1$ indicate more ventilation than perfusion, a situation that occurs during heavy exercise (figure 11.10). $\dot{V}_A/\dot{Q}c$ ratios >1, while showing poor matching, generally result in good gas exchange during exercise because blood perfusing the lung has a high probability of interfacing with well-ventilated alveoli. $\dot{V}_A/\dot{Q}c$ ratios <1 indicate considerable areas of the lung with less ventilation than perfusion, which result in low arterial PO_2 values. At rest, small areas of mismatching of $\dot{V}_A$ and $\dot{Q}c$ cause the drop in PO_2 observed from capillary to the arterial blood (figure 11.13).

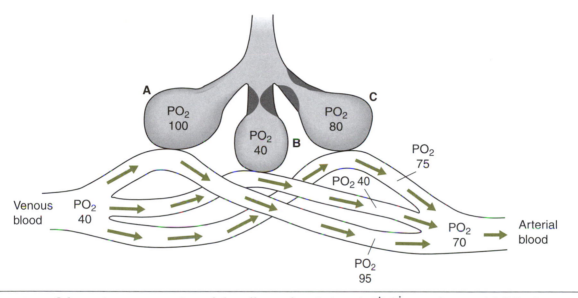

Figure 11.12 Schematic representation of the effects of variations in $\dot{V}_A/\dot{Q}c$ on the arterial PO_2. See text for explanation.

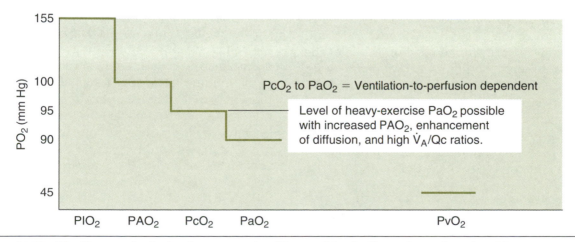

Figure 11.13 The O_2 cascade during heavy exercise. The combined effects of exercise-related improvements in PAO_2, diffusion, and $\dot{V}_A/\dot{Q}c$ are shown by the black line.

During upright exercise, the increased cardiac output and outflow pressures result in a more even distribution of blood flow in the lung (figure 11.10), ensuring that blood flowing through the capillaries surrounding the alveoli is more evenly distributed than at rest. The more even distribution of $\dot{Q}c$ during exercise ensures good matching with the $\dot{V}_A$, which also is more evenly distributed.

The $\dot{V}_A/\dot{Q}c$ ratio, reflecting the matching of lung air and blood, is dramatically improved during exercise (figure 11.10). An improved $\dot{V}_A/\dot{Q}c$ is a direct result of the exercise-related redistribution of both $\dot{V}_A$ and $\dot{Q}c$ in such a way that a close interface between the two is maintained or increased.

Figure 11.10 shows that not only is the $\dot{V}_A/\dot{Q}c$ improved throughout the lung; the absolute value is also greatly increased. The $\dot{V}_A/\dot{Q}c$ during exercise is elevated because the increase in $\dot{V}_A$ (from a resting value of 6 L/min to a peak exercise value of >100 L/min) is significantly greater than the increase in $\dot{Q}c$ (from a resting value of 6 L/min to a peak exercise value of 25 L/min). The greatly increased $\dot{V}_A$ ensures that virtually all of the blood entering the alveoli will be exposed to fresh, O_2-rich alveolar gas.

An improved $\dot{V}_A/\dot{Q}c$ is a direct result of the exercise-related redistribution of both $\dot{V}_A$ and $\dot{Q}c$ in such a way that a close interface between the two is maintained or increased.

Blood flow redistribution in the supine position also increases venous return to the heart, raising stroke volume to near-maximal levels. The optimization of stroke volume, $\dot{V}_A$, $\dot{Q}c$, and $\dot{V}_A/\dot{Q}c$ ratio in the supine position is one of the reasons young swimmers can compete with adults. The relatively small heart volume (and therefore stroke volume) and smaller lung surface area in children, as compared to adults, are less of a limiting factor during exercise in the supine position than in the upright position.

PaO_2 is maintained or increased during exercise

1. by an increased $\dot{V}_E$, with increases in Vt such that "wasted" ventilation is minimized;
2. by an improvement in the pressure gradient and area available for diffusion; and
3. by an improvement in overall lung $\dot{V}_A/\dot{Q}c$.

PaO_2 is maintained or increased during exercise by an increased $\dot{V}_E$, with increases in Vt such that "wasted" ventilation is minimized; by an improvement in the pressure gradient and area available for diffusion; and by an improvement in overall lung $\dot{V}_A/\dot{Q}c$.

Respiratory Mechanics

The recruitment of respiratory muscles to increase ventilation during exercise results in an increase in the energy expended to breathe. At the end of a normal exhalation during quiet breathing, the respiratory system is in a balanced state such that little or no energy is being expended for breathing. An increase or a decrease in volume away from this end-expiratory volume requires muscle recruitment and causes an increase in the expenditure of energy for breathing. In this section we review the mechanical factors that contribute to energy expenditure by the respiratory system and show how this expenditure is minimized during exercise.

If the chest were opened, the lungs would tend to collapse and the chest wall would tend to flare out. In the closed chest at the end of a normal, relaxed exhalation, the inward-drawing force of the lungs and the outward drawing of the chest wall are balanced. If you took a series of inhalations and then relaxed your respiratory muscles for exhalation, the respiratory system would consistently return to this state of balance between the two forces. This point of relaxation, termed the relaxation volume of the lung (Vrel), represents the lung volume indicated by **functional residual capacity** (FRC) (figure 11.14). The FRC represents the volume of air left in the lungs at the end of a normal relaxed exhalation. **Residual volume** (RV) is the volume left in the lungs at the end of a maximal exhalation. It is critical to remember that FRC represents the point at which the pulmonary system expends no energy to maintain volume, and that movement away from this point is accomplished only with the expenditure of energy. To conserve energy, the respiratory system tends to remain as close to FRC as possible.

Elastic Forces

Increases in lung volume away from FRC require a change in pressure in the pulmonary system.

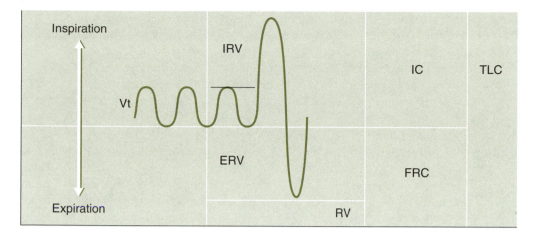

Figure 11.14   Standard division of the lung volume. TLC = total lung capacity; IRV = inspiratory reserve volume; Vt = tidal volume; ERV = expiratory reserve volume; RV = residual volume; IC = inspiratory capacity; EC = expiratory capacity; FRC = functional residual capacity.

The pressure change measured in the lung is used as an estimate of the muscle force necessary to produce that change in volume. The pressure required to move the elastic structures of the respiratory system is measured by having the person inhale to a certain volume, and then measuring the pressure in the mouth while the person relaxes the system at that volume against an occluded mouth and nose. The pressure measured in the lung at that point is positive, reflecting the tendency of the system to want to collapse back to the relaxed FRC volume. You can feel this by inhaling to different volumes and holding your hand over your mouth while relaxing. The positive pressure you feel against your hand is the pressure (force) that was required to change the volume of the elastic tissues of the system; the higher the volume, the greater the pressure. Pressures measured during occlusion at the end of an exhalation below FRC are negative, reflecting the tendency of the system to expand back to the relaxed FRC volume. Again, you can feel this by exhaling below resting volume and holding your hand against your mouth. You will feel a negative pressure against your hand. The generation of inspiratory and expiratory forces represents work done by the respiratory muscles against the elastic structures of the respiratory system and represents the elastic **"work"** done by the system. During quiet breathing, exhalation is a passive process in which the elastic structures recoil back to FRC after being stretched during inspiration. During exercise the elastic work of each inspiration and exhalation is minimized by breathing as close to FRC as possible.

The pressure required to increase lung volume is the same over much of the inspiratory volume, but it increases dramatically at very high lung volumes or very low lung volumes. The energy required to increase, or decrease, volume at this point also increases dramatically. Energy expenditure to move the elastic structures of the lung is minimized when we avoid breathing at very high or very low lung volumes.

Imagine breathing on the horizontal line starting at FRC (figure 11.14). As the ventilatory demand increases during exercise, Vt also increases until high lung volumes are reached and the energy to breathe is greatly increased. At this point the pressure, and therefore the energy required to breathe in, becomes quite large. To avoid this increase in energy, further increases in Vt are accomplished by decreasing lung volumes below FRC, or by breathing below the FRC line on figure 11.14. Since this movement below FRC is still close to FRC, it takes place with the minimal use of energy. Of course, a decrease in volume below FRC requires the recruitment of expiratory muscles. Once the end-inspiratory and end-expiratory volumes approach the point where further increases in Vt could take place only with great increases in energy expenditure, the energy cost to increase volume further is extremely high, and little further increase in Vt occurs. At maximal exercise, only about 60% of our vital capacity is used for Vt because of the high energy requirement to increase volume further.

> Energy expenditure to move the elastic structures of the lung is minimized when we avoid breathing at very high or very low lung volumes.

Diseases that stiffen either the chest wall (e.g., severe scoliosis) or the lungs (e.g., sarcoidosis) increase the elastic force necessary for breathing. To avoid these forces, patients with these diseases assume a shallow, rapid breathing pattern. We consider these topics in more detail in the next chapter.

Flow–Resistive Forces

Muscular force is used not only to move the elastic structures of the respiratory system, but also to move the air through the airways. You can get a sense of these forces if you breathe through a straw. An increased pressure (force) is required to move air through the small tube.

The forces required to move air through the airways are the **flow-resistive forces**, which represent the **flow-resistive "work."** The faster you move air through the airways, the greater this flow-resistive work. During exercise, the rate of inspiration and exhalation increases, thus increasing the flow-resistive forces. To generate the higher flow rate during exercise, both inspiratory and expiratory muscles are activated to increasingly higher degrees.

Diseases that decrease the lumen of the airways necessitate greater muscle activation than normal to overcome the increased flow-resistive forces. These obstructive diseases include chronic bronchitis, emphysema, and cystic fibrosis (see chapter 12).

> The forces required to move air through the airways are the flow-resistive forces, which represent the flow-resistive "work."

Breathing Economy

The increase in ventilation during exercise is accomplished through increases in both Vt and Fb (figure 11.3). At low exercise intensities, increases

in Vt are primarily responsible for the increase in ventilation, whereas increases in Fb account for the increased ventilation at higher exercise intensities. The initial rise in Vt allows ventilation to increase with moderate increases in elastic work, since these avoid the extremes of high or low lung volume. Of course, this initial rise in Vt also minimizes the influence of Vd.

Optimizing the elastic forces and the flow-resistive forces minimizes the work of breathing. Figure 11.15 shows that as Vt is increased, the elastic work increases, and that as Fb increases, the flow-resistive work increases. To minimize the combined, total, work, the best combination of Vt and Fb is found—not because of a conscious decision, of course, but as the result of feedback to the respiratory control center regarding pressure and flow. Actual measures of the work of breathing show that Vt and Fb are optimized such that we breathe in the trough of the total-work-of-breathing curve in figure 11.15. The minimization of work of breathing is a powerful controller of how we breathe. It is extremely difficult to alter breathing pattern from the one that occurs naturally, since this alteration typically requires expending more energy—something the system resists. In some cases, respiration is controlled to minimize the work of breathing at the expense of maintaining blood gases. For instance, obesity increases elastic forces by increasing the load on the chest wall. In some

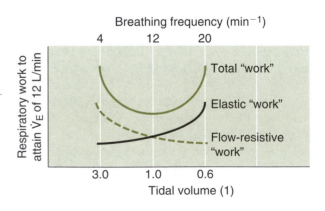

Figure 11.15 The total respiratory work, the work associated with moving the elastic structures of the respiratory system, and the work associated with moving air through the airways during breathing at a $\dot{V}_E$ of 12 L/min.

persons who are obese, $PaCO_2$ is "allowed" to increase because the work to move the chest wall becomes very large. The interaction between disease and work of breathing is discussed in the next chapter.

At low exercise intensities, the increase in $\dot{V}_E$ is accomplished primarily by increases in Vt, which plateaus at moderate exercise intensities (figure 11.3). Most of the increase in $\dot{V}_E$ at higher exercise intensities is accomplished by increases in Fb. Moderate exercise-stimulated increases in ventilation are achieved with increases in Vt rather than with increases in Fb because this is very economical. Increases in Vt beyond this level—that is, at very high or low lung volumes—cost more and more energy. Because flow-resistive work increases with any increase in Fb, increases in Fb are avoided until high Vt is reached. Control of changes in Vt and Fb is accomplished in such a way that the energy cost of breathing is minimized.

During resting quiet breathing, the energy required to ventilate is on the order of 0.5 to 1.0 ml O_2/L ventilation. This represents less than 5% of the total resting energy expenditure. This relative economy is achieved by keeping volume close to FRC and by taking advantage of elastic recoil during exhalation so that no energy is needed. At maximal exercise, the energy cost of ventilation increases to 2 to 3 ml O_2/L $\dot{V}_E$ or, assuming a $\dot{V}O_2$max of 5 L and a $\dot{V}_E$ of 100 L/min, slightly greater than 5% of the total energy expenditure. In spite of increased expiratory muscle energy requirement and increased energy needed to create more airflow, the cost of ventilation remains a relatively small percentage of the total energy expenditure. This remarkable economy of breathing is achieved, in part, by staying as close to FRC as possible (thereby minimizing the elastic forces) and by increasing the diameter of the airways (thereby reducing airway resistance and flow-resistive forces).

In summary, Vt is regulated to minimize the energy required to move the elastic structures of the respiratory system; Fb is regulated to minimize the energy needed to move air through the airways. At rest and during exercise, the combination of Vt and Fb is regulated to ensure that the required $\dot{V}_E$ is attained with use of the minimum amount of energy. It is very difficult to alter this naturally controlled breathing pattern.

> Vt is regulated to minimize the energy required to move the elastic structures of the respiratory system; Fb is regulated to minimize the energy needed to move air through the airways.

Case Study

You have been assigned to work with a track and field club. After participating in several practices, you observe the coach telling several runners that they need to breathe deeper while they run so that they don't get so out of breath. The coach explains to you her assumption that the runners were having difficulty breathing because they "lacked oxygen." She reasoned that their rather rapid breathing pattern prevented good gas exchange, and she was trying to rectify the inefficient breathing pattern.

You need to explain to the coach that the breathing pattern adopted by a person is usually the most efficient pattern—and that attempts to breathe in a different pattern will increase the energy used to breathe, potentially "stealing" energy from the working muscles.

Control of Ventilation

How and how much we ventilate play a critical role in ensuring that PaO_2 is maintained or increased during exercise with minimal energy expended. The magnitude of change in $\dot{V}_E$ and its components, Vt and Fb, is regulated by a central (medullary) controller that receives primary inputs from peripheral O_2 and central CO_2 (pH) sensors, as well as joint and lung receptors. These inputs are

- neural in origin,
- blood-borne (humoral), and
- temperature related.

This control system matches increases in $\dot{V}_E$ with increases in $\dot{V}O_2$ very closely at mild to moderate work levels (segment A, figure 11.16). At higher levels of $\dot{V}O_2$, $\dot{V}_E$ increases disproportionately to the increases in $\dot{V}O_2$ (segment B, figure 11.16).

Neural Control of Ventilation

Ventilation increases with the first breath on initiating movement, indicating a neural connection

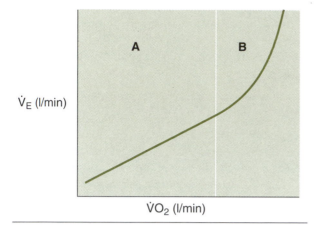

Figure 11.16   Increase in minute ventilation. As energy demand ($\dot{V}O_2$) increases, minute ventilation ($\dot{V}_E$) increases. At lower demands (segment A) the increase in $\dot{V}_E$ is linear, whereas at higher demands (segment B) the $\dot{V}_E$ increases more than $\dot{V}O_2$ does.

from the periphery to the central controller. The increase in $\dot{V}_E$ during passive movement of the limbs supports the notion that the most likely source of the neural input is joint receptors, but muscle **proprioceptors** also provide feedback to the respiratory controller. This initial input into the central controller is sometimes called the "gross" controller: it stimulates an increase in $\dot{V}_E$ that is pretty close to the increase needed to remove the increased production of CO_2 and maintain PaO_2, but it may require some fine-tuning. Proprioceptors in the joints that sense motion, as well as those in the muscles that sense stretch, motion, and tension, provide the neural input to the central controller.

In addition to the muscle and joint proprioceptors, there are **mechanoreceptors** (receptors which respond to mechanical pressure or distoration) in the lung that provide feedback to the central controller. These mechanoreceptors (located in the lung and chest wall) provide feedback to the medullary centers regarding the absolute lung volume, changes in the lung volume, and the rate of change in the volume. These receptors are important in controlling Vt and the duration of inspiration and expiration.

Humoral Control of Ventilation

The circulating levels of O_2 and CO_2 are important sources of feedback to the medullary center. At rest, decreases in O_2 and increases in CO_2 stimu-

late increases in $\dot{V}_E$ (figure 11.17). The primary receptors for changes in O_2 are in the carotid body; those for CO_2 are in the medulla. Note that inspired O_2 (F_IO_2) decreases from 100% to below 15% before a significant increase in $\dot{V}_E$ is observed. In contrast, small increases in inspired CO_2 (F_ICO_2) result in immediate and significant increases in $\dot{V}_E$. The medullary center is exquisitely sensitive to small changes in blood CO_2. Studies have shown that the medullary "CO_2 receptors" are really sensitive to changes in pH. The reaction relating CO_2 and pH is as follows:

$$H_2O + CO_2 \longleftrightarrow H_2CO_3 \longleftrightarrow HCO_3^- + H^+$$

where H_2CO_3 is carbonic acid and HCO_3^- is bicarbonate. You can see that as CO_2 levels increase, the equation will move to the right and increase H^+, thereby lowering the pH. The medullary center senses this H^+ and increases ventilation so that CO_2 is breathed out, thereby lowering the level of H^+ and returning pH toward its original level. Conversely, an increase in H^+, such as occurs during the metabolic production of lactic acid, also stimulates an increase in ventilation, so that CO_2 is blown off and pH is returned toward its original level. Through this mechanism, ventilation becomes an important part of acid-base regulation of the blood and the rest of the body.

▶ The medullary center is a part of the medulla oblongata, the most caudal part of the brainstem, where a major part of respiratory control takes place.

During exercise, the increased CO_2 production raises blood CO_2 and lowers blood pH. Sensing of this shift in pH by the medullary center stimulates an increase in $\dot{V}_E$. This chemical control of ventilation is sometimes called the "fine-tuning" of ventilation. The combination of neural input and chemical control results in the increases in ventilation relative to increases in $\dot{V}O_2$ in segment A of figure 11.16. The disproportionate increase in $\dot{V}_E$, relative to increases in $\dot{V}O_2$ in segment B, results from the accumulation of metabolic acids at these higher intensities of exercise. As noted earlier, the lactic acid that accumulates dissociates into lactate and H^+, which stimulates the respiratory center. Although the point at which this increase in $\dot{V}_E$ occurs is sometimes referred to as the **anaerobic threshold**, this term is avoided in this text.

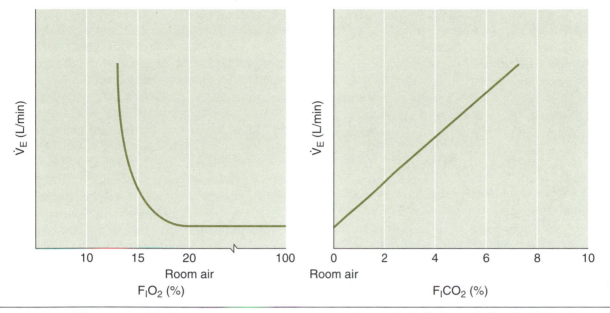

Figure 11.17 The relationship between changes in the levels of inspired O_2 (F_IO_2) and CO_2 (F_ICO_2) and minute ventilation ($\dot{V}_E$).

Figure 11.18 Arm exercise demanding more ventilation than leg exercise. When patients will be returning to jobs that demand heavy arm work, you must take into account the heavy load this work will place on their pulmonary systems.

© Eric Berndt/The Image Finders

Ventilation during arm exercise is higher at any $\dot{V}O_2$ than during leg exercise and results in hyperventilation. You must consider this when evaluating any exercise involving arm work, particularly in patients with pulmonary disease. The higher $\dot{V}_E$ will increase the ventilatory cost of exercise and can contribute to ventilatory limitations in these patients. The hyperventilation associated with arm work can lead to sensations of hypocapnia (e.g., light-headedness, tingling sensations in the fingers), which may interfere with the demands of work (figure 11.18). Such interference has been noted, for instance, in equipment inspectors who must operate in confined spaces while manipulating valves and other objects with the arms. The sensations of hypocapnia may prevent the operators from performing their tasks optimally.

Temperature Related

The third factor that stimulates increases in ventilation during exercise is elevated body temperature, particularly that associated with higher exercise intensities or with environmental load. The increased body temperature appears to stimulate the medullary control center directly.

In summary, ventilation is controlled by signals from peripheral proprioceptors and **chemoreceptors**. The medullary central controller processes this feedback, establishing the absolute amount of ventilation ($\dot{V}_E$) as well as the depth and frequency of breathing. During exercise, the control system increases $\dot{V}_E$ with the combination of Vt and Fb to maintain, or improve, blood gases using the minimum energy expenditure.

Ventilation is controlled by signals from peripheral proprioceptors and chemoreceptors.

Oxygen Transport

O_2 is transported in the blood both in a dissolved form and as carried on Hb. The Hb molecule consists of a central globin surrounded by four iron-associated **hemes** (figure 11.19). Each of these hemes can carry one O_2 molecule. One gram of Hb can carry 1.34 ml O_2 if it is fully saturated. Assuming a normal blood Hb level of 15 g Hb/100 ml blood and assuming that each Hb is fully saturated, we could carry 20.1 ml O_2/100 ml blood attached to the Hb. If we add the amount of O_2 dissolved in the blood, the total concentration of O_2 is approximately 20.4 ml O_2/100 ml blood. The attachment of O_2 to Hb allows the transport of high amounts of O_2 with a small cardiac output. To transport the same amount of O_2 without Hb would require a cardiac output of over 80 L/min at rest and over 1500 L/min during exercise!

The amount of O_2 on the Hb is dependent on the O_2 pressure in the blood. If the O_2 pressure is low, it may be that only one heme is associated with an O_2, making the Hb 25% saturated (SO_2 = 25%). As blood PO_2 rises, two hemes may associate, making the Hb 50% saturated (SO_2 = 50%), and so on. The relationship between PO_2 and the saturation level is plotted in figure 11.20. This relationship is not linear: this means that O_2 is readily released from the Hb on the steep part of the curve and the Hb becomes **desaturated**. On the flat part of the curve the O_2 is tightly held by the Hb as it becomes more saturated. Functionally, at the PO_2 levels associated with the steep part of the curve, the blood is in the peripheral capillaries where you would expect easier offloading of the O_2 so that it is available to the working muscle. Conversely, at PO_2 levels associated with the flat part of the curve, the blood is in the pulmonary capillary where you would expect the O_2 to be held tightly to the Hb. The PO_2 on the abscissa in figure 11.20 indicates arterial blood PO_2 during breathing of pure O_2 at about 250 mm Hg. Note that the Hb is 100% saturated at this PO_2. At a normal PaO_2 of 90 mm Hg, the Hb is about 96% saturated. This small change in SaO_2 with a rather large change in PO_2 indicates how tightly the O_2 attaches to the Hb at PO_2s >80 mm Hg. As the blood traverses through the muscle capillary, O_2 is transferred into the muscle tissue, thereby decreasing PO_2 and the SO_2% such that at a normal resting PvO_2 of 40 mm Hg, the saturation is <80%.

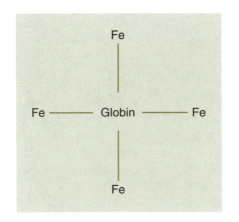

Figure 11.19 Simple structure of the hemoglobin molecule.

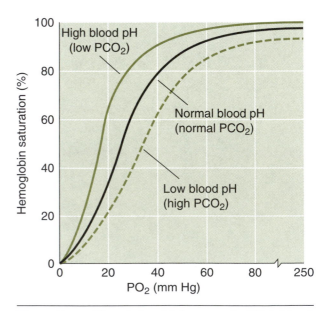

Figure 11.20 The relationship between the PO_2 in blood and the amount of O_2 carried on hemoglobin, expressed as a percentage of the total possible saturation.
Reprinted from Wilmore and Costill 1999.

During exercise, several factors affect the O_2 dissociation curve, enhancing O_2 delivery to the working muscles. Increases in blood CO_2, decreases in pH, and increases in blood temperature shift the curve to the right (dashed, dotted curve; inset, figure 11.20). The significance of this shift is illustrated using a PO_2 of 40 as a reference point. With the curve in its normal position, SO_2 will be approximately 75%, whereas with the rightward-shifted curve it will be closer to 65%. This means that the Hb is holding less O_2 at the same PO_2 when the curve is shifted to the right; the "extra" off-loaded O_2 is available to the working muscles. The extra off-loading of O_2 does not come at the expense of on-loading at the lung, since the shift does not affect the plateau at a normal PO_2 seen in lung capillaries ($PO_2 = 90$ mm Hg).

Hemoglobin allows the transport of far more O_2 in the blood than would be possible if the O_2 were simply carried in solution. This transport is optimized during exercise through alteration of the relationship between blood PO_2 and the amount of O_2 on the Hb.

Hemoglobin allows the transport of far more O_2 in the blood than would be possible if the O_2 were simply carried in solution.

Case Study

You have recently been assigned to work with a U.S. high school athletic program. At the first football game you observe team members breathing O_2 from tanks on the sideline. When you question the coach regarding his rationale, he tells you that the O_2 increases the amount of O_2 to the working muscles when the player returns to the field. What is wrong with this rationale?

During breathing of ambient air, the Hb is nearly 97% saturated with O_2. Using pure O_2 to increase the blood PO_2 will increase the saturation by only 3%. This extra 3% and the small amount dissolved in the blood would have a minimal effect on O_2-carrying capacity.

If saturated to 100% under ambient conditions, Hb can carry 1.34 ml O_2; but at a normal SaO_2 of 97%, this is reduced to 1.3 ml O_2 (1.37 $\times$ .97), making carrying capacity 19.5 ml O_2/ 100 ml blood (assuming a normal Hb of 15 ml O_2/100 ml blood). Breathing pure O_2 and increasing SaO_2 to 100% increases the carry-

ing capacity to 20.1 ml O_2/100 ml blood, a difference of only 0.5 ml of O_2/100 ml blood. Of course this small increase in O_2 would be quickly eliminated as soon as the breathing mask was removed and the player began moving.

Your advice to the coach is that the use of pure O_2 is likely to have a psychological benefit, if any, and that he could save the cost of purchasing the O_2.

Carbon Dioxide Transport

CO_2 is transported via three mechanisms. Approximately 75% of the CO_2 is transported as carbonic acid in the plasma and the red blood cell. Between 5% and 10% of the CO_2 is dissolved in the plasma and in the red blood cell. The remaining CO_2 is transported attached to Hb (carbaminohemoglobin) and attached to plasma proteins. At higher PO_2 levels, Hb carries less CO_2; at low PO_2, the Hb carries more CO_2. This effect of PO_2 on CO_2Hb, called the Haldane effect, enhances CO_2 transport away from the muscles, where the CO_2 is produced. The low PvO_2 during exercise allows ready removal of the exercise-related CO_2 production by this mechanism.

Special Considerations

Extreme changes in the respiratory pressures are reflected in the cardiovascular system. Trying to forcefully exhale while keeping the mouth and nose closed is an example of this situation. The following section addresses this phenomenon, called the Valsalva maneuver, as well as several other respiratory system-related phenomena that you should be familiar with.

• **Hyperventilation**. Attempts to prolong underwater breath-hold times are frequently preceded by a period of deep breathing sufficient to produce large decreases in $PaCO_2$ (**hyperventilation**) (figure 11.21). The hyperventilation also produces increases in PaO_2. At the onset of the breath-hold and underwater swimming, PaO_2 decreases and $PaCO_2$ increases. The ensuing relative hypoxia reaches a point where it induces loss of consciousness before $PaCO_2$ can increase to the extent that it stimulates the urge to breathe and a need to surface. Under these circumstances, drowning may occur.

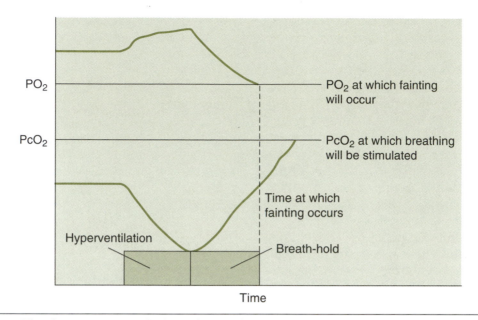

Figure 11.21 The changes in arterial PO$_2$ and PcO$_2$ after a period of hyperventilation and breath-hold.

• **Valsalva.** Many lifting or pushing actions are done while the upper body is fixed. In some cases this fixation is associated with glottal closure and a breath-hold, an attempt to provide further stabilization of the thorax. This maneuver, called the Valsalva maneuver, can be associated with increases in blood pressure. Patients doing resistance training or hoisting heavy loads by hand should understand that they must not hold their breath (figure 11.22). Direct measurements of blood pressure during performance of a leg press with a Valsalva maneuver show systolic pressures of over 300 mm Hg (chapter 13, pages 224-225). This is equivalent to the pressure exerted by water at a depth of about 4 m (13 ft). Anyone who has dived to the bottom of a 2.75 m (9 ft) swimming pool will recognize the effects that this pressure has on the whole body. Pressure of 300 mm Hg exerted on the vascular system can be deadly if it results in the bursting of cerebral vessels. Blood vessels of patients with diabetes are particularly susceptible to hemorrhage, with the ocular vasculature being particularly vulnerable.

• **Second wind.** **Second wind** is a little-understood phenomenon that, in spite of the terminology, is likely not associated with changes in the respiratory system. Because of the inconsistency of occurrence, second wind is difficult to study. Some data suggest that second wind occurs when the cardiovascular adjustments to exercise are complete.

Figure 11.22 The Valsalva maneuver. An increase in blood pressure, associated with a breath-hold, can be fatal for people with vascular problems.

• **Stitch in the side.** As with second wind, the stitch in the side is a phenomenon that is not understood. There is some evidence that it is associated with changes in diaphragm recruitment pattern suggestive of fatigue, but this idea has been difficult to verify. Because the description of a stitch is so subjective and because the side stitch occurs only occasionally, the mechanism will be difficult to determine.

• **"Anaerobic threshold."** The anaerobic threshold has been classically defined as the exercise intensity at which energy production shifts from primarily aerobic to anaerobic (non-oxidative) metabolism. Because of this shift to anaerobic metabolism, lactate begins to form or accumulate, stimulating increases in $\dot{V}_E$ and altering the chemical environment of the cells. This "threshold" can be determined as the point at which $\dot{V}_E$ increases out of proportion to the increases in $\dot{V}O_2$ (line separating segment A from segment B, figure 11.16). As discussed in chapter 4, lactic acid is formed at all work levels. This lactic acid is used as a fuel in the working muscle as well as in nonworking muscle such that little, if any, accumulates in the blood at lower work levels. At higher work levels, however, the ability of the body to use the increased lactic acid production becomes increasingly limited, and the acid begins to accumulate in the blood. Healthy sedentary people reach this point at about 60% of the $\dot{V}O_2$max, whereas athletes may not reach this point until 80% or 90% of $\dot{V}O_2$max. The point where lactic acid begins to accumulate stimulates the increase in $\dot{V}_E$ and other changes associated with the anaerobic threshold. Although there is a level of exercise intensity associated with increased blood lactic acid levels, this should not be interpreted as a threshold point indicating the shift from aerobic to anaerobic metabolism. Terms such as "ventilatory threshold" and "lactate threshold" are currently being used to refer to the phenomenon.

What You Need to Know From Chapter 11

Key Terms

alveolar ventilation

anaerobic threshold

breathing frequency

chemoreceptors

dead space

desaturated

elastic "work"

flow-resistive forces

flow-resistive "work"

functional residual capacity

heme

hemoglobin

hyperpnea

hyperventilation

hypoventilation

minute ventilation

mechanoreceptors

proprioceptors

residual volume

second wind

stitch in the side

tidal volume

transit time

Valsalva

ventilation-to-perfusion matching

Key Concepts

1. You should understand the O_2 cascade and know how the pulmonary system responds at each level of the cascade to ensure that gas exchange is adequate during exercise.

2. You should understand the roles that elastic forces and flow-resistive forces play in determining how we breathe.

3. You should know the role of humoral, or blood-borne, factors and proprioceptors in controlling exercise ventilation.

4. You should understand the O_2 dissociation curve and how changes in the curve might affect O_2-carrying capacity.

(continued)

5. You should be aware of the possible consequences of special respiratory maneuvers such as hyperventilation prior to swimming and the Valsalva maneuver.

Review Questions

1. A worker lying in the prone position may have the movement of the chest wall restricted. What are the likely consequences of this on the way this person will breathe and on gas exchange?

2. Why will a low PAO_2, such as is experienced at altitude, result in a reduced diffusion of O_2 into the blood?

3. Why is $\dot{V}_A/\dot{Q}c$ matching better in the supine position?

4. A patient who is obese will breathe at a high frequency and, in spite of an apparent high ventilation, may show high $PaCO_2$ and low PaO_2. Why?

5. Why might a worker doing arm exercise and leg exercise against the same loads describe sensations of dizziness with the arm work but not the leg work?

6. Why might breathing high levels of O_2 help a pulmonary patient, but not a healthy person, perform activities of daily living?

7. Why would the ventilatory or lactate threshold occur at a higher percentage of the $\dot{V}O_2max$ in a trained athlete than in a sedentary person?

Bibliography

Dempsey JA. Is the lung built for exercise? *Med Sci Sports Exerc* 1986, 18:143–155.

Dempsey JA and Fregosi RF. Adaptability of the pulmonary system to changing metabolic requirements. *Am J Cardiol* 1985, 55:59D–67D.

Dempsey JA, Mitchell AS, and Smith CA. Exercise and chemoreception. *Am Rev Respir Dis* 1984, 129 (suppl):s31–s34.

Dempsey JA and Rankin J. Physiologic adaptation of gas transport system to muscular work in health and disease. *Am J Phys Med* 1974, 46:582–647.

Gallagher CG, Brown E, and Younes M. Breathing pattern during maximal exercise and during submaximal exercise with hypercapnia. *J Appl Physiol* 1987, 63:238–244.

Grimby G. Respiration in exercise. *Med Sci Sport* 1969, 1:9–14.

Henke KG, Sharratt M, Pegelow D, and Dempsey JA. Regulation of end-expiratory lung volume during exercise. *J Appl Physiol* 1988, 64:135–146.

Leith DE and Bradley M. Ventilatory muscle strength and endurance training. *J Appl Physiol* 1976, 41:508–516.

Miyachi M and Tabata I. Relationship between arterial oxygen desaturation and ventilation during maximal exercise. *J Appl Physiol* 1992, 73:2588–2591.

O'Kroy JA, Loy RA, and Coast JR. Pulmonary function changes following exercise. *Med Sci Sports Exerc* 1992, 24:1359–1364.

Robinson EP and Geldgaard JM. Improvement in ventilatory muscle function with running. *J Appl Physiol* 1982, 52:1400–1406.

Scharf SM, Bark H, Heimer D, Cohn A, and Macklem PT. "Second wind" during inspiratory loading. *Med Sci Sports Exerc* 1984, 16:87–91.

Symposium on ventilatory control during exercise. *Med Sci Sport* 1979, 11:190–226.

Thoden JS, Dempsey JA, Reddan WG, Birnbaum ML, Forster HV, Grover RF, and Rankin J. Ventilatory work during steady state response to exercise. *Fed Proc* 1969, 28:1316–1321.

Thompson JM, Dempsey JA, Chosy LW, Shahidi NT, and Reddan W. Oxygen transport and oxyhemoglobin dissociation during prolonged muscular work. *J Appl Physiol* 1974, 37:658–664.

Warren GL, Cureton KJ, Middendorf WF, Ray CA, and Warren JA. Red blood cell pulmonary capillary transit time during exercise in athletes. *Med Sci Sports Exerc* 1991, 23:1353–1361.

Whipp B. Ventilatory control during exercise in humans. *Ann Rev Physiol* 1983, 45:393–413.

Williams JH, Powers SK, and Stuart MK. Hemoglobin desaturation in highly trained athletes during heavy exercise. *Med Sci Sports Exerc* 1986, 18:168–173.

Pulmonary Diseases

12

This chapter will discuss

➤ the etiology and pathophysiology of common pulmonary diseases and the effects these diseases have on the response to exercise, and

➤ the effects of rehabilitation and regular exercise on patients with pulmonary disease.

Pulmonary diseases are divided into two general categories: **obstructive** and **restrictive**. These categories are not mutually exclusive; many patients with an obstructive disease such as emphysema also may have a restrictive component. Diseases that obstruct the airways include **asthma**, **bronchitis**, **emphysema**, **chronic obstructive pulmonary disease (COPD)**, and **cystic fibrosis**. Diseases that restrict the ability to expand or contract the lung or chest wall include obesity, interstitial fibrosis, and lupus. Many of these diseases involve the destruction and permanent loss of lung function. In these cases, rehabilitation in the sense of restoration of lung function is not possible. These patients can, however, benefit from a rehabilitation program that may allow them to better cope with their disease-related disability.

Some abbreviations used in this chapter are defined in Abbreviations Used in Pulmonary Physiology in chapter 11, page 177.

Obstructive

In general, obstructive pulmonary diseases are caused by irritation of the airway resulting in an inflammatory response that may become permanent. Chronic inflammation causes tissue injury and progressive destruction. For instance, bronchitis, generally caused by cigarette smoking that irritates the airway, may progress from an intermittent problem to a chronic bronchitis that in turn may progress to emphysema. Symptomatology reflects the progressive nature of the disease process, starting with intermittent cough and developing to chronic shortness of breath and eventual hypoxemia (i.e., low partial pressure of O_2 in the arterial blood [(PaO_2)]).

Asthma

Asthma is a disease characterized by airways that are abnormally responsive to a variety of stimuli, or triggers.

In patients with severe asthma, the obstruction caused by highly responsive airways may be chronic; but in most patients the obstruction is intermittent. Common triggers of asthma include dust, animal dander, pollen, proteins associated with specific food items, and exercise. Not all patients respond to all allergens, and a particular patient may not respond to a given allergen all the time. It is difficult to estimate the incidence of

197

asthma because much of it is not reported, but estimates suggest that more than 10% of the population may have the disease. Incidence is increasing as a consequence of greater recognition by the medical community and increases in environmental allergens. Although we hear that children "outgrow" asthma, this is probably not the case. As children's airways increase in size, the obstruction may become less noticeable; and in many who seem to outgrow their asthma, it returns later in adulthood, usually in association with an upper respiratory infection.

Asthma is a disease characterized by airways that are abnormally responsive to a variety of stimuli, or triggers.

Asthma and Exercise-Induced Bronchospasm

Asthma obstructs, or narrows, the airways by

- contracting the airway smooth muscle, either directly through neural pathways or indirectly through the local release of histamine;
- causing swelling of the airway mucosa; and
- stimulating the production of mucus secretions.

The primary exception is **exercise-induced asthma (EIA)**, in the medical community more commonly called **exercise-induced bronchospasm (EIB)**, which involves only the contraction of airway smooth muscle. In this condition, the airway becomes narrowed and the resistance to airflow is increased.

Exercise-induced bronchospasm is defined as a narrowing of the airways that is

- acute,
- reversible,
- usually self-terminating,
- triggered by exercise or hyperpnea, and
- accentuated by breathing dry and/or cold air in known, atypical, or latent asthmatics.

This definition highlights differences between EIB and asthma that is initiated by other triggers. Exercise-induced bronchospasm is acute and does not involve the inflammation and associated mucus formation and swelling that prolong asthma in other cases. Exercise-induced broncho-

spasm also is usually self-terminating, which means that at cessation of exercise the bronchoconstriction reverses itself within 30 min (figure 12.1). When people experience EIB it is important to recognize this fact and to reassure them that relief will occur within a short period if they stop exercising and relax. The basis for the self-termination is the fact that the bronchoconstriction is mediated neurally or by local agents and is not complicated by inflammation. Exercise-induced bronchospasm is not triggered by exercise per se but by loss of heat or water, or both, from the airway. Thus an increase in ventilation, or hyperpnea, in the absence of exercise can also result in airway heat or water loss and trigger the bronchospasm. Finally, EIB occurs in 80% to 90% of those diagnosed with asthma induced by other triggers. It is only in very unusual cases that exercise is the sole trigger of asthma. The high incidence of EIB in known asthmatics makes an exercise test to induce EIB useful in diagnosing asthma (see chapter 18). The incidence of EIB is higher in males than in females.

Exercise-induced bronchospasm is defined as a narrowing of the airways that is acute, reversible, usually self-terminating, triggered by exercise or hyperpnea, and accentuated by breathing dry and/or cold air in known, atypical, or latent asthmatics.

Treatment of Asthma and Exercise-Induced Bronchospasm

Treatment for asthma includes the use of bronchodilators to relieve the smooth muscle contraction, drugs to stabilize the cells that release histamine, and anti-inflammatory drugs including corticosteroids. Once the process of airway inflammation begins, treatment becomes more difficult because swelling and mucus plugging prevent inhaled medications from reaching the affected sites.

➤ A histamine is a powerful stimulator of the smooth muscle in the airways, causing airway constriction. Histamine is stored in pulmonary mast cells.

Treatment of EIB starts with good control of the patient's asthma as already described. In patients who are receiving chronic treatment but in whom

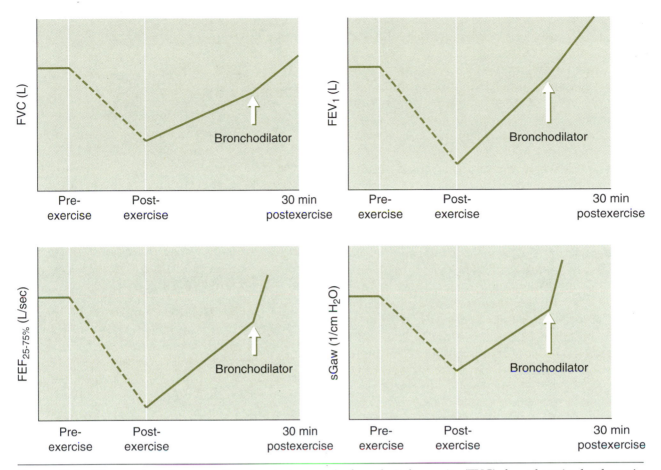

Figure 12.1   Changes in pulmonary function tests, namely, forced vital capacity (FVC), forced expired volume in one second (FEV$_1$), forced expired flow between 25% and 75% of FVC (FEF$_{25-75\%}$), and specific airway conductance (sGaw), after exercise in a patient with exercise-induced bronchospasm. The effect of a bronchodilator after 25 min of recovery is also shown.

this treatment may not prevent EIB entirely, bronchodilators may be administered before exercise to open the airway. Pre-exercise treatment with a mast cell stabilizer, to prevent histamine release, also is effective in preventing EIB. If prevention is unsuccessful, postexercise treatment with bronchodilators always reverses the bronchoconstriction (figure 12.1). Mast cell stabilizers are not effective once EIB has occurred. The ease of prevention and treatment of EIB is important to recognize because it indicates that EIB need not be an excuse to avoid exercise. Following strenuous exercise, patients are protected from EIB in subsequent bouts of exercise for 1 to 2 hr. The mechanism of protection is unknown, but it may be related to the loss of histamine from the mast cells during the original bout of exercise.

Exercise lasting less than 2 min appears to be too short to trigger EIB. On the other hand, exercise lasting more than 10 to 12 min is associated with a decreased incidence of EIB. The decreased incidence and/or severity of EIB during exercise bouts <2 min is one reason why patients with asthma tolerate "stop-and-go" or intermittent activities better than continuous activities. The decreased EIB with longer-duration exercise is what is happening when we say that someone can "run through" his asthma. Sympathetic stimulation associated with exercise is a bronchodilator and contributes to the ability to avoid severe EIB during exercise of longer duration.

> The ease of prevention and treatment of EIB is important to recognize because it indicates that EIB need not be an excuse to avoid exercise.

Exercise in warm, humid environments offers some protection from EIB. For this reason, swimming is a preferred exercise for these patients (figure 12.2). Exercise in cold, dry environments exacerbates EIB and should be undertaken only when medication can be optimized or when the inhaled air can be conditioned through the use of a mask. Because nose breathing enhances the warming and humidification of inhaled air, nasal breathing should be encouraged in patients with EIB. Caution is in order when one places these patients in competitive situations in which they may be under pressure to ignore signs of EIB. Literature to educate coaches and others involved with athletics is available from the American Lung Association.

Proper warm-up, gentle and intermittent, can minimize or attenuate EIB. The incidence of EIB is higher when patients are forced to exercise at high levels without warming up. We have successfully used a protocol of a series of 2 min of exercise followed by 2 min rest for 6 to 8 min, with the exercise periods becoming progressively more intense. The ability of warm-up to prevent EIB is

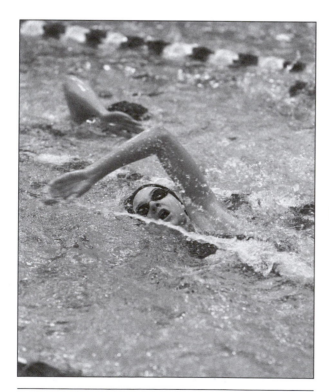

Figure 12.2 Exercise in a warm, humid environment. Swimming is an ideal exercise for a person with exercise-induced bronchospasm because it rarely involves the loss of heat or water from the airways, as do many other forms of exercise.

related to the effect of previous exercise in preventing the triggering of EIB.

Chronic Obstructive Pulmonary Diseases

Chronic obstructive pulmonary diseases include chronic bronchitis and emphysema. Emphysema has become virtually synonymous with COPD. Since the primary cause of emphysema is cigarette smoking, this is a largely self-inflicted, preventable disease.

Characteristics of Emphysema

Emphysema is characterized by airway obstruction, alveolar septa breakdown, increased lung compliance, and hyperinflation.

• Airway obstruction is caused by chronic inflammation that stimulates increases in mucus production and thickening of the airway walls. The degree of obstruction is monitored through use of **pulmonary function** tests. The most common tests are done by (a) spirometry, including **forced vital capacity (FVC), forced expired volume in one second (FEV$_1$)**, and **forced expired flow in the middle 50% of FVC (FEF$_{25-75\%}$)**; and (b) body plethysmography, including functional residual capacity (FRC), **airway resistance (Raw)**, and **specific airway conductance** (sGaw; "G" is a term from physics designating conductance). Chronic obstructive pulmonary disease results in progressive decreases in FVC and flow rates due to increasing Raw and decreasing sGaw. In addition, **hyperinflation** is reflected in increases in FRC.

➤ Spirometry is taking measurements of pulmonary function with a device that measures air volume and flow.

• The breakdown of the alveolar septa results in large balloon-like air spaces with high compliance, that is, like a floppy balloon. The loss of septal surface area is reflected in a loss of surface area for gas exchange. This process would be equivalent to converting the alveolus in figure 11.1b to that in figure 11.1a. The loss of alveolar area in conjunction with the loss of associated capillaries results in a dramatic decrease in the capacity to diffuse gases from the alveolus into the capillary (see chapter 11). The disruption of normal alveolar-to-capillary interface results in gross

$\dot{V}_A/\dot{Q}c$ inhomogeneities and increases in Vd. The increase in Vd increases the "wasted" ventilation (Vd/Vt), with a consequent reduction in $\dot{V}_A$. The attempt to maintain $\dot{V}_A$ causes $\dot{V}_E$ to rise. Of course this increased $\dot{V}_E$ is accomplished against narrowed, obstructed airways, increasing the flow-resistive forces that must be overcome in order for ventilation to occur.

➤ Septa are thin segments of the alveoli that create a large alveolar surface area (see figure 11.1).

Emphysema results in a disruption of normal gas exchange, forcing an increase in $\dot{V}_E$ to overcome large increases in Vd.

Figure 12.3 Climbing stairs with COPD. Because of the disruption of septal walls, COPD patients lose much of the gas exchange area in their lungs. For such patients, ordinary activities such as climbing stairs become an ordeal.

• The increase in lung compliance lessens the tendency of the lung to collapse such that the tendency of the chest wall to flare out can cause the lung volume to expand. This larger FRC results in a condition called hyperinflation, which gives these patients their characteristic barrel-chested look and which is measured by an increase in FRC (see figure 11.14).

• Hyperinflation causes a shortening and flattening of the diaphragm, putting it at a mechanical disadvantage as an inspiratory muscle so that accessory muscles play a greater role in inspiration. Patients breathe at higher lung volumes, leading to a chronically increased work of breathing due to the greater effort required to overcome the elastic forces of the respiratory system.

Emphysema results in hyperinflation and consequent increases in the work of breathing.

Exercise Response in Patients With Chronic Obstructive Pulmonary Disease

The reduced gas-exchange capacity and mechanical disadvantage induced by COPD eventually reduce exercise capacity. Exercise limitations are not evident immediately when the process of lung destruction begins.

In a healthy person, the capacity of the pulmonary system to ventilate and exchange gases is never exceeded, even at maximal exercise. Even during exercise above maximum $\dot{V}O_2$, only 80% of the capacity of the pulmonary system is used, leaving a 20% reserve (figure 12.4). As the disease process begins, exercise is not limited because of this reserve; but as the disease progresses, the reserve is reduced, and eventually exercise limitation is noticeable. Of course, if the person does not exercise strenuously, the limitation may not be evident until the disease process has destroyed enough pulmonary capacity to affect activities of daily living (figures 12.3 and 12.4).

The reduced gas-exchange capacity and mechanical disadvantage induced by COPD eventually reduce exercise capacity.

The alterations in the pulmonary response to exercise in patients with COPD are illustrated in figure 12.5. The disease-associated increase in Vd/Vt noted at rest becomes greater during exercise because of exercise-induced increases in $\dot{V}_A/\dot{Q}c$

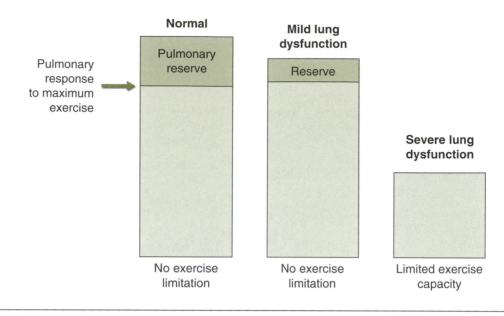

Figure 12.4 Schematic depicting the change in functional lung capacity with increasing lung dysfunction related to chronic obstructive pulmonary disease.
Adapted from Haas and Axen 1991.

inhomogeneities. In a healthy person, Vd/Vt is about 30% at rest and decreases to less than 10% at maximal exercise (figure 11.4). Moderate COPD results in moderate increases in Vd/Vt, and Vd/Vt may rise slightly during exercise. In the patient with severe COPD, Vd/Vt may be as high as 50% at rest and rise during exercise. Consequently, $\dot{V}_E$ is higher at all exercise levels in the patient with COPD. Remember that the increase in $\dot{V}_E$ is an attempt to move sufficient air to overcome the greater amount of "wasted" air. In the patient with severe lung dysfunction, this increase in $\dot{V}_E$ is insufficient to overcome the greater Vd, and gas exchange is compromised—resulting in a decrease in PaO_2 and possible increases in $PaCO_2$ (figure 12.5).

Case Study

Mr. Robert Schoenfeld is a 67-year-old, 170 cm (5 ft 7 in.), 80 kg (176 lb) former carpenter with a diagnosis of emphysema. He has been a pack-a-day smoker for the past 50 years. He has a 35-year history of coughing and mucus production with frequent upper respiratory infections. He has not worked for the past six years because he became unable to meet the demands of the job. More recently he has been short of breath when performing low-level activities such as walking. His most recent pulmonary function tests show severe lung dysfunction.

Test	Result	% Predicted
FVC	3.61 L	92
FEV$_1$	1.31 L	48
FEF$_{25-75\%}$	0.73 L/s	28
FRC	6.26 L	182
sGaw	0.023 L/cm H$_2$O/s	12

His physician has requested a progressive exercise test with full monitoring of ventilatory and cardiovascular variables. You select a modified Balke protocol (see chapter 18) with treadmill speed at 3.0 mph and 2% grade increments every 2 min.

Mr. Schoenfeld reaches a grade of 6% (5.5 METs) at a heart rate of 121 beats/min. His ventilatory results are shown in figure 12.5 (severe COPD patient). Because of the likely increase in Vd, his increased $\dot{V}_E$ is not unexpected. However, on the basis of his airway obstruction you would have predicted a much greater increase in Vt than in Fb. Your rationale was that the best strategy to minimize the increased flow resistance would have been to minimize increases in flow by increasing Vt. You seek an answer from the attending pulmonologist.

Dr. Redman explains that your assumptions are correct but that you forgot to consider the second aspect of respiratory work, the elastic forces. Because Mr. Schoenfeld is

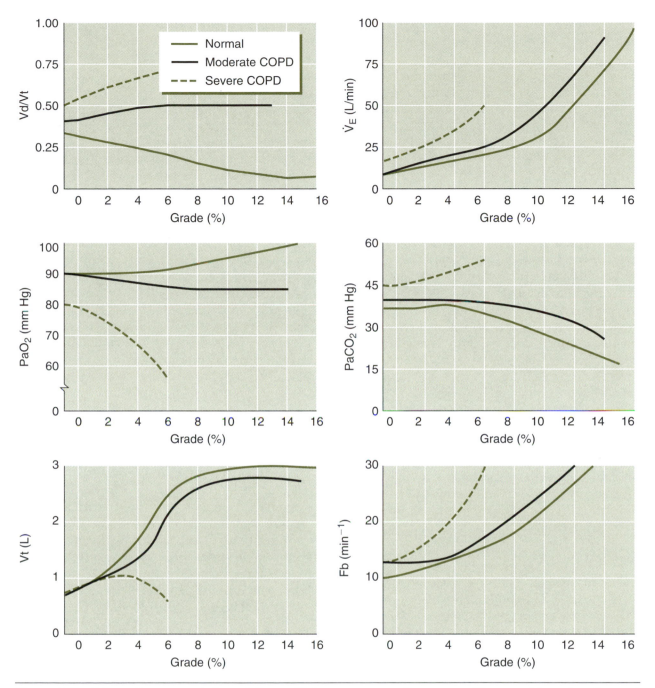

Figure 12.5   Schematic representation of the ventilatory responses of a person with normal lung function and those of chronic obstructive pulmonary disease patients with moderate or severe lung dysfunction.

so hyperinflated, it would be extremely difficult for him to increase Vt by going to higher lung volumes. As a result, Mr. Schoenfeld's only mechanism to increase $\dot{V}_E$ is to increase Fb. This, of course, is ineffective in increasing $\dot{V}_A$ because an increase in Fb exacerbates the Vd problem (see pages 174-180).

Chronic obstructive pulmonary disease alters the mechanical response of ventilation during exercise. You will remember that the combination of Vt and Fb to attain any level of $\dot{V}_E$ is the one that minimizes the flow-resistive and elastic forces of breathing. The patient with COPD must contend with airway obstruction, which increases flow-resistive forces, and also hyperinflation, which increases elastic forces. The combination of Vt and Fb measured in any given patient depends on the degree of obstruction balanced against the degree of hyperinflation. In general, however, the more severe the lung dysfunction, the greater the hyperinflation. Up to a point, the work required to increase Vt at high volumes is far greater than the work required to move air at a higher velocity through obstructed airways, so patients assume a breathing pattern with a high breathing frequency. Again, this may seem contradictory because the result is an increased work to move the air through the obstructed airways—but that work is still less than the work to inflate a hyperinflated pulmonary system. Figure 12.5 shows schematically the progression of ventilatory changes from normal lung function to moderate and severe lung dysfunction. Patients with moderate lung function show moderate alterations in ventilatory response to exercise, with little change in exercise capacity. Patients with severe lung dysfunction show extreme alterations in the ventilatory response to exercise, with notable changes in exercise capacity. The high Fb in patients with severe dysfunction seems paradoxical, especially considering that an increase in Fb exacerbates the existing increased Vd. The decrease in PaO_2 and increase in $PaCO_2$ indicate that the patient's ventilatory responses to exercise are inadequate.

Cystic Fibrosis

Cystic fibrosis (CF) is the most common lethal genetic disease among children, occurring in about 1 in every 2000 live Caucasian births. The defect affects exocrine gland function by altering chloride transport. The sweat glands produce sweat with high concentrations of Na^+, Cl^- and K^+, making possible the diagnostic sweat test for the disease. Pancreatic insufficiency results in digestive absorption deficiencies, which affect the ability to digest fat and may result in growth failure unless the pancreatic enzymes are replaced orally. The primary cause of death is from respiratory failure due to the thick tenacious mucus secreted into the airways. These secretions plug the airways and promote chronic infections, leading to greater mucus secretion. In spite of its name, CF is rarely associated with fibrotic changes in the lung. Treatment includes replacement of pancreatic enzymes, oral and intravenous antibiotics,

and chest physical therapy to promote mucus mobilization and expectoration. With aggressive treatment the mean survival age has increased from 12 to over 30 years in the last two decades.

The increased mucus production and plugging of the airways lead to airway obstruction and severe lung dysfunction similar to that found in emphysema. Flow rates (FEV_1, $FEF_{25-75\%}$) are reduced and FRC is increased, reflecting hyperinflation. The airway obstruction increases $\dot{V}_A/\dot{Q}c$ inhomogeneities, which in turn increase Vd and gas-exchange abnormalities. As a consequence of the gas-exchange and mechanical disruption, the exercise response is similar to that described for emphysema (figure 12.5). As in COPD, these pulmonary abnormalities increase ventilatory demand, but reduce the ability of the system to respond to the increased demand. This tension between increased demand and reduced response ultimately leads to an inability to maintain arterial blood gases during exercise (figure 12.5, severe patient).

Case Study

Colleen is a 13-year-old CF patient. She was diagnosed at age 6 months and has been on regular antibiotic and pancreatic enzyme therapy since. In addition, she receives chest physical therapy, including postural drainage and percussion and vibration, two to three times daily. Colleen has been admitted to the hospital for an acute exacerbation of her lung disease.

Her pulmonary functions at admission show severe lung dysfunction.

Test	Result	% Predicted
FVC	1.55 L	59
FEV_1	0.87 L	36
$FEF_{25-75\%}$	0.40 L/s	12
FRC	2.89 L	188
sGaw	0.041 L/cm H_2O/s	19

Colleen's pulmonologist has asked that she receive appropriate exercise therapy twice daily for 10 to 15 min, or less initially if she can't tolerate 10 to 15 min.

To determine Colleen's appropriate exercise level, you perform a graded, progressive exercise test to find out whether she desaturates (that is, decreases SaO_2) during exercise, and if so, at what level. You need this information so that you can set her exercise intensity below the level at which her arterial O_2 decreases. At the time of the test, Colleen's resting SaO_2 is 90%. During free-wheeling on the cycle ergometer, her SaO_2 decreases to 89%. At 120 W, saturation is 85%; at 180 W, her saturation drops below 80%, and you terminate the test.

Since your facility has a policy of not having CF patients exercise when SaO_2 decreases below 90%, you are not sure what you should do.

You have possible two solutions. You can find an exercise—other than on the cycle ergometer—that offers a lower exercise intensity, or you can supplement with O_2 to keep the SaO_2 above 90%. Since Colleen prefers heavier-intensity exercise, you begin her exercise program on the cycle ergometer at 180 W and use 0.5 L O_2 supplementation by nasal cannula, which is sufficient to keep SaO_2 >90%.

Patients with CF are able to maintain a normal sweat response to exercise, but the increased NaCl concentration in their sweat increases the loss of Na^+ and Cl^-. The normal sweat response allows these patients to maintain normal core temperatures and heart rates during exercise. In addition, patients are able to acclimate normally to heat, as measured by exercise heart rate and temperature responses, but are unable to alter sweat electrolyte concentrations. Nevertheless, making table salt available and encouraging fluid ingestion ensure maintenance of normal body electrolyte balance and hydration in patients who may be exposed to high sweat loss.

Restrictive Diseases

Restrictive lung diseases affect the lung or chest wall tissues in such a way that they become stiffer. Sarcoidosis is a systemic disease that results in fibrotic changes in the alveolar walls, causing increased stiffness of the lungs. Diffuse interstitial fibrosis results in a thickening, and therefore increased stiffness, of the alveolar interstitium. Obstructive diseases that include a significant component of lung restriction due to fibrosis include silicosis, asbestosis, and related diseases. The causes of sarcoidosis and diffuse interstitial fibrosis are unknown; therefore treatment is lim-

ited to controlling inflammation through the administration of corticosteroids. The treatment of mixed obstructive and restrictive diseases is oriented to the use of bronchodilators and anti-inflammatory drugs.

The increased stiffness, or decreased elasticity, in these diseases makes it difficult to expand the lung. The energy requirement to overcome the high elastic forces increases the work of breathing. In contrast to patients who adopt a slow, deep breathing pattern to breathe through narrow, obstructed airways, patients with restrictive diseases breathe at high frequencies with shallow tidal volumes to minimize work against elastic forces. The higher breathing frequencies and lower tidal volumes increase the Vd/Vt, potentially decreasing gas exchange. The increased Vd reduces gas exchange in patients with mixed obstructive and restrictive disease. In patients with primarily restrictive disease, ventilation is exaggerated such that they hyperventilate, with normal quiet breathing $PaCO_2$ below 35 mm Hg (figure 12.6).

The rapid, shallow breathing pattern observed at rest is exaggerated during exercise (figure 12.6). At the onset of exercise there is an immediate rise in Fb; there is then a further, dramatic rise as exercise intensity increases. This Fb response is reflected in a parallel decrease in $PaCO_2$, which drops precipitously as exercise becomes more intense. Note also the decrease in PaO_2, reflecting the high Vd/Vt in these patients.

Rehabilitation

A pulmonary rehabilitation program must be considered as a part of a comprehensive treatment program for patients with pulmonary diseases. This comprehensive program must include

- medical management of the disease,
- patient and family education,
- psychosocial counseling, and
- physical reconditioning.

One must consider each of these components in concert with the other goals of rehabilitation. The primary goals of pulmonary rehabilitation (page 206) are based on a 1981 statement of the American Thoracic Society. As part of the comprehensive rehabilitation program, exercise therapy will directly impact many of these goals (a, d, h, i, j) and indirectly impact others (b, c, g).

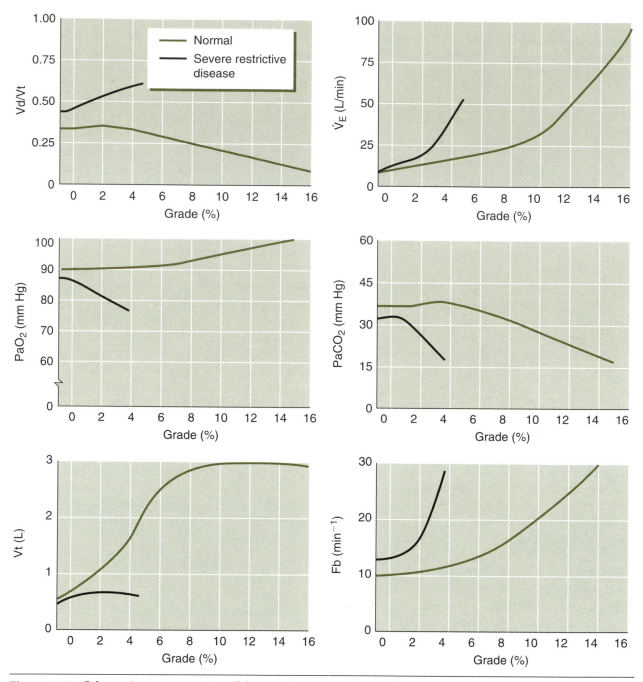

Figure 12.6 Schematic representation of the ventilatory responses of a person with normal lung function and a patient with severe restrictive lung disease.

Primary Goals for Pulmonary Rehabilitation

a. An improvement in cardiopulmonary function.

b. The prevention and treatment of complications.

c. The recognition and treatment of stress and depression, which so often accompany severe pulmonary disease and interfere with a healthy adaptation to it.

d. An active rather than passive lifestyle, with a reversal of withdrawal from society.

e. The promotion of patient acceptance of and compliance with optimum medical care, wherein the patient assumes increasing responsibility for his or her own care and well-being.

f. An increased understanding of the disease and the disease process, so that the patient and family can confront it realistically without fear of the unknown.

g. A reduction in numbers of exacerbations, emergency room visits, and hospitalizations.

h. A return to work and/or a more active, productive, emotionally satisfying life.

i. The provision of tools with which the patient may regain independence and cope with his or her sense of loss, including loss of control of personal and social relationships, self-esteem, and self-worth.

j. Improvement in the quality of life for both the patient and his or her family.

Benefits and Goals of Pulmonary Rehabilitation

Comprehensive pulmonary rehabilitation programs have been shown to reduce hospitalizations and days in the hospital, as well as disease-related psychological disorders, and also to increase survival, employment, and exercise tolerance.

Most of the scientific evidence supporting exercise training has been obtained in COPD patients; few studies have examined patients with restrictive diseases.

> Comprehensive pulmonary rehabilitation programs have been shown to reduce hospitalizations, days in the hospital, and disease-related psychological disorders and also to increase survival, employment, and exercise tolerance.

Exercise training in the patient with COPD produces the same physiological effects as in a healthy person. Benefits include improved cardiorespiratory endurance and muscle strength and endurance. Patients have a reduced heart rate and increased stroke volume at any given workload after training. Resting blood pressure is also lowered with training, particularly in patients who may be hypertensive. Improved metabolic function is reflected in lower blood lactic acid levels at a given exercise level as shown in pre- and posttraining comparisons. In addition, patients with lung disease show an improvement in respiratory muscle function after exercise training. All

these exercise training-related changes translate into improved exercise capacity and greater ease in performing activities of daily living.

The specific goals of a pulmonary exercise therapy program are listed below. There is general agreement that exercise training in pulmonary patients should emphasize aerobic activities. Resistance exercise, as a supplement to the aerobic conditioning program, can enhance the effects of the aerobic program on specific muscle strength and endurance (figure 12.7). Upper-body exercises, including rowing-like activities—which provide a combination of aerobic and resistive exercise—can improve the strength and endurance of the respiratory muscles.

The selection of specific activities is also critical for the long-term compliance to, and therefore success of, a rehabilitation program. As is true for the general population, the more enjoyable the activity, the more likely it is that people will continue to participate in rehabilitation over time.

> The more enjoyable the activity, the more likely it is that people will continue to participate in rehabilitation over time.

Specific Goals of a Pulmonary Exercise Training Program

General

- Increase cardiorespiratory fitness
- Increase neuromuscular efficiency
- Strengthen respiratory muscles
- Help patient become more comfortable with their ability to participate in physical activity and to build a healthy self-image

Cystic fibrosis patients

- Enhance airway clearance
- Improve gas exchange

Guidelines for Pulmonary Rehabilitation

Pulmonary rehabilitation programs generally follow the same model that is used for rehabilitation of cardiac patients (see chapter 14).

The rehabilitation model defines a progression of rehabilitation from the acute, hospitalization phase to the time when the patient is able to engage

Figure 12.7   Mild exercise. Patients with COPD can benefit from exercise as light as the activity pictured here.

in independent activity. There are differences between the pulmonary and cardiac patients that one must consider when developing a physical rehabilitation program. In most cases, the cardiac patient experiences a sudden event that becomes a defined point for the onset of rehabilitation. The disease process in most pulmonary patients, however, is more insidious, with no such well-defined starting point. Patients begin to experience episodes of dyspnea that become evident at progressively lower work levels. Typically they do not seek advice regarding exercise until the dyspnea begins to interfere with activities of daily living and severe lung dysfunction is present. Often patients express this concern during a hospitalization for an acute exacerbation of their disease; but in other cases they seek assistance well before the disease has reached this stage. Pulmonary rehabilitation that is started early in the disease process is more effective than that initiated after the disease has begun to limit daily function.

Pulmonary rehabilitation programs generally follow the same model that is used for rehabilitation of cardiac patients.

Patients who are admitted to the hospital should begin physical rehabilitation during the hospitalization. This is an excellent opportunity to begin the education that is part of the rehabilitation process. Ongoing physical rehabilitation need not be interrupted by admission to the hospital for an acute exacerbation of the disease. For patients unable to ambulate to the rehabilitation facility but able to move about the room, elastic straps attached to the bed frame can provide both low-intensity aerobic and resistive exercise. Likewise, ankle weights used during leg-extension exercise can help to maintain or improve muscle strength while providing some aerobic activity. Intensity must be low enough that SaO_2% does not fall below 90. Portable oximeters allow bedside monitoring of SaO_2 during the activities and may indicate the need for O_2 supplementation. Patients tolerate four or five periods of short-duration (5-10 min) exercise of this type better than attempts to exercise for longer periods of time. Patients should perform these levels of light activity five to six days a week, or more, depending on tolerance.

Once patients are able to ambulate, they can perform exercise in a rehabilitation or laboratory setting where monitoring of SaO_2, as well as heart rate, if desired, is easier. The duration of the exercise sessions should increase progressively until the patient can tolerate low levels of exercise for 10 to 15 min. As soon as possible, ambulatory hospitalized patients should be encouraged to perform informal exercise such as walking the halls or other types of activities, perhaps related to their occupation, at least five to six days a week. Patients should attempt this type of activity only after having performed monitored exercise to determine the exercise intensity at which desaturation may occur. Once this level has been identified, the patient should be taught which activities are safe and should be instructed to recognize signs of hypoxemia, such as shortness of breath or increased anxiety. Use of the modified **Borg rating of perceived exertion (RPE) scale** (figure 18.8, page 303) can aid in teaching patients self-monitoring of exercise intensity. Individual RPEs closely match exercise intensity targets determined by heart rate, and adult patients can learn to identify an exercise intensity that is associated with a specific RPE. Until late adolescence, though, RPE is an unreliable indicator of exercise intensity.

Case Study

Ms. Brown is a 61-year-old smoker with emphysema. Her pulmonary function tests show mild lung dysfunction (FVC = 73% predicted; FEV_1 = 65% predicted; $FEF_{25-75\%}$ = 59% predicted; $SaO_2\%$ = 92). She works as a secretary for the local school district and would like to work for another two years to qualify for retirement benefits, but she has fallen into a sedentary lifestyle that, combined with her disease, makes even light work difficult. She has come to you for an exercise program. Her insurance won't cover the cost of a health club, so her program must include activities that she can do on her own. You perform a treadmill exercise test and find that she desaturates to <85% at a heart rate of 148. Using the RPE scale, Ms. Brown rates the activity at that level a 5 out of 10. You need to identify activities of low to moderate intensity that she can carry out at this level of perceived exertion.

Activities that the patient prefers, and that she can perform at an intensity that will keep her heart rate below 145, include walking, recreational tennis, and cycling. As her exercise capacity improves, she hopes to be able to engage in exercises of higher, but still moderate, levels, including canoeing and skating. You need to determine the best way for Ms. Brown to monitor her exercise intensity to minimize the possibility of desaturation. Although a heart rate monitor is a possibility, you suggest that Ms. Brown become familiar with rating her perceived exertion and use these ratings to monitor exercise intensity. After several treadmill exercise sessions, she is able to adjust her exercise intensity so that her rating falls between 4 and 5 when her heart rate is above 135 beats/min but below 145 beats/min.

Chapter 18 includes more specific exercise recommendations.

Patients who are not hospitalized also should be encouraged to participate in regular activity. An important component of any activity program is the social aspect. On an individual basis, patients should be counseled to recruit an exercise partner who can provide support and encouragement on those days when exercise may be difficult. An organized program at a health care or fitness facility is ideal for providing such support and may include educational programs designed for this patient population (figure 12.8). If exercise intensity cannot be established through an exercise test with monitoring of SaO_2, the patient should be started with low-intensity activities and observed for signs of hypoxemia. Likewise, progress should be slow, and patients should not be pressured to exert themselves beyond comfortable levels. There is a risk that patients may develop exertional hypoxemia sufficient to aggravate subclinical cor pulmonale. Cor pulmonale is a dangerous condition in which hypoxemia causes elevated pulmonary arterial pressures leading to right heart failure, and for which exercise is contraindicated. It is important also to regulate duration of exercise carefully so that patients do not exercise beyond their level of comfort. The relative vagueness of this recommendation emphasizes the need for a monitored exercise evaluation before one establishes safe exercise limits in pulmonary patients.

One of the goals of physical rehabilitation is to return the patient to his or her occupation, or to establish the physical capability to perform in an alternative occupation. The goal of such

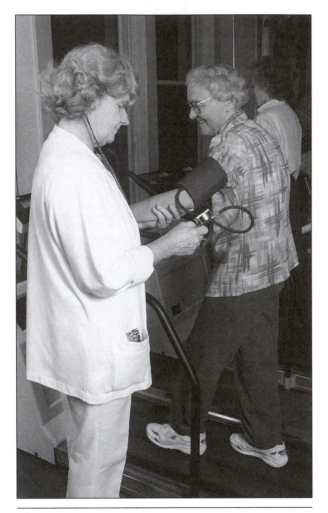

Figure 12.8 Exercise class for pulmonary patients. Both supervision and exercising with others can help motivate COPD and other pulmonary patients to comply with their activity prescriptions.

occupational conditioning is to ensure that patients can perform the tasks required in their job. An occupational program, however, should start with general conditioning principles to improve functional aerobic capacity and general body strength. This conditioning will provide the basis for the more specific conditioning or "work hardening" to follow. As with all physical rehabilitation in pulmonary patients, one should establish exercise intensity after conducting a monitored exercise test (see chapter 18).

Case Study

Mr. Licht has been a regular participant in your exercise rehabilitation program for six months. He is a 64-year-old ex-smoker with a diagnosis of emphysema. His resting blood gases show a PaO_2 of 90 mm Hg ($SaO_2\% = 90$) and a $PaCO_2$ of 46 mm Hg. His FVC is 60% predicted; his FEV_1 is 40% predicted; and his $FEF_{25-75\%}$ is 23% predicted. Pulmonary functions have improved only marginally over the six months, but Mr. Licht has significantly improved his exercise capacity from less than 3 METs to over 3.5 METs. His enthusiasm for exercise is contagious.

Over the past several days, Mr. Licht has been complaining of increasing shortness of breath and reduced exercise tolerance. His pulmonary functions haven't changed, and his resting SaO_2 is still 90 mm Hg. Much to your surprise, his resting $\dot{V}_E$ is over 20% higher than before, and his ventilatory response to a brief, two-stage exercise test on the treadmill is exaggerated, lending credence to his complaints or shortness of breath.

As part of your discussion of these results with the patient, you ask whether he has recently changed any of his habits. The only thing he can think of is that, since reading in a popular running magazine that a high-carbohydrate diet could lead to better exercise performance, he has been eating considerably more carbohydrates than he used to. You can't think of any immediate reason why this should increase $\dot{V}_E$ and cause shortness of breath.

Upon reflection, however, an idea occurs to you. You remember that when carbohydrates are burned, the amount of CO_2 produced—relative to the $\dot{V}O_2$ (expressed as the respiratory exchange ratio; see chapter 4)—is greater than that produced when lipids are burned. By consuming a high-carbohydrate diet, Mr. Licht is increasing the ventilatory requirement to eliminate CO_2. The increased requirement for ventilation, in combination with his reduced capacity to ventilate, leads to Mr. Licht's dyspnea. Your recommendation is that he return to his previous diet.

What You Need to Know From Chapter 12

Key Terms

airway resistance (Raw)

asthma

bronchitis

chronic obstructive pulmonary diseases (COPD)

cystic fibrosis

emphysema

exercise-induced asthma

exercise-induced bronchospasm (EIB)

forced expired flow in mid-vital capacity ($FEF_{25-75\%}$)

forced expired volume in one second (FEV_1)

forced vital capacity (FVC)

hyperinflation

obstructive pulmonary diseases

pulmonary function

rating of perceived exertion (RPE, Borg scale)

restrictive pulmonary diseases

specific airway conductance (sGaw)

Key Concepts

1. You should know the causes of lung diseases including asthma, COPD, and restrictive diseases; should understand the differences between obstructive and restrictive lung diseases; and should know how pulmonary function changes with each.

2. You should understand how obstructive and restrictive lung diseases affect the responses to exercise.

3. You should know what the basic elements of a pulmonary rehabilitation program are and how to institute an exercise program in patients with lung disease.

Review Questions

1. Consider the following pulmonary function test results and determine whether the patient has obstructive or restrictive disease.

FVC	78% predicted
FEV_1	82% predicted
FEV_1/FVC	95%

2. A young athlete wants to cross-country ski but has EIB. What is the potential problem with this, and what advice would you give him?

3. Your friend, a 26-year-old smoker, keeps telling you that she can be just as active now as she was 10 years ago. What can you tell her that will help her understand that she is being lulled into a false sense of security?

4. A patient with COPD breathes with a high frequency and low tidal volume. This seems paradoxical, because you would expect the patient to try to reduce the flow-resistive forces by breathing at low frequencies and higher tidal volumes. Explain this paradox.

5. You are proud that you have just spent an hour teaching a pulmonary patient to breathe more effectively; but when you see him a short time later he is breathing the way he was before your wonderful lesson, and his blood gases (PaO_2 and $PaCO_2$) haven't improved. Why shouldn't you be surprised?

Bibliography

Alpert JS, Bass H, Szucs MM, Banas JS, Dalem JE, and Dexter L. Effects of physical training on hemodynamics and pulmonary function at rest and during exercise in patients with chronic obstructive pulmonary disease. *Chest* 1974, 66:647–651.

American Thoracic Society. Pulmonary rehabilitation: Official American Thoracic Society position statement. *Am Rev Respir Dis* 1981, 124:663-666.

Babb TG and Rodarte JR. Exercise capacity and breathing mechanics in patients with airflow limitation. *Med Sci Sports Exerc* 1992, 24:967–974.

Bellemare F and Grassino A. Force reserve of the diaphragm in patients with chronic obstructive pulmonary disease. *J Appl Physiol* 1983, 55:8–15.

Casaburi R, Patessio A, Ioli F, Zanaboni S, Donner C, and Wasserman K. Reductions in exercise lactic acidosis and ventilation as a result of exercise training in patients with obstructive lung disease. *Am Rev Respir Dis* 1991, 143:9–18.

Cerny FJ. Relative effects of bronchial drainage and exercise for in-hospital care of patients with cystic fibrosis. *Phys Ther* 1989, 69:633–639.

Cerny FJ and Armitage L. Exercise and cystic fibrosis. *Ped Exerc Sci* 1989, 1:116–126.

Cerny FJ, Pullano TP, and Cropp GJA. Cardiorespiratory adaptations to exercise in cystic fibrosis. *Am Rev Respir Dis* 1982, 126:217–220.

Haas F, Pineda H, Axen K, Gaudino D, and Haas A. Effects of physical fitness on expiratory airflow in exercising asthmatic people. *Med Sci Sports Exerc* 1985, 17:585–592.

Levison H and Cherniack RM. Ventilatory cost of exercise in chronic obstructive pulmonary disease. *J Appl Physiol* 1968, 25:21–27.

Longo AM, Moser KM, and Luschinger PC. The role of oxygen therapy in the rehabilitation of patients with chronic obstructive pulmonary disease. *Amer Rev Respir Dis* 1971, 103:690–697.

McFadden ER. Exercise performance in the asthmatic. *Am Rev Respir Dis* 1984, 129 (suppl):584–587.

Cardiovascular System

This chapter will discuss

➤ the anatomy and function of the cardio-vascular system, including the anatomy and physiology of the heart and circulatory systems,

➤ basic electrocardiography,

➤ blood pressure responses to exercise, and

➤ volume versus pressure overload of the heart.

Of all the body systems that alter function in response to exercise, the cardiovascular (CV) system responds in the greatest magnitude. Resting blood flow is very low in skeletal muscle (3-5 ml/100 g/min) but can increase by 50 or 60 times that at maximum exercise! This means that regulation of the CV system must be very precise, but at the same time must operate over a wide range of functional capacities. Our first key principle is based on the ability to redistribute blood within the body to areas that have an increase in metabolic demand. The CV system responds to the increased demand by lowering blood flow to "nonessential" organs during exercise, while increasing flow to the working muscle in direct proportion to the increased metabolic needs. This precise redistribution of blood is especially critical for long-term, steady-state exercise. You should become familiar with terms common to CV physiology to understand concepts and principles, so before continuing, refer to the list below.

Common Terms in Cardiovascular Physiology

heart rate (HR)—Beats/min

stroke volume (SV)—Blood ejected/beat in ml

cardiac output ($\dot{Q}$)—SV × HR

systole—Heart is contracting

diastole—Heart is relaxed, fills with blood

end-diastolic volume (EDV)—Amount of blood in ventricle at end of diastole

end-systolic volume (ESV)—Amount of blood in ventricle at end of systole

total peripheral resistance (TPR)—Resistance to blood flow in entire systemic circulation

blood pressure (BP)—$\dot{Q}$ × TPR

systolic BP—Represents work of the heart

diastolic BP—Reflects peripheral resistance

Cardiac Output

Cardiac output (Q̇) at rest and at V̇O₂max for three groups of people is shown in figure 13.1. The group with lowest maximum cardiac output has a mechanical filling deficiency of the left ventricle. This deficiency is caused by mitral stenosis, a narrowing of the opening between the left atrium and left ventricle. The middle group is called "normally active" (NA)—people in this group participate in low-level physical activity on a regular basis but are not conditioned. The group with the highest V̇O₂max comprises highly trained athletes (ATH). The point of showing these widely differing responses to exercise is that almost everyone in the general population falls between the lowest and highest cardiac outputs illustrated here. Note that these values are for males; females have values about 25% lower.

You should note the following key points as you look at this figure:

• All three of our groups are the same age and sex and have approximately the same body mass and composition. Thus, resting cardiac output in all three is approximately the same. Note from the figure that about 20% of Q̇ at rest is directed toward skeletal muscle, which makes up about 40% of the total body mass. This demonstrates that resting blood flow in skeletal muscle is very low.

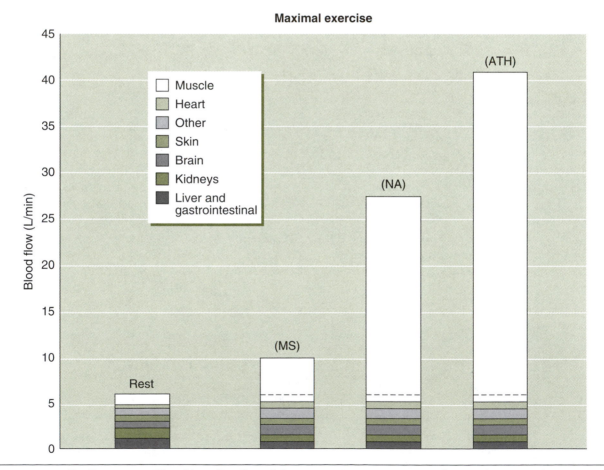

Figure 13.1 Cardiac output. Blood is redistributed during maximum exercise so that working muscle receives the majority of cardiac output. Blood flow to nonessential organs decreases or remains the same. People with mitral stenosis (MS) have an extremely low maximum cardiac output, while the cardiac output of elite endurance athletes (ATH) is quite high. Cardiac output of normally active people (NA) falls between the two extremes. Because these groups have approximately the same maximum heart rate, this graph illustrates the limitations of stroke volume on cardiovascular function.

Adapted from Rowell 1986.

• Blood flow is redistributed during exercise so that blood vessels in the working muscle are dilating and those in nonessential organs (splanchnic area, kidney, etc.) are constricting (figure 13.2). We are assuming here that body temperature is not rising, so there is a decrease, not an increase, in skin blood flow. Note that blood flow is decreased even in nonactive muscle, underlining the precision with which blood flow is regulated during exercise.

• All three groups have the same relative response of the CV system to exercise. That is, they all increase flow to the working muscles and decrease flow to nonessential areas.

• The percentage of total cardiac output delivered to the working muscles varies among the three groups: it is about 50% for the MS and 75% for the NA, but about 90% for the ATH. In the ATH, this represents an amazing capability to direct blood to working muscles at the expense of other areas of the body—in essence the formation of a closed loop between the heart and the working muscles, with very little blood going elsewhere.

• Because all groups are the same age and sex, maximum heart rates are very similar. This means that the biggest limitation on CV function, in terms of $\dot{V}O_2$max, is **stroke volume**. The major goal of a CV training program, then, should be to increase stroke volume.

We'll look at the best way to increase the pumping capabilities of the heart later in the chapter. First let's briefly examine basic anatomy and function of the heart and vascular system.

Since the biggest limitation on CV function, in terms of $\dot{V}O_2$max, is stroke volume, the major goal of a CV training program should be to increase stroke volume.

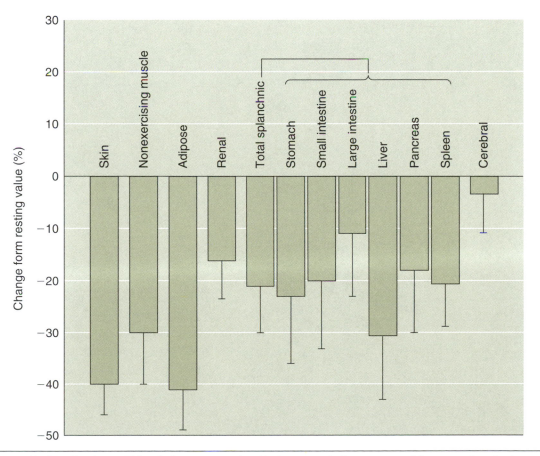

Figure 13.2   Redistribution of blood during exercise. To increase delivery to working muscles during exercise, blood flow is reduced to nonessential organs, including nonexercising muscle. Because skin blood flow decreased, it is assumed that exercise is in a thermoneutral environment.
Reprinted from Rowell 1986.

Anatomy and Function of the Heart

Just as with other body systems we have already covered, there is an intricate relationship between structure and function in the CV system. This is especially true of the heart. Did you know that the heart beats approximately 100,000 times a day, over 36 million times in one year, and about 2.9 billion times in the lifetime of someone who lives to be 80? The importance of educating patients and clients on how to be "heart healthy" is obvious.

Anatomy

The four main chambers and valves of the heart are illustrated in figure 13.3. O_2-poor venous blood returns to the right side of the heart and is pumped to the lungs for oxygenation. The pulmonary circulation is a low-pressure system (approximately 25 mmHg systolic blood pressure and 15 diastolic blood pressure), resulting in low resistance for emptying the right ventricle. Oxygenated blood that returns from the lungs enters the left side of the heart and is pumped to the rest of the body. Resistance to flow leaving the left ventricle is much higher than resistance on the right side, so the left ventricle has to work harder, resulting in higher pressures on the left, compared to the right side of the heart.

➤ Systolic blood pressure is measured as the peak pressure during the pumping phase (systole) of the heart, while diastolic blood pressure is measured during the heart filling phase (diastole). Blood pressure is usually expressed as the systolic over the diastolic, for example, 120/80, and abbreviated as BP.

Wall Thickness

Because the left ventricle must overcome a much higher resistance than the right to expel blood, the left ventricular wall is much thicker than the right wall. What would happen to the mass of the right ventricle if resistance to flow was increased in the pulmonary circulation (such as occurs in chronic pulmonary edema)? The right ventricle

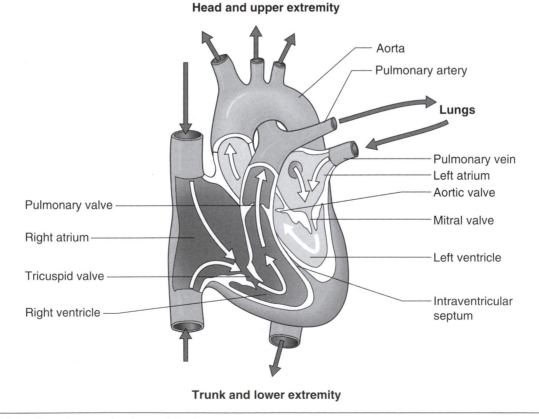

Figure 13.3   Chambers and valves of the heart.
Adapted from Guyton 1991.

would hypertrophy and could even reach or exceed the mass of the left ventricle. The key point is that changes in ventricular mass are due largely to changes in resistance. The heart muscle responds much the same as skeletal muscle when it is overloaded—it hypertrophies. Think of the stress the heart undergoes in someone with hypertension. The left ventricle must pump against a high resistance 24 hr a day! Considering the average person's heart beats over 100,000 times a day, this is quite an overload. It's no wonder that the heart becomes "worn out" in a short time if hypertension is untreated.

Changes in ventricular mass are due largely to changes in resistance to flow.

Coronary Blood Flow

The major coronary vessels travel along the surface of the heart, then bifurcate and dive into the myocardium (figure 13.4). Two major mechanisms control blood flow through the heart muscle:

• **Metabolic.** Increases and decreases in coronary blood flow are closely linked to metabolism. Coronary arterioles are very sensitive to metabolic by-products (especially **adenosine**) that are released from muscle cells and diffuse to arteriolar walls. Here they cause the vessel to relax (**vasodilation**), and blood flow is matched almost perfectly with the metabolic needs of the heart muscle. Heart muscle cells are highly oxidative and rely on an efficient O_2 delivery system.

Increases and decreases in coronary blood flow are closely linked to metabolism.

Heart muscle cells are highly oxidative and rely on an efficient O_2 delivery system.

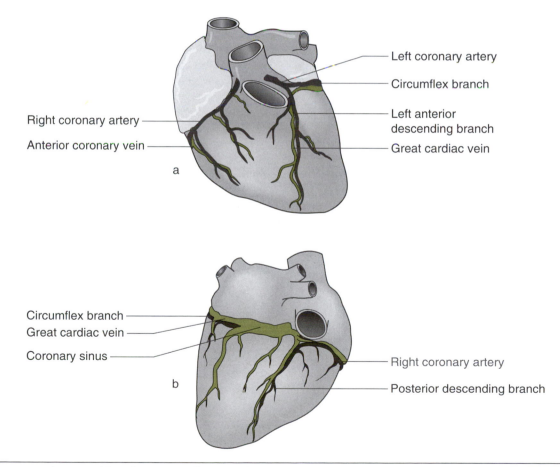

Figure 13.4   Anterior (*a*) and posterior (*b*) view of the major coronary blood vessels. Major vessels travel along the surface of the heart and then branch and enter myocardium. Blood flow through coronary vessels within the heart muscle is interrupted during the isovolumic contraction phase of the cardiac cycle.
Reprinted from Little 1977.

• **Mechanical.** The force generated when the heart contracts collapses arterioles (especially in the innermost portion of the heart, or the **endocardium**), so that blood flow is interrupted. This is especially true when the heart is contracting isometrically—when all the valves are closed at the beginning of systole (called the **isovolumic stage**). Normally this period of diminished blood flow lasts for a short time, and the heart function is not compromised. But when peripheral resistance is high (as occurs with hypertension), and coronary blood vessels are narrowed (as occurs with atherosclerosis), and the demand is high (as occurs during exercise), the heart may not receive as much blood as it needs to function properly. We will encounter more about this problem later in the chapter.

> During the isovolumic stage of systole, vessels within the myocardium collapse because of force of contraction, and blood flow is interrupted.

Function: The Heart as a Pump

Two main factors affect function of the heart as a pump: **preload** and **afterload** (figure 13.5). Before exploring the roles played by afterload and preload on heart function, we must examine the vascular system that delivers blood to the body's tissues and returns blood to the heart.

Vascular System

Examine figure 13.5 and notice that the arteries, capillaries, and veins form a closed system through which blood travels. Large arteries have a large elastic component, which facilitates propulsion of the blood through the system by elastic recoil. These larger vessels act more or less as conduits. The larger arteries have few sympathetic nerves innervating them and do not actively constrict or dilate much. Smaller arteries and **arterioles** (the smallest arteries) have less elasticity, and the smooth muscle that surrounds them is more richly innervated by the sympathetic nervous system. We can deduce, then, that the major function of the arterioles is to regulate blood flow through specific organs by constricting or dilating. As discussed earlier, metabolic demand determines whether these vessels are dilated or constricted.

Feed arteries are those that enter directly into the muscle and thus regulate total flow to the

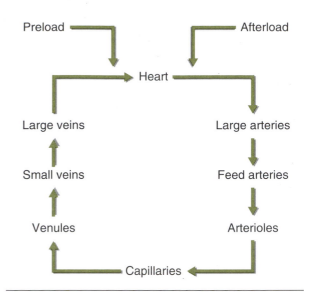

Figure 13.5   The heart as a pump. The pumping function of the heart is affected by preload (ventricular filling pressure) and afterload (resistance to flow from the heart). Large arteries have a large elastic component that propels blood through the vessels. Feed arteries regulate flow into the muscle, whereas arterioles, which are very reactive, distribute blood to active areas within the muscle. Constriction of arterioles increases afterload and work of the heart. During exercise, constriction of veins and muscle pumping action assist venous return and increase preload, resulting in a larger stroke volume.

muscle. Arterioles within the muscle distribute the blood to the areas where it is most needed by dilating in active areas of the muscle and constricting in nonactive areas, as already discussed. Arteries and arterioles are under the control of the sympathetic nervous system, which in general constricts these vessels. How, then, can we increase blood flow only to active areas of the muscle while most other areas of the body are vasoconstricted? Arterioles within skeletal muscle, much the same as coronary arterioles, are very sensitive to substances released from contracting skeletal muscle fibers (e.g., adenosine, potassium, inorganic phosphate, hydrogen ions). The signal from a sympathetic nerve to constrict the vessel is blocked, or overcome, by these local vasodilator substances, resulting in dilation of vessels only in the active areas of the muscle. This dilator response is propagated backward along the vascular tree to the feed arteries (which are outside the muscle) (figure 13.6). The

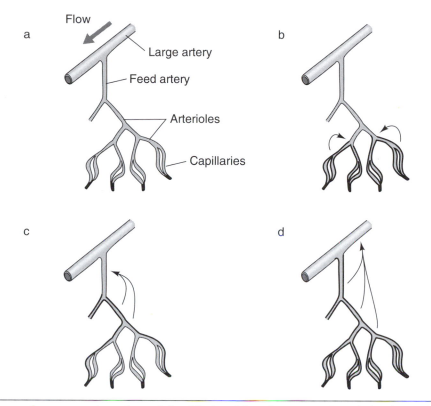

Figure 13.6   Retrograde propagation of arteriolar dilation. Vascular tree within skeletal muscle at rest (*a*). Small arterioles vasodilate in response to vasoactive substances released from contracting fibers (*b*). Vasodilation (indicated by dark lines) is propagated along the vessels back up the vascular tree (*c*) until feed artery dilates enough to provide needed blood to the muscle (*d*). The signal also slightly dilates the larger conduit artery.
Adapted from Segal 1992.

feed arteries dilate enough to divert exactly as much blood as is needed into the muscle. Since the vessels inside the muscle are dilated (from largest down to the smallest, in turn) only in the active areas, blood is distributed to the active fibers and not inactive fibers. In effect, a communication link exists between the small arterioles within the muscle and the larger feed arteries that control blood flow into the muscle.

A communication link exists between the small arterioles within the muscle and the larger feed arteries that control blood flow into the muscle.

Capillaries, the smallest blood vessels, have a diameter of 5 to 8 μm. Capillary cell walls are only one cell thick (endothelial cells form capillaries), which facilitates diffusion of O_2, CO_2, nutrients, waste products, and so on. Blood flow into the capillaries is controlled by small arterioles immediately proximal to them.

Blood flowing out of the capillaries is collected by the venules, which branch into successively larger veins until the blood returns to the heart. The pressure in veins is very low, so there is not much of a driving force to help return blood to the heart. When we are in the upright position, it is especially difficult to get blood back to the heart. Because of the tendency for blood to pool in the lower extremities during standing, veins have a unique feature—one-way valves that help keep the blood moving in the right direction.

Veins have much thinner cell walls than arteries, mostly because they have less smooth muscle surrounding them. They still have the capability of constricting, though, and they do so during exercise. At rest, about 65% of the total blood in our body is located in the veins. During endurance exercise, the veins constrict (a process called **venoconstriction**), and blood is expelled from the veins back to the heart. This increased venous return helps to fill the heart more, increasing ventricular preload.

Venoconstriction is important during exercise to prevent blood from collecting in the venous system, which would lower the amount available for the working muscles.

Venoconstriction is important during exercise to prevent blood from collecting in the venous system, which would lower the amount available for the working muscles.

Ventricular Preload and Afterload

As mentioned, venoconstriction increases venous return and loads the heart with blood. By deduction, then, **ventricular preload** refers to ventricular filling pressure (the amount of blood filling the heart). Changes in preload affect the magnitude of the stroke volume: if preload is high, plenty of blood is filling the ventricles, and stroke volume is high. If the filling pressure is low, stroke volume is low (the heart can pump only as much blood as it receives).

At rest, preload is affected mostly by body position. Think about what happens to preload when you suddenly stand from a supine position. Blood tends to pool in the lower extremities, and preload decreases abruptly. This lowers stroke volume, and the body attempts to stabilize BP and maintain cardiac output by increasing heart rate.

If preload is high, plenty of blood is filling the ventricles, and stroke volume is high. If the filling pressure is low, stroke volume is low.

During exercise, three major factors increase preload:

1. Venoconstriction, as already discussed
2. The **muscle pump**, which is very effective in moving blood through veins back to the heart with rhythmic muscular contractions (the muscle pump is made more effective by one-way valves)
3. Deeper and more rhythmic breathing, which helps decrease pressure inside the thoracic cavity, thereby facilitating venous return to the heart (keep in mind that the pressure is very low in the venous system, but that pressure within the thoracic cavity becomes more negative when you inhale deeply, increasing blood flow toward the heart)

Figure 13.7 Rhythmic exercise. Skating and other rhythmic/dynamic exercise promote venoconstriction; consistent work of the muscle pump to return blood to the heart; and deep, rhythmic breathing—all of which increase preload.

These three factors are involved more with rhythmic/dynamic exercise like running, swimming, cycling, and rowing (figure 13.7) than with stop-and-start exercises like baseball and basketball.

Afterload is the other major force that affects the function of the heart: it is the resistance to flow from the heart. Resistance to flow is largely a function of the most active blood vessels, the arterioles. If we add together relative constriction or dilation of all arteries in the body (except the lungs), we get **total peripheral resistance (TPR)**. This is the pressure that holds the **aortic valve** closed (figure 13.3). Remember, afterload is reflected in diastolic blood pressure (DBP)—high DBP means a high resistance to overcome—and vice versa, which affects the work of the heart. For example, hypertension is characterized by increased resistance to flow from the heart. Consequently, the left ventricle must generate more force with each beat to overcome the pressure that is holding the aortic valve closed. To maintain stroke volume under this condition, systolic blood pressure must be elevated. Thus, someone with hypertension has high systolic and diastolic BP. Unlike preload, which is affected mostly by rhythmic/dynamic exercise, afterload is affected mostly by heavy-resistance or static exercise.

Afterload is reflected in diastolic blood pressure (DBP)—high DBP means a high resistance to overcome—and vice versa, which affects the work of the heart.

Unlike preload, which is affected mostly by rhythmic/dynamic exercise, afterload is affected mostly by heavy-resistance or static exercise.

Basic Electrocardiography

The electrocardiogram (ECG) is a measure of the voltage change between two points on the surface of the body as a result of electrical events of the cardiac cycle. Thus, in looking at an ECG we are viewing the heart between one negative and one positive electrode at any one time. In essence the ECG gives us a graphic representation of the depolarization and repolarization of different portions of the heart muscle. A pen deflection on an ECG strip (or the beam deflection if you are looking at an oscilloscope) represents the average magnitude (measured by the amplitude of the deflection) and direction—that is, whether the deflection is positive (above the baseline) or negative (below the baseline) in relation to the current flow within the heart. The time between electrical events gives information about the conduction properties of the signal.

Electric Axis

By convention, the tracings on an ECG recording are expressed relative to the positive electrode; that is, a wave of depolarization traveling toward a positive electrode produces a positive deflection. If you look at the ECG recording in figure 13.8, you will notice that both the P and the R waves are positive. Why? This means that the average current flow for depolarization of both the atria (P wave) and ventricles (R wave) is traveling toward the positive electrode. If you place two electrodes on your chest, with the negative electrode at the sternoclavicular junction (or right arm) and the positive electrode at the fifth

intercostal space to the left and below the heart (or left arm), this view of the heart produces the positive P and R waves just described (the third lead is used as a ground). This electrode positioning creates the V5 lead, which produces the most recognizable ECG recording, the one in figure 13.8. Thus, the average current flow of both atrial and ventricular depolarization can be determined by drawing a line between the positive and negative electrodes. For ventricular depolarization, the magnitude and direction of this average flow are represented by a vector called the **electric axis** (figure 13.9). Notice that the direction of the current flow is biased toward the left ventricle—a function of the larger mass of the left compared to the right ventricle. The electric axis remains the same with each heartbeat, but if we place a positive electrode in position A, B, and C, respectively, on the figure, then during ventricular depolarization we get the three different types of tracings illustrated.

A vector called the electric axis represents the average current flow during depolarization of the ventricles. The direction is biased toward the left side because the left ventricle has a greater mass than the right.

The last wave in the ECG recording, the T wave, represents ventricular repolarization, and is also in a positive (+ve) direction in the V5 lead just described. The principle is a little different here from that for depolarization, in that a wave

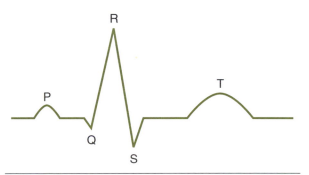

Figure 13.8   A typical electrocardiogram recording from the V5 lead. The P wave represents atrial depolarization; the Q wave is depolarization of the intraventricular septum; the R wave is ventricular depolarization; the T wave is ventricular repolarization.

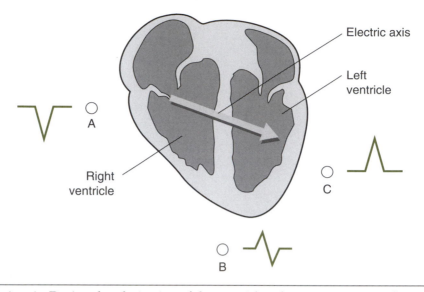

Figure 13.9 Electric axis. During depolarization of the ventricles, the average current flow is represented by a vector called the electric axis. Because the left ventricle has a greater mass than the right, the direction is biased toward the left side. A positive electrode placed in position A produces a negative deflection because the current flow is away from the electrode. Current flow is toward, then away, from the electrode in position B, producing a positive, then a negative, recording. If placed in position C, the electrode produces a positive deflection. Reprinted from Thaler 1999.

of repolarization traveling toward a positive electrode always gives a negative (–ve) deflection. What does this tell you about the direction of the wave of ventricular repolarization? Since the deflection is +ve in this lead, the direction of the repolarizing current must be away from the +ve electrode. Indeed, the last cells to be depolarized are the first to be repolarized, so the wave of repolarization travels in the direction opposite that of the wave of depolarization. To review, a wave of depolarization traveling toward a +ve electrode produces an upward deflection on the ECG tracing, but the opposite is true for a wave of repolarization traveling toward a +ve electrode—the deflection is down (–ve).

A wave of depolarization traveling toward a +ve electrode produces an upward deflection on the ECG tracing, but the opposite is true for a wave of repolarization traveling toward a +ve electrode—the deflection is down (–ve).

Twelve-Lead Electrocardiogram

The 12-lead ECG offers a comprehensive record of the electrical activity in various parts of the heart. The six limb leads provide a view of the heart on the frontal plane, and electrodes for these leads are placed on the wrists and the ankles. Each lead is variably designated as +ve or –ve to provide different angles of orientation (remember that each ECG tracing represents the potential difference between only two points on the body at any one time). The six remaining leads, the precordial leads, are placed across the chest to provide anterior and posterior views of the heart.

Limb Leads

The three standard limb leads are created by making left arm +ve and right arm –ve (lead I); legs +ve and right arm –ve (lead II); and legs +ve and left arm –ve (lead III). Again, the view of the heart, or angle of orientation, is achieved by drawing a line from the –ve to the +ve electrodes (figure 13.10).

Which of the three standard limb leads provides the best representation of the average current flow during depolarization of the ventricles, the electric axis? Note from figure 13.10 that lead II gives the best representation of the electric axis. The average current flow during depolarization of both ventricles is almost directly toward the +ve electrode. Therefore, the amplitude of the R wave is the highest in lead II.

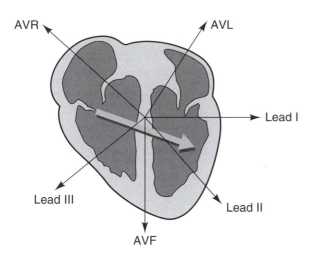

Figure 13.10 Orientation of standard and augmented limb leads. AVL, AVR, and AVF are augmented leads and are created by making one lead positive and all others negative. In AVL, the left arm is positive; in AVR, the right arm is positive; and in AVF, the left leg is positive.
Reprinted from Thaler 1999.

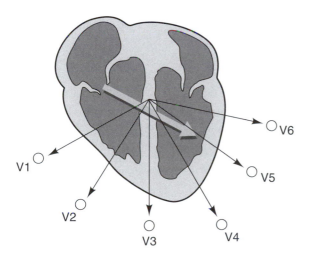

Figure 13.11 Orientation of precordial leads.
Reprinted from Thaler 1999.

The remaining three limb leads are called **augmented** (simply because the tracings must be amplified to get an adequate recording) and are achieved by making one electrode +ve and all the others –ve. The three augmented limb leads, AVL, AVR, and AVF, are created by making the left arm, right arm, and legs +ve, respectively.

Precordial Leads

The six precordial leads are **unipolar**, meaning that each one is made +ve in turn and the whole body is used as the ground. As illustrated in figure 13.11, leads V1 and V2 lie over the right ventricle; leads V3 and V4 lie over the interventricular septum; and leads V5 and V6 lie over the left ventricle. Look at the placement of these electrodes and visualize how the R (+ve) and S (–ve) waves would change as we move around the heart from V1 to V6. V1 would have the smallest R and the largest S wave, while V5 would have the largest R wave. These wave configurations are related to the position of the electrodes relative to the electric axis. The electric axis is traveling mostly away from lead V1, resulting in a small +ve deflection, but is traveling directly toward V5, resulting in a large +ve deflection.

Case Study

Recall that the electric axis represents the average direction and magnitude of depolarization of the ventricles. It is biased toward the left ventricle because of the difference in mass between the left and the right ventricle. What would you expect to happen to the electric axis over time in a patient with chronic pulmonary edema? Why? How would this be reflected in a change in the R and S waves in the precordial leads?

Chronic pulmonary edema causes increased pressure within the lungs, thereby increasing resistance to blood flow. The right ventricle has to work harder to overcome this resistance and pump blood through the lungs. Over time, the right ventricle will hypertrophy. Now the mass of the right ventricle is approaching (or may exceed) the mass of the left ventricle, so the electric axis shifts to the right, making the R wave larger in precordial leads on the right side of the heart (V1, V2, and V3). Lead V3 or V4 may now have the largest R-wave amplitude. This phenomenon is called a right axis deviation.

Blood Pressure Responses

To this point we have examined regulation of the heart and the redistribution of blood flow within the body during exercise. We also learned that the two major factors affecting function of the heart are preload and afterload. Recall that afterload is the resistance to flow from the heart and is reflected in DBP. Now we will examine BP responses to two different types of exercise—heavy resistance or static and rhythmic/dynamic.

Blood Pressure Response to Heavy-Resistance Exercise

Blood pressure responses to heavy-resistance exercise, taken from a study by MacDougall et al., are represented in figure 13.12. These are among the highest BPs ever recorded. The BPs are very accurate because they were recorded from a small pressure transducer placed directly in the brachial artery. Note the large increase in both diastolic and systolic BPs with a single arm curl! Because the increases in BP are relatively small when muscle mass is added (1 and 2 legs), we can deduce that the increase in BP is dependent, but only to a small extent, on muscle mass. The BP does increase as mass increases, but not to the extent you would expect if there were a direct relation between muscle mass and BP.

The largest increase in BP is seen with two legs at 90% of maximum to failure. The subjects were highly trained weight lifters who could perform 10 to 15 repetitions of 90% of their maximum voluntary contraction. At the end of these repetitions the BP had risen, in some cases as high as 400/300! So, BP rises most dramatically during repetitious heavy lifts rather than during a single maximum lift. The higher BP during repetitive lifts, according to MacDougall et al., may relate to recruitment of additional motor units and to increased involvement of accessory muscles as the primary muscle fatigues.

An important point to remember when prescribing resistance training for sedentary people with CV risk factors is that trained people such as those who participated in this study actually have an attenuated BP response to resistance exercise. This means that the BP response to heavy lifting in sedentary people is relatively higher compared with trained people. Thus, caution is in order when prescribing heavy-resistance exercise for persons who have a high risk for CV disease.

> Since the BP response to heavy lifting in sedentary people is relatively higher compared with trained people, caution is in order prescribing heavy-resistance exercise for persons who have CV disease.

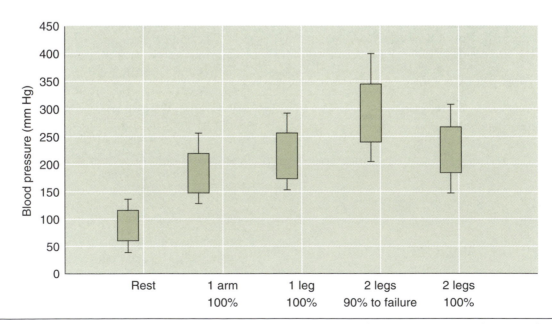

Figure 13.12 Blood pressure response to heavy resistance exercise. Blood pressure rises dramatically during single lifts at maximum effort (100% MVC), but especially during repeated lifts (90% of MVC to failure). MVC = maximum voluntary contraction.
Adapted from McDougall 1985.

Problems Associated With This Blood Pressure Response

Obviously, young healthy adults can withstand the high BPs associated with heavy lifting, but let's consider some of the potential occurrences in persons who are elderly, persons who have diseased arteries, and those with weak hearts.

Case Study

Hank, 75 years old, has coronary artery disease and a weak heart. He is a patient in your rehabilitation program. He has been a farmer all his life and refuses to reduce his work on the farm. Much of the work he is able to do consists of bending over and lifting heavy objects. The legs and arms are involved, and he usually braces himself by contracting the abdominal muscles as well. This represents a large functioning muscle mass in a heavy or static lift. Hank has been complaining to his family about dizzy spells, and they have expressed concern to you. What do you tell them?

You need to explain that under these conditions there is a lot of vascular constriction that makes the heart pump against a high resistance. The afterload on the heart increases dramatically. If the heart is not strong enough to overcome the afterload, then the **pulse pressure** (difference between systolic and diastolic pressure) decreases, and stroke volume is lowered. If the resistance to flow is high enough, stroke volume falls dangerously low. In this case the person may become dizzy, or faint.

Another factor leading to Hank's spells is the very high diastolic pressure that increases the length of the isovolumic stage of the heart (the time the heart is without blood flow). The reduced blood flow can be exacerbated if he holds his breath and contracts the abdomen and chest wall muscles. In a person with diseased arteries, blood flow to the heart may already be compromised. Creating a high demand while at the same time shutting off blood flow leads to prolonged ischemia and possible arrhythmias, among other things. Lastly, the high pressures create a high shear force on vessel walls and may burst an aneurysm (weakened vessel wall) or loosen a piece of plaque, which in turn may cause blockage of a blood vessel in the heart or brain. Hank and his family need to be made aware of these risks.

It is worth noting here that resistance exercise can be, and is, used to rehabilitate cardiac patients. Low to moderate resistance training is advantageous because raising the DBP moderately theoretically increases the perfusion pressure of the heart muscle. Indeed, cardiac patients have reported less angina with resistance exercise (lasting for a relatively short time) than with longer-lasting aerobic exercise that increases the demand for coronary blood flow and may cause a relative ischemia. Still, patients at risk need to guard against heavy lifting, especially when it is repeated (e.g., snow shoveling).

➤ Angina is chest pain caused by ischemia of the heart muscle.

➤ Ischemia is inadequate blood flow resulting in relative O_2 deprivation.

Mechanisms

Three mechanisms are responsible for the high BPs associated with heavy-resistance exercise:

• **Valsalva maneuver.** Most people hold their breath when lifting a heavy weight. The Valsalva maneuver increases afterload on the heart and increases BP momentarily.

• **Total peripheral resistance.** Blood vessels in each muscle participating in the lift are mechanically closed off by the force of the muscle contracting. This increases TPR in direct proportion to the amount of muscle mass that is contracting.

• **Pressor response.** Perhaps the largest contributor to the increased BP is a reflex called the pressor response. Blood pressure increases in response to the powerful contraction in an attempt to perfuse the ischemic tissue. This response is mediated initially by mechanical and then by chemical receptors in the muscle. A signal is sent from the muscle to the CV control center, and reflex increase in BP results.

Blood Pressure Response to Aerobic or Rhythmic/Dynamic Exercise

The BP response to aerobic exercise for both a small and a large muscle mass is illustrated in figure 13.13. Notice that the BPs during aerobic exercise are much lower than the response to heavy-resistance exercise. In fact, the leg exercise (large mass) produces a small drop in DBP at the low exercise level, and then DBP is elevated only slightly as the intensity of the exercise increases. Even at $\dot{V}O_2$max, DBP is at or lower than resting values. This means that the afterload on the heart changes very little, if at all, even at high workloads. This type of exercise elicits active vasodilation in the working muscle, thereby reducing the TPR. Systolic pressure increases as workload increases—but remember this reflects the force of contraction of the heart, which is increasing as workload increases.

Now go back to figure 13.4 and examine the coronary circulation. Notice that the larger vessels all lie on the surface of the heart. Smaller vessels come off these large surface vessels and bifurcate many times, getting smaller and smaller as they move toward the endocardium (inner surface of the heart). During the isovolumic stage of the heart cycle, the small vessels inside the heart are actually closed off during this isometric contraction, making the heart ischemic.

➤ To bifurcate means to divide into two branches.

Case Study
You have a patient with moderate hypertension (BP of 150/100), whom you are monitoring during a graded exercise test on a treadmill. What do you expect to happen to DBP as the exercise intensity increases (compare to normal response in figure 13.13)? As a result of this, what happens to the isovolumic stage of the cardiac cycle in your patient? What are the dangers associated with this phenomenon?

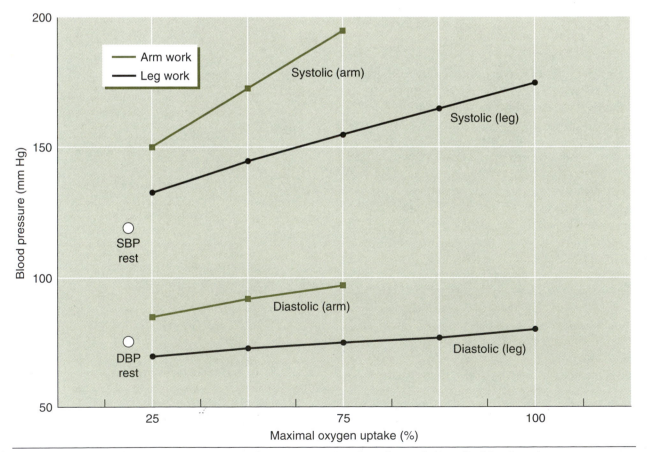

Figure 13.13 Blood pressure response to dynamic exercise. Systolic and diastolic blood pressure response to exercise using small (arm) and large (leg) muscle mass. As exercise intensity increases from rest, blood pressure during arm work increases more than with leg work.
Adapted from Astrand and Rodahl 1986.

Hypertension is characterized by a relative vasoconstriction in the periphery, which increases resistance to blood flow. Because patients with hypertension have a decreased ability to vasodilate, you would observe a slow rise in DBP as exercise intensity increases. This differs from the small rise, or absence of any rise, in DBP in healthy people. Because the peripheral resistance is higher, the pressure holding the aortic valve closed is greater; this extends the isovolumic stage of the cardiac cycle. (The heart has to generate more force to open the aortic valve, so the time spent in an isometric contraction is longer and blood flow is compromised.) The increased resistance causes the heart to work harder; and as a consequence of this, coupled with a probable coronary artery disease, the myocardium becomes O_2 deprived. The patient may experience angina, which is usually accompanied by a depressed ST segment on the ECG tracing. (A depressed ST segment is associated with myocardial ischemia.)

The BP response for arm work (a small mass) during rhythmic/dynamic exercise is different than for a large mass (figure 13.13). Both systolic and diastolic pressures are actually higher for arm work than for leg work at the same absolute workload. The primary mechanism responsible for this difference is relative vasodilation. That is, a small muscle mass has a small area of vasodilation leaving a larger relative area in the rest of the body for vasoconstriction compared to a larger muscle mass; the larger muscle mass has a larger area for vasodilation, which results in a lower TPR.

Aerobic exercise puts much less stress on the heart than resistance exercise (figure 13.4). Stress on the heart with resistance exercise relates to the large increase in afterload or resistance to flow from the heart. This type of training, then, produces a **pressure overload** of the heart. Aerobic, or endurance, exercise is associated with large increases in cardiac output and a large volume of blood returning to the heart. This type of training results in a **volume overload** of the heart and produces a much different adaptation.

Aerobic exercise puts much less stress on the heart than resistance exercise.

Figure 13.14   Heavy resistance training. This kind of demanding resistance exercise makes much more stressful demands of the cardiovascular system than does aerobic training.

Volume Versus Pressure Overload of the Heart

As mentioned previously, one of the prime goals of CV conditioning is to increase stroke volume. This can be accomplished by increasing the strength of the heart so that more blood is ejected with each beat, or by increasing the volume of the heart so that it can accept more blood between beats.

Resistance or Aerobic Training?

When you are considering resistance versus aerobic training for a patient, you must think about acute pressure overload, chronic pressure overload, and volume overload.

• **Acute pressure overload.** Resistance training produces an increased afterload on the heart, which requires the heart to contract more forcefully with each beat. This acute pressure overload produces an increase in wall thickness of the left ventricle over time, but little if any change in the size of the left ventricular (LV) chamber (or volume). The acute training stimulus makes the heart stronger and able to increase stroke volume by reducing **end-systolic volume (ESV)**. The increased stroke volume is much smaller than that observed with aerobic, or endurance, training.

• **Chronic pressure overload.** Hypertension produces a chronic pressure overload, which results in detrimental changes to the heart over time. The left ventricle of someone with untreated systemic hypertension has a very thick wall (which reduces the size of the LV chamber) and also has a large amount of connective tissue (which increases stiffness). Consequently, the LV cannot expand much during diastole, reducing end-diastolic volume (EDV); this in turn results in a very low stroke volume. The contractility of the heart also declines, further reducing the ability to eject blood.

• **Volume overload.** One can increase the volume of the ventricular chamber by increasing the compliance of the walls and the pericardium. The best way to do this is by stretching the heart repeatedly. The large venous return associated with aerobic exercise produces a volume overload, leading to an increased size of the ventricular chamber with only a small increase in the thickness of the wall. This training stimulus produces large increases in EDV, greatly enhancing stroke volume. Increases in EDV observed with endurance training are also related to an increase in plasma volume in these trained people. The increased volume of blood provides greater filling pressure for the ventricles. Table 13.1 presents a hypothetical example of the differences in changes in stroke volume between resistance and endurance trainers.

This table illustrates the benefits of aerobic over resistance training in terms of increasing the heart's ability to deliver blood to the working muscle. Resistance training produces only small increases in stroke volume. This effect relates directly to the training stimulus: resistance training produces a pressure overload and results in a stronger heart, but little or no increase in compliance or volume of the heart. Even with use of a series of resistance exercises having high repetitions and low intensities, separated by short rest periods (circuit training), increases in $\dot{V}O_2$max are limited (approximately 5-8%). In contrast, aerobic exercise produces a small increase in the strength of the heart but a large increase in volume of the LV chamber accompanied by an expansion of plasma volume. This training produces much larger increases in $\dot{V}O_2$max. The utility of this information for prescribing different types of exercise for patients should be obvious. Resistance training is useful for improving muscle strength and muscle endurance whereas endurance, or aerobic, training is best for CV improvements. The key is understanding the adaptations from the different exercise stimuli.

Resistance training is useful for improving muscle strength and muscle endurance whereas aerobic training is best for CV improvements.

Exercise Limitations of Cardiac Output

The performance limitations imposed by a low maximum cardiac output are illustrated in

Table 13.1 Volume Versus Pressure Overload

	End-diastolic volume (ml)	End-systolic volume (ml)	Stroke volume (ml)
Resting	110	40	70
Maximal exercise	140	20	120
Resistance-trained maximal exercise	145	5	140
Endurance-trained maximal exercise	190	10	180

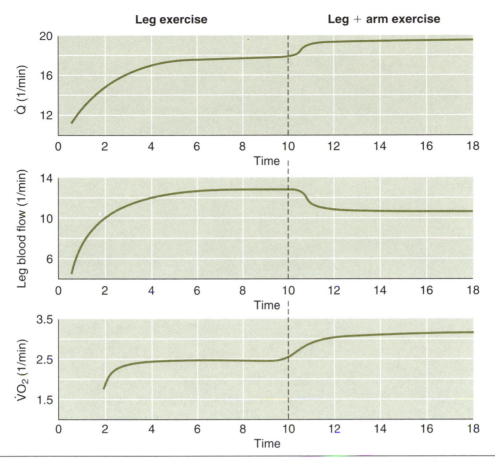

Figure 13.15   Performance limitations of low maximum cardiac output. During leg work in an unconditioned person, cardiac output, leg blood flow, and $\dot{V}O_2$ increase until demand and delivery are matched. When arm exercise is added, the exerciser cannot increase cardiac output enough to supply both the arms and the legs. Arterioles in the working muscle of the leg constrict, and blood flow is lowered. Now delivery is lower than demand, and non-oxidative energy sources must be used. Under these conditions, the onset of muscle fatigue is hastened.
Reprinted from Secher et al. 1977.

figure 13.15. Initially, the person is exercising with the legs. The cardiac output, leg blood flow, and $\dot{V}O_2$ all increase linearly, then level off when the person reaches a steady state. When arm work is added to the leg work, however, the person is limited in the amount he or she can increase cardiac output. How do we know this? We know because cardiac output increases slightly, but leg blood flow decreases. This means that the heart cannot supply both areas at the same time, so this person is probably unconditioned. Blood pressure drops as the arm vessels dilate, and the CV control center responds by vasoconstricting very powerfully in the work-

ing legs. Now we have a condition in which supply does not meet demand and some of the energy for the leg work must come from non-oxidative sources. This of course produces a more rapid fatigue, and the person cannot continue the work. If we put this person on an endurance-training program, what would be the difference in the response when we add arm work to leg work? Endurance training would increase the capacity of the heart to pump blood, thereby increasing maximum cardiac output. Now, under the same conditions, the person has more reserve and can handle the extra demand for blood when arms are added to leg work.

What You Need to Know From Chapter 13

Key Terms

adenosine

afterload

aortic valve

arterioles

augmented

cardiac output

diastole

electric axis

end-diastolic volume (EDV)

endocardium

end-systolic volume (ESV)

isovolumic stage

muscle pump

preload

pressure overload

pulse pressure

stroke volume

systole

total peripheral resistance (TPR)

unipolar

vasodilation

venoconstriction

ventricular preload

volume overload

Key Concepts

1. You should be able to describe how blood flow is redistributed from resting to exercise conditions.

2. You should be able to describe how coronary and skeletal muscle blood flow is regulated during exercise.

3. You should be able to relate how changes in ventricular preload and afterload affect the function of the heart.

4. You should be able to describe the basic principles of electrocardiography and the location of the 12 ECG leads.

5. You should be able to compare and contrast the blood pressure responses and adaptations to heavy-resistance and rhythmic/dynamic exercise.

6. You should be able to describe conditioning responses to exercises that produce volume and pressure overload of the heart.

7. You should know the principles that underlie performance limitations imposed by a low cardiac output.

Review Questions

1. A trained endurance athlete has about 20% of cardiac output directed toward skeletal muscle under resting conditions, but as much as 90% during maximum exercise. Describe how this is accomplished.

2. Describe the differences in structure and innervation between large and small arteries and describe how these differences relate to their respective function.

3. Describe the mechanisms responsible for the difference in mass between the left and right ventricle.

4. Precise regulation of muscle blood flow is needed for optimum endurance performance. Describe how muscle arterioles direct blood only to active areas and how they communicate with feed arteries to ensure adequate flow into the muscle.

5. Describe the three factors that increase ventricular preload during rhythmic/dynamic exercise (like running).

6. Three females aged 26, 24, and 25, with similar body mass and composition and height, have widely different $\dot{V}O_2$max. Person #1 has mitral stenosis (MS) ($\dot{V}O_2$max = 1.3 L/min); person #2 is sedentary (SED) ($\dot{V}O_2$max = 1.8 L/min); and person #3 is moderately active (ACT) ($\dot{V}O_2$max = 2.7 L/min). These women have come to you for advice on endurance fitness training and want to know what improvements they can expect.

a. Describe any differences in cardiac output among the three at rest and at maximum exercise. Include differences in heart rate and stroke volume.

b. In terms of percentage increase in $\dot{V}O_2$max, in which woman would you expect the greatest adaptation to an endurance-training program lasting six months?

7. A sedentary adult asks you for advice about CV conditioning. Like most people, he has heard that resistance training gives a good CV workout because the heart rate is elevated during and after the exercise. Do you agree? Why or why not? Would you recommend another type of exercise for good CV benefits? Why or why not?

Bibliography

Beniamini Y, Rubenstein JJ, Zaichkowsky LD, and Crim M. Effects of high-intensity strength training on quality of life parameters in cardiac rehabilitation patients. *Am J Cardiol* 1997, 80 (7):841–846.

Bertovic DA, Waddell TK, Gatzka CD, Cameron JD, Dart AM, and Kingwell BA. Muscular strength training is associated with low arterial compliance and high pulse pressure. *Hypertension* 1999, 33 (6):1385–1391.

Fleck SJ. Cardiovascular adaptations to resistance training. *Med Sci Sports Exerc* 1988, 20 (5):S146–S151.

Hare DL, Ryan TM, Selig SE, Pellizzer A, Wrigley TV, and Krum H. Resistance exercise training increases muscle strength, endurance, and blood flow in patients with chronic heart failure. *Am J Cardiol* 1999, 83 (12):1674–1677.

Jennings G, Dart A, Meredith I, Korner P, Laufer E, and Dewar E. Effects of exercise and other nonpharmacological measures on blood pressure and cardiac hypertrophy. *J Cardiol Pharm* 1991, 17 (suppl. 2):S70–S74.

MacDougall JD, Tuxen D, Sale DG, Moroz JR, and Sutton JR. Arterial blood pressure response to heavy resistance exercise. *J Appl Physiol* 1985, 58:785–790.

McCartney N, McKelvie RS, Haslam DRS, and Jones NL. Usefulness of weightlifting training in improving strength and maximal power output in coronary artery disease. *Am J Cardiol* 1991, 67:939–945.

Moore RL and Palmer BM. Exercise training and cellular adaptations of normal and diseased hearts. *Exerc and Sport Sci Rev* 1999, 27:285–305.

Peterson DF, Armstrong RB, and Laughlin MH. Sympathetic neural influence on muscle blood flow in rats during submaximal exercise. *J Appl Physiol* 1988, 65 (1):434–440.

Rerych SK, Scholz PM, Sabiston, DC Jr., and Jones RH. Effects of exercise training on left ventricular function in normal subjects: A longitudinal study by radionuclide angiography. *Am J Cardiol* 1980, 45:244–252.

Seals DR, Victor RG, and Mark AL. Plasma norepinephrine and muscle sympathetic discharge during rhythmic exercise in humans. *J Appl Phsyiol* 1988, 65 (2):940–944.

Secher NH, Clausen JP, Noer I, and Trap-Jensen J. Central and regional circulatory effects of adding arm exercise to leg exercise. *Acta Physiol Scand* 1977, 100:288–297.

Segal SS. Communication among endothelial and smooth muscle cells coordinates blood flow control during exercise. *News Physiol Sci* 1992, 7:152–156.

Shepard RJ and Balady GJ. Exercise as cardiovascular therapy. *Circulation* 1999, 99 (7):963–972.

Turpeinen AK, Kuikka JT, Vanninen E, Vainio P, Vanninen R, Litmanen H, Koivisto VA, Bergstrom K, and Uusitupa MI. Athletic heart: A metabolic, anatomical, and functional study. *Med Sci Sports Exerc* 1996, 28:33–40.

Cardiovascular Diseases

This chapter will discuss

➤ the etiology and pathophysiology of common cardiovascular disesases,

➤ the effects these diseases have on the response to exercise, and

➤ the effects of rehabilitation and regular exercise on patients with cardiovascular disease.

In spite of a decline in their incidence over the past 15 years, cardiovascular diseases, including **coronary heart disease** (CHD) or **coronary artery disease** (CAD) and other cardiovascular diseases such as **stroke**, are still the leading cause of death in Western societies. This remarkably high incidence is seen in the face of considerable scientific evidence that regular exercise and attention to the diet can significantly reduce the risk factors for the development of these diseases. Health professionals should understand the need for prevention as well as for rehabilitation. This chapter presents material to help you learn about the basis for both approaches.

Risk Factors

A multitude of factors can increase the risk of cardiovascular disease. Some of these are uncontrollable, whereas others can be controlled through lifestyle changes. It is important to recognize these **risk factors** and to be prepared to intervene with patients who are putting themselves at risk. This section discusses the risk factors and the ways in which they might be involved in the pathogenesis of cardiovascular disease.

Risk Factors for Cardiovascular Disease

- Family history
- Smoking
- Increased lipids/cholesterol
- Hypertension
- Obesity
- Sedentary lifestyle
- Age
- Sex
- Personality type
- Stress

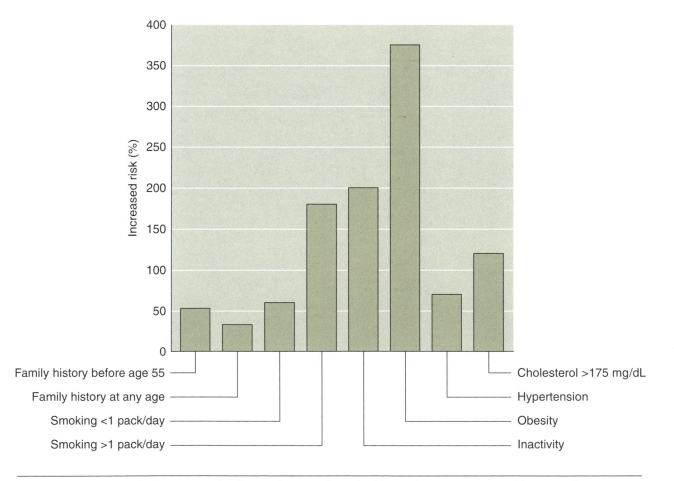234 Exercise Physiology for Health Care Professionals

• **Family history.** A family history of cardiovascular disease, including **hypertension** (increased blood pressure), CHD, stroke, or diabetes, represents the greatest risk factor for developing cardiovascular diseases.

If parents or siblings of a person had disease before the age of 55, the risk for the person is over 50%; and if they had disease at any age, the risk is 33% higher than for someone who has no family history. The next three major risk factors—smoking, increased cholesterol, and hypertension—are similar to one another in their contribution to risk. If you have one of these factors, the risk of having a heart attack or stroke is doubled; if you have two, it is quadrupled; and if you have all three, it is double that (eight times the risk for someone with none of these factors) (figure 14.1).

A family history of cardiovascular disease, including hypertension, CHD, stroke, or diabetes, represents the greatest risk factor for developing cardiovascular diseases.

• **Smoking.** The precise relation between cigarette smoking and cardiovascular disease is unclear. It appears likely that chemicals in cigarette smoke have a direct effect to constrict the vascular smooth muscle, contributing to hypertension and to injury of the vascular wall. This injury then becomes a site for formation of **plaque** (areas of vascular smooth-muscle deformation) as well as narrowing **(atherosclerosis)** and hardening **(arteriosclerosis)** of the arteries. Atherosclerosis and arteriosclerosis further contribute to increases in blood pressure (hypertension). Smoking increases your risk of not only cardiovascular disease, but

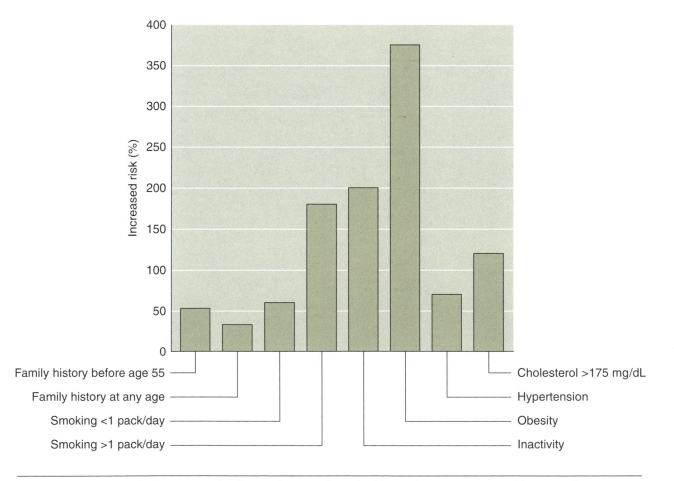

Figure 14.1 Factors that increase risk of heart disease. The risk of heart disease is increased by many factors, most of which are controllable. The risk is further increased when two or more of these factors are combined.

also lung disease. As the lung disease (emphysema or lung cancer) progresses, work of the heart also increases because of lowered O_2 levels. The combined effects of smoking-related hypertension, vascular disease, and low O_2 levels may lead to **heart failure**, an inability of the heart to pump blood to the periphery.

• **Blood lipids.** High blood lipid levels in general, and cholesterol levels in particular, contribute to the development of atherosclerosis. The lipoproteins that carry cholesterol and other lipids in the blood are classified into high-density (HDL), low-density (LDL), and very low density (VLDL) lipoproteins, based on the amount of lipid carried (see chapter 2). Since lipids have a relatively low density, HDL carries fewer associated lipids than the LDLs. Because the HDL carries fewer lipids, it may play a role in protecting against heart disease. It is clear that the high lipid levels of the LDLs increase the risk of heart disease. In this regard, the total amount of lipoprotein in the blood is less important than the relative proportions: a high HDL relative to LDL is associated with a lower risk of cardiovascular disease.

• **Hypertension.** Hypertension is another important risk factor for cardiovascular disease. Hypertension is present in over 90% of stroke patients. The relationship between risk and increasing systolic or diastolic pressure, or both, is linear. Hypertension is higher in people who are obese, who smoke, or who are under a lot of stress, suggesting that control of these factors can reduce the risk of disease. A growing body of evidence indicates that regular exercise may moderate blood pressure in people who are hypertensive. Currently available medications can be used to control blood pressure and significantly reduce the associated risks.

The combined effects of smoking-related hypertension, vascular disease, and low O_2 levels may lead to heart failure, an inability of the heart to pump blood to the periphery.

High-fat diets are associated with elevated LDL levels, whereas regular exercise is associated with increased HDL levels; both regular exercise and higher HDL levels decrease the risks for development of cardiovascular diseases.

Cardiac Diseases

Cardiac disease is the major cause of mortality in North America. Among the population with cardiac diseases, people with CHD and chronic heart failure are the ones most commonly referred for exercise rehabilitation.

Common Cardiovascular Diseases

Cardiac diseases

- Coronary heart disease
- Chronic heart failure
- Myocarditis
- Heart defects
- Cardiomyopathies

Vascular diseases

- Atherosclerosis
- Hypertension
- Venous insufficiency
- Claudication
- Aneurysms

Coronary Heart Disease

Coronary heart disease is characterized by a reduced perfusion of the heart muscle due to a narrowing, or a complete occlusion, of one or more of the coronary vessels. In addition to atherosclerosis, thrombi and spasm of the coronary vessels can reduce myocardial perfusion. Atherosclerosis is caused by a sequence of events starting with infiltration of the endothelium of the vascular smooth muscle by **monocytes** (mononuclear leukocytes—phagocytes). This infiltration apparently can be stimulated by injury to the vascular wall through the chronic effects of hypertension and is accelerated by the presence of high circulating LDL cholesterol. The monocytes respond by ingesting and stimulating the oxidation of the LDL, eventually resulting in fatty streaks in the smooth-muscle wall. This process may further stimulate an inflammatory response, contributing to the injury and to the narrowing of the vessel. Disruption of the resulting plaque may lead to a **thrombosis**, or clotting within the vessel, which will contribute to the narrowing or—in extreme cases—occlusion of the vessel. Finally, should elements of the plaque break off, they may cause occlusion of a vessel at some distant site, leading to coronary vascular occlusion (heart attack) or cerebral vascular occlusion (stroke—a brain attack).

➤ Phagocytes are cells capable of ingesting foreign particles. They include leukocytes and macrophages.

Incomplete perfusion of the heart muscle results in symptoms of **angina pectoris**, including uncomfortable pressure or pain in the center of the chest that spreads to the shoulders, neck, or arms. Complete occlusion of a coronary vessel is known as a **myocardial infarction** (MI).

Infarctions are classified as left ventricular (anterior, inferior, posterior, or lateral) or right ventricular, depending on the site of the occlusion. An MI affecting a large area of the myocardium can result in sudden death. A less severe MI results in an area of myocardial death proportionate to the size of the vessels occluded. Immediate treatment with "clot-busting" drugs can minimize the damage caused by an MI. In addition, regular cardiovascular exercise, which stimulates the formation of **collateral circulation** (additional blood vessels) in the myocardium, can help to minimize the damaged area should an MI occur.

Incomplete perfusion of the heart muscle results in symptoms of angina pectoris. Complete occlusion of a coronary vessel is known as a myocardial infarction.

The size of the area of the myocardium affected by an MI is predictive of subsequent mortality. Damage to >12% of the myocardium increases two-year mortality to 7% as compared to 0% for damage to <12% of the myocardium. An MI involving >35% of the left ventricle is predictive of a high short-term mortality.

Case Study
Mr. Cassotti has been referred for rehabilitation. You visit him in his room to introduce yourself and to begin the process of orientation and education. The chart describes him as a 76-year-old male weighing 77 kg (170 lb) with a height of 173 cm (5 ft 8 in.). Unfortunately, Mr. Cassotti's chart isn't complete, and you are forced to obtain as much of the history as you can from the patient. In re-

sponse to your question about the site of his heart attack, he says it was in either the right or the left coronary artery. Why is it important for you to know whether it was the right or the left?

In over 50% of the population, the right coronary artery carries the majority of the blood flow to the myocardium. If the MI was in the right coronary artery, there is a likelihood that damage was greater than if the MI had been in the left coronary artery. In addition to knowing which vessel was occluded, it may be important to know the results of radionuclide imaging of the myocardium. This technique can identify the extent of myocardial damage.

Chronic Heart Failure

Chronic heart failure affects more than 5 million persons in the United States with a prevalence of over 10% for those 80 years of age and above. Hypertension and chronic myocardial ischemia can initiate a sequence of events that result in left ventricular dysfunction, including a loss of compliance of the left ventricle. The loss of compliance decreases diastolic filling (diastolic dysfunction) and emptying (systolic dysfunction) and reduces stroke volume. In a normal heart, the **ejection fraction** (end diastolic volume – end systolic volume) is 60% to 70%, while in severe left ventricular dysfunction, or heart failure, it can be <30%. Once left ventricular function is severely compromised, prognosis is poor, with mortality approaching 80% within six months. The associated loss of cardiac output results in severe exercise limitations.

Vascular Diseases

Diseases of the circulation include those affecting both the myocardial circulation and the peripheral vasculature (e.g., atherosclerosis), as well as those affecting only the peripheral vasculature. With the exception of aneurysms, the vascular diseases listed on page 235) can be positively affected by regular exercise.

Atherosclerosis

We have already looked at the pathology of atherosclerosis as it relates to the coronary circulation. Atherosclerosis of the peripheral vasculature presents as **arteriosclerosis obliterans (ASO)**. Since the vessels of the pelvis and lower extremities are affected primarily, the symptoms of ASO are experienced in the buttocks and legs. The major symptom, similar to that associated with narrowing of the arteries in the myocardium (i.e., angina pectoris), feels like a muscle cramp. These occasional ischemic events are called **intermittent claudication**. Intermittent claudication occurs during exercise when the demand for increased blood flow cannot be met because of the narrowed vasculature. The resultant ischemia causes pain in the affected areas. Regular, low-level exercise may improve the condition.

Hypertension

Hypertension is typically defined as a systolic pressure greater than 140 mm Hg or a diastolic pressure greater than 90 mm Hg. Since blood pressure normally increases with age, this definition is somewhat age dependent.

Primary, or **essential**, **hypertension** is associated with no known etiology whereas **secondary hypertension** occurs secondary to another disorder—namely, pregnancy, drug use, sleep apnea, or renal disease. The causes of hypertension include atherosclerosis, hormonal- or drug-related increases in vascular resistance, and genetics. In people who are elderly, systolic hypertension is most prevalent because of the stiffening of the vascular system with age. Hypertension accelerates the atherosclerotic process, increasing the risk of CHD, stroke, or other cardiovascular diseases. Reduction of risk factors and, in those who are particularly susceptible to hypertension, regular exercise can reduce the risk and consequences of hypertension.

Hypertension is defined as a systolic pressure greater than 140 mm Hg or a diastolic pressure greater than 90 mm Hg at rest.

Venous Insufficiency

The most common form of venous insufficiency is **varicose veins**. Venous varicosities are associated with genetic weakness of the venous walls and can be related to pregnancy, a lack of exercise, obesity, and smoking. The weakness of the venous wall apparently causes the valves of the system to become incompetent, leading to venous pooling—

particularly in vessels that are not supported by the muscles, such as peripheral veins of the legs. Treatment of mild varicose veins includes regular exercise, elastic support, and elevation of the affected limb. Treatment of severe peripheral varicose veins involves the surgical removal (stripping) of the affected vessels.

Aneurysms

A weakened area of the vascular wall may develop a localized dilation or bulging called an **aneurysm**. The most common type of aneurysm occurs in the large arteries such as the abdominal aorta. The weak sites may be congenital or may be a result of the arteriosclerotic process. There is no evidence that exercise might prevent an aneurysm other than through the reduction of arteriosclerotic risk factors.

Prevention and Rehabilitation

Of course, the primary goal of clinical practice is the prevention of disease. Smoking cessation, lowering blood pressure, and instituting an exercise program can significantly reduce the risk of heart attacks (figure 14.2). Unfortunately, too many health professionals are involved in the consequences of inadequate prevention and must deal

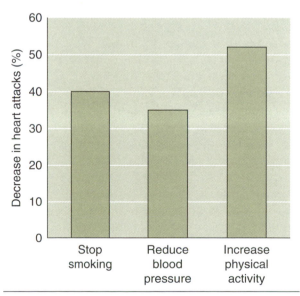

Figure 14.2 Changes that reduce risk of cardiovascular disease. A change in lifestyle and medical modification of blood pressure can significantly reduce cardiovascular disease risk factors. Combining any of these factors will have an additive effect.

with the rehabilitation of the patient. It would be wrong, however, to consider prevention a failure and therefore to ignore preventative efforts at the rehabilitation stage. It is important to incorporate preventative strategies as part of a rehabilitation program. This section deals with these two aspects of cardiovascular disease.

Prevention and Risk Reduction

As you can imagine, prevention of cardiovascular diseases involves reducing the controllable risk factors—smoking, high levels of lipids/cholesterol, hypertension, obesity, sedentary lifestyle, stress, and some personality characteristics. Changing one's lifestyle can significantly reduce the risk of cardiovascular disease. Stopping smoking and becoming less sedentary directly reduce risk. Indirectly these factors will also reduce hypertension, obesity, diabetes, and stress—further lowering the total risk by over 400%. It is clear that prevention of cardiovascular disease is possible and that it should be a priority of health professionals.

• **Cigarette smoking.** Cigarette smoke carries chemicals that create an addiction to cigarettes. Nicotine is the most important chemical responsible for establishing and maintaining a tobacco addiction. Nicotine initially acts as a stimulant, making the smoker more alert; but as the effect wears off, the person feels the need for another boost through another cigarette. After a short time, this cycle creates an intense dependency on cigarettes. This addiction is what makes it so difficult to stop smoking. Other important factors in smoking dependency are the behavioral and psychological aspects. The most effective smoking-cessation programs address chemical, behavioral, and psychological dependencies. Regardless of how long a person has smoked, the risks associated with smoking decrease upon cessation. And they continue to decrease—to the point that some people who have smoked 20 years or less can reduce their risks to normal levels within 10 years after smoking their last cigarette.

• **Cholesterol level.** The American Medical Association currently recommends that adults strive to maintain a cholesterol level below 200 mg/dl blood. Since the risk of CAD is increased in those with high levels of LDL and decreased in those with high levels of HDL, prevention of CAD should focus on lowering cholesterol below 200 mg/dl and HDL and LDL levels below 35 mg/dl and 130 mg/dl, respectively. A diet consisting of <30% fat in general, and low in saturated fat and cholesterol in particular, will reduce cholesterol and LDL levels. If dietary strategies are inadequate, cholesterol-lowering medication may be prescribed. Regular exercise, even of low to moderate intensity, is an effective mechanism to increase HDL levels and can do so regardless of changes in body fatness. A combination of appropriate dietary changes and regular exercise can significantly alter risk factors relating to blood cholesterol levels.

• **Hypertension.** Elevated blood pressure is a dangerous threat to cardiovascular health because it may be present with no outward signs. Early treatment is critical to preventing the complications of high blood pressure. The evidence regarding the possible effect of regular exercise on hypertension is unclear. It does appear that many people can moderate or even reduce blood pressure with regular exercise. Other research has suggested that a lifetime of regular exercise can prevent high blood pressure in people who may be susceptible on the basis of family history. In addition to regular exercise, lowering stress and losing weight can moderate blood pressure. People who snore or who experience episodes of sleep apnea have a high incidence of hypertension. Treatment of these disorders effectively reduces the hypertension.

• **Diet and exercise.** Prevention or treatment of obesity will reduce the incidence of diabetes, which is independently associated with a higher risk of cardiovascular disease; these measures can also moderate hypertension and improve the blood lipid profile. It is clear that treatment of obesity must include both dietary restriction and exercise (see chapter 3). That appropriate dietary modifications and increased activity levels can significantly reduce the threat of cardiovascular diseases also is clear. Dietary recommendations are discussed in chapter 2. The traditional recommendation concerning regular exercise has been to perform a minimum of 20 to 30 min of continuous, moderate-intensity exercise at least three days a week for cardiovascular health. The recommended intensity was between 65% and 85% of heart rate reserve (HRR = maximum HR − resting HR; see chapter 18). Recent evidence suggests that this recommendation could be revised—that the 20 to 30 min of exercise could be accumulated over the course of the day and still have a significant positive impact on cardiovascular health. Many people had difficulty finding a 20 to 30 min block of time and consequently did not attempt regular exercise. According to the current recommendation, simple measures such as parking farther away from the destination and taking the stairs instead of the elevator can fulfill the activity requirement. Chapter 18 contains specific recommendations.

It is clear that prevention of cardiovascular diseases must involve continual public education regarding the risk factors; clearly, too, all health care workers must focus more strongly on encouraging the public to address these risk factors.

> It is clear that prevention of cardiovascular diseases must involve continual public education regarding the risk factors.

Rehabilitation

Rehabilitation of a patient who has experienced an MI has changed dramatically since the 1960s. Patients used to maintain strict bed rest and reduced activity, including turning over, for several weeks (figure 14.3). Because strict bed rest can lead to 1% to 5% losses of strength and aerobic capacity per day, and because enforced inactivity results in a 15% decrease in stroke volume over several weeks, this regime made rehabilitation more difficult. Cardiac rehabilitation programs

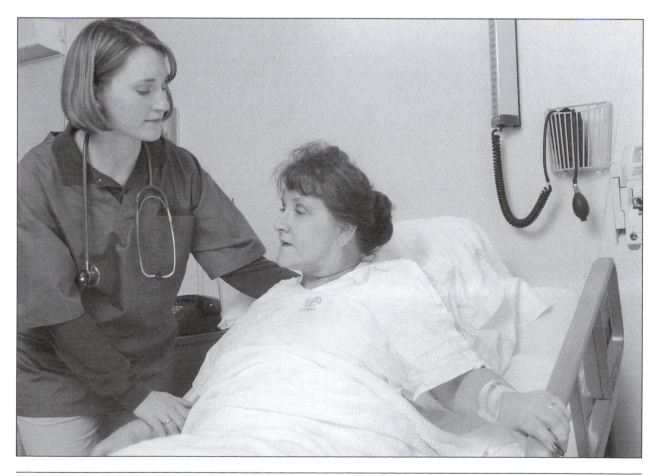

Figure 14.3 A post-operative patient being assisted to sit up. It is no longer the case that post-operative heart patients are restricted to total bed rest for long periods. Now they are usually urged to sit up as soon as within 24-48 hours of their surgery.

now begin within days of an event. In addition, rehabilitation now includes patients who have had **coronary artery bypass graft surgery** (CABGS) or coronary artery stent implantation and those who have chronic heart failure.

Exercise Guidelines

General guidelines for exercise rehabilitation programs can be established according to an initial classification of the patient based on the New York Heart Association classification system (table 14.1). On the basis of a description of the patient's symptoms and comfort level during exercise, one can estimate his or her likely exercise capacity and even establish reasonable targets for exercise during rehabilitation. Beyond the general guidelines shown in table 14.1, the rehabilitation specialist must attempt to obtain more specific information about the patient's functional capacity through a graded exercise test (GXT), described in chapter 18.

Case Study

Mr. Cassotti (see case study on page 236) tells you that when he first came to the hospital he was unable to move around without some chest discomfort. He describes this as angina. Now, three days later, he tells you that he is feeling much better and can go to the washroom with just a little discomfort. What is your best guess for starting points for his exercise program?

From Mr. Cassotti's description you would classify him as status III—he has symptoms with minimal work. His likely exercise capacity, then, would be about 3 METs. The most he is likely to tolerate while exercising continuously is about 2 METs—if short exercise periods are interspersed with short rest periods, about 3 METs. His electrocardiogram (ECG), of course, should be monitored while he is performing this exercise.

Table 14.1   NYHA Classifications With Expected Exercise Capacities

Status	Characteristics	Maximal capacity (METs)	Maximal permissible workload (Cal/min and METs*)	
			Continuous	Intermittent
I	Can walk without symptoms or limitation; can do most light-effort activities. 0–15% impairment.	6.5	4.0/3.2	6.0/4.9
II	Has symptoms with light work; slight limitation of physical activity. Comfortable at rest. Ordinary physical activity results in fatigue, palpitations, dyspnea, or angina. 15–30% impairment.	>4.5 <6.5	3.0/2.5	4.0/3.2
III	Has symptoms with minimal effort; marked limitation of physical activity. Comfortable at rest. Less-than-ordinary physical activity results in fatigue, palpitations, dyspnea, or angina. 30–70% impairment.	3.0	2.0/1.6	3.0/2.5
IV	Is unable to carry on any physical activity without discomfort; discomfort increases with exercise. Symptoms may be present at rest. >70% impairment.	1.5	1.0/1.0	2.0/1.6

*METs calculated assuming 5 Cal/L O_2 and a 70 kg person.

Adapted from American Heart Association.

On the basis of what you learned in chapter 1, you should be able to establish his exercise levels on the treadmill and cycle ergometer. Let's assume you wish him to exercise on the treadmill and that you want him to exercise with no elevation (grade). Calculate the $\dot{V}O_2$:

$$2 \text{ METs} \times 3.5 \text{ ml } O_2/\text{kg}/\text{min} = 7 \text{ ml } O_2/\text{kg}/\text{min}$$

Calculate the speed:

$$7 \text{ ml } O_2/\text{kg}/\text{min} = 0.1 \text{ (speed in m/min)} + 3.5$$

$$3.5 = 0.1 \text{ (speed in m/min)}$$

$$\text{speed} = 35 \text{ m/min or about 1.3 mph}$$

Should you wish to have him exercise on the cycle ergometer, the calculations would be as follows:

The $\dot{V}O_2$ must be expressed in absolute terms:

$$7 \text{ ml } O_2/\text{kg}/\text{min} \times 77 \text{ kg} = 539 \text{ ml/min or 0.539 L/min}$$

Calculate the load on the ergometer:

$$539 \text{ L/min} = 1.8 \text{ (work rate)} + (3.5 \times 77 \text{ kg})$$

$$539 = 1.8 \text{ (work rate)} + 269.5$$

$$269.5 = 1.8 \text{ (work rate)}$$

$$149.7 \text{ kg/m/min} = \text{work rate}$$

$$149.7/6.1 = 25.5 \text{ W}$$

Rehabilitation Programming

Patients who have congestive heart failure, angina with exertion, dangerous arrhythmias, myocarditis, or emboli should exercise with extreme caution, if at all. In general, these conditions are absolute contraindications for exercise. Relative contraindications include consistent ectopic activity of the heart, uncontrolled or high-rate supraventricular dysrhythmias, moderate aortic stenosis, and pulmonary hypertension. The decision to exercise is made by the patient's physician and is based on an evaluation of the patient's potentially disqualifying condition in conjunction with all other factors.

➤ Supraventricular dysrhythmia is an arrhythmia of the heart that is observed in the atria.

➤ A stenosis is a narrowing. Aortic stenosis impedes blood flow from the left ventricle.

The Phases of Rehabilitation Programs

Cardiac rehabilitation programs are divided into four phases, with phase I starting after the initial event, usually while the patient is still in the hospital (figure 14.4). As the patient's condition improves, he or she progresses into a monitored, well-controlled phase II and then into a transitional phase III. From this phase the patient graduates into the long-term-maintenance phase IV program. The progression from one phase to the next is dependent on the conditions of the patient at the onset, medication, compliance, and many other factors. Figure 14.4 provides a general guide. The health care team must monitor each patient's progress and determine progression on an individual basis.

• **Phase I.** Phase I is the inpatient phase of rehabilitation, typically lasting from four days to two weeks. Rehabilitation may begin in the cardiac care unit or cardiac ward with education and some passive range-of-motion exercises, progressing to increased activity as soon as possible. The rate of progression from passive to more active and resistive exercises is dependent on the patient's status. Patients with a mild incident who have been exercising regularly will progress faster than patients who have had a more severe incident or who had been sedentary before the event. These previously sedentary patients typically have a low exercise tolerance at discharge. The goal of phase I is to prepare the patient for living at home with as little as possible assistance after discharge. The rehabilitation specialist should be aware of the great potential for anxiety and depression during the initial stages of rehabilitation. A GXT may or may not be performed at the time of discharge, or soon after, depending on the facility.

• **Phase II.** Phase II, the first phase of outpatient care, usually lasts from one to three months and can start as soon as four or five days after the event. Phase II programs are held in a hospital or other clinical setting where medical supervision is available and monitoring of ECG is possible. Monitoring should include heart rate, blood pressure, ECG, and symptoms (angina, dyspnea).

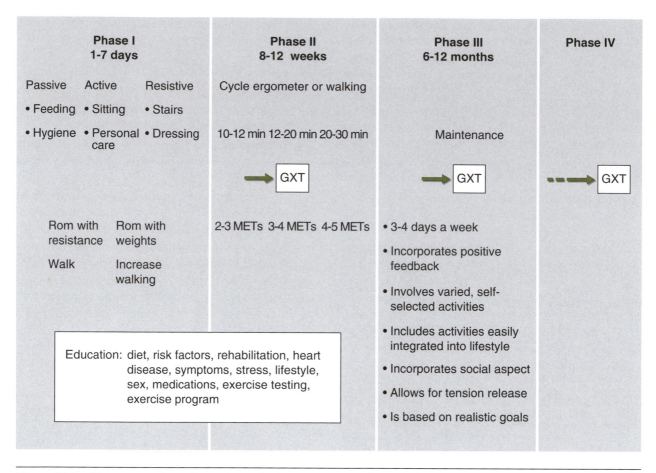

Figure 14.4 The progression of exercise and education in a cardiac rehabilitation program. Phase I starts as soon as possible after admission. The progression through the phases and the duration of each phase depends on patient condition and response to rehabilitation.

Exercise capacity at entry is usually symptom limited. The patient's symptoms, such as angina and shortness of breath, determine the end point of the exercise test and define the limits of tolerable, safe exercise for phase II. The supervised exercise progresses in both duration and intensity until the person is able to tolerate 20 to 30 min of exercise at 60% to 65% of his or her peak work capacity. A GXT is performed at the end of phase II rehabilitation.

• **Phase III.** Phase III, or the maintenance phase, lasts from 6 to 12 months. The purpose is to maintain and improve on the gains made during phase II while weaning the patient from the comfort of the supervised setting. The primary goal of phase III is to enable the person to exercise in his or her chosen setting. It is important to determine the activity likes and dislikes of each patient and to design the long-term program accordingly.

Promoting a social aspect of the activity is critical to the success of rehabilitation; long-term compliance to exercise is greater when the exerciser has a support group.

• **Phase IV.** Once a regular, positive exercise pattern has been established, patients graduate to the phase IV program, in which they are treated along with the non-involved population.

Resistance Exercise and Rehabilitation

For many years, resistance exercise (weight lifting) within the rehabilitation program was discouraged. Rehabilitation specialists feared that blood pressure changes with lifting could put excess stress on a compromised myocardium. Recent evidence indicates that, when implemented properly, resistance exercise can be a valuable part of rehabilitation. Patients who have undergone bypass surgery, in particular, may benefit from range-of-motion resistance exercises of the upper extremity. Among patients who should not participate in resistance-exercise training programs are those who are also excluded from the aerobic portion of rehabilitation programs (e.g., patients with congestive heart failure, embolism, or myocarditis.) Mild resistance exercise can be done as early as phase I, starting with patient resistance and progressing to light external resistance. During the early phase of rehabilitation, exercises should be performed for 10 to 15 repetitions before fatigue so that lighter weights are used. As the program progresses, repetitions can be reduced to 8 to 12 repetitions with heavier weights to fatigue. Patients should be instructed in proper lifting techniques and in how to avoid the Valsalva maneuver.

➤ An embolism is a blockage of a vessel by a detached thrombus or foreign body.

What You Need to Know From Chapter 14

Key Terms

aneurysm

angina pectoris

arteriosclerosis

arteriosclerosis obliterans (ASO)

atherosclerosis

collateral circulation

coronary artery bypass graft surgery

coronary artery disease

coronary heart disease (CHD)

ejection fraction

heart failure

hypertension

intermittent claudication

monocyte

myocardial infarction

plaque

primary (essential) hypertension

risk factors

secondary hypertension

stroke

thrombosis

varicose veins

Key Concepts

1. You should understand the factors that increase the risks of CHD and know how these factors interact to enhance or reduce the risk of disease.

2. You should know the etiology of cardiovascular diseases.

3. Given an understanding of the risk factors and the etiologies, you should be able to develop programs designed to prevent cardiovascular diseases.

4. You should be able to identify the basic elements of cardiac rehabilitation programs.

Review Questions

1. Many people become confused about the relationships between smoking and cardiovascular diseases. Can you come up with a simple explanation that will tie these together?

2. How can smoking lead to chronic heart failure?

3. You have a cycle ergometer, a stepping device, and a treadmill available. Which of these would be least desirable for a patient with intermittent claudication and why?

4. You have the results of an exercise stress test from a patient referred for an exercise program. Resting heart rate was 78/min, and blood pressure was 128/92. At the first exercise level (5 METs), the heart rate was 139/min and the blood pressure 142/92. At this level some heart arrhythmias were noted, but the test was continued. Exercise was terminated after 30 min at the next level because the arrhythmias became worse.

(continued)

The patient is a homemaker with a husband and three teenagers at home. Her cholesterol is >250. Her physical activity has been limited to the occasional weekend walk with her husband. You have been asked to provide an exercise program for this patient.

a. What is her peak exercise capacity in METs for the purposes of establishing a safe exercise recommendation?

b. What is her cardiovascular functional status?

c. The information you have would characterize this patient as starting in which phase of cardiac rehabilitation?

d. What would you recommend for specific activities that she could safely do in the rehabilitation facility and at home?

e. How and when would you monitor this patient?

f. What would you include in the first two scheduled 10 min educational sessions?

Bibliography

American Association of Cardiovascular and Pulmonary Rehabilitation. *Guidelines for cardiac rehabilitation and secondary prevention programs.* 3rd ed. Champaign, IL: Human Kinetics, 1999.

American College of Sports Medicine. *ACSM's guidelines for exercise testing and prescription.* 6th ed. Philadelphia: Lippincott Williams & Wilkins, 2000.

Cardiac Rehabilitation. Clinical Practice Guidelines. Publication no. 96-0672. Agency for Health Care Research and Quality, 1995.

Dazu VJ. Pathobiology of atherosclerosis and plaque complications. *Am Heart J* 1994, 128:1300–1304.

Dressendorfer RH, Franklin BA, Cameron JL, Trahan KJ, Gordon S, Timmis GC. Exercise training frequency in early post infarction cardiac rehabilitation: Influence on aerobic conditioning. *J Cardiopul Rehab* 1995, 15:269–276.

Franklin BA, Bonzheim K, Gordon S, et al. Resistance training in cardiac rehabilitation. *J Cardiopul Rehab* 1991, 11:99–107.

Hahn RA, Teutsch SM, Rothenberg RB, and Marks JS. Excess deaths from nine chronic diseases in the United States. *JAMA* 1990, 264:2654–2659.

Haskell WL, Alderman EL, Fair JM, et al. Effects of intensive multiple risk factor reduction on coronary atherosclerosis and clinical cardiac events in men and women with coronary artery disease: The Stanford Coronary Risk Intervention Project (SCRIP). *Circulation* 1994, 89:975–990.

Pollock ML, Gaesser GA, and Butcher JD. The recommended quantity and quality of exercise for developing and maintaining cardiovascular and muscular fitness and flexibility in healthy adults. *Med Sci Sports Exerc* 1998, 30:975–991.

Pu CT and Nelson ME. Aging, function, and exercise. In *Exercise in rehabilitation medicine,* ed. WR Frontera, DM Dawson, and DM Slovik. Champaign, IL: Human Kinetics, 1999.

Roberts, WC. Preventing and arresting coronary atherosclerosis. *Am Heart J* 1995, 130 (1 Pt 3):580–600.

Rochmis P and Blackburn H. Exercise tests: A survey of procedures, safety, and litigation experience in approximately 170,000 tests. *JAMA* 1971, 217:1061–1066.

Wenger NK, Froelicher ES, Smith LK, et al. *Cardiac rehabilitation.* Clinical Practice Guidelines no. 17. Rockville, MD: U.S. Department of Health and Human Services, Agency for Health Care Policy and Research and National Heart, Lung and Blood Institute.

Wilson PF, D'Agostino RB, Levy D, et al. Prediction of coronary heart disease using risk factor categories. *Circulation* 1998, 97:1837–1847.

Pediatrics

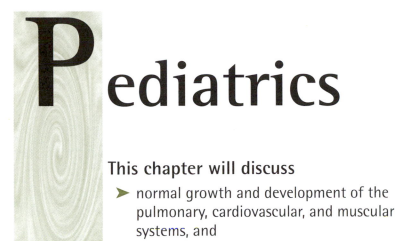

This chapter will discuss

➤ normal growth and development of the pulmonary, cardiovascular, and muscular systems, and

➤ the influence of these systems on exercise response and exercise capacity in the pediatric population.

This chapter consists of two sections. The first section reviews the development of the pulmonary, cardiovascular, and muscular systems. As you read this section, try to apply some of the concepts you learned in previous chapters to predict the influence that each developmental factor might have on the exercise response in the pediatric age group. The second section of the chapter is an opportunity for you to check your predictions against a description of the responses to exercise in this age group. It is difficult to obtain measurements during exercise in children below the age of 6. For this reason the responses to exercise are described primarily from age 6 to adulthood (18 years).

Growth and Development

The response to exercise is governed by the function of the pulmonary, cardiovascular, and muscle/metabolic systems. An understanding of the response to exercise in the pediatric population must therefore be based on consideration of the underlying patterns of development of these systems. In some cases it is also necessary to consider sex differences in growth and development. You might be tempted to assume that the

material in previous chapters can simply be applied to children. However, children are not just small adults. There are important physiological distinctions between young children and adults, and these are reflected in differences in their exercise responses.

We must also consider the significant variability in childhood development when dealing with exercise responses in children. An examination of standard growth charts (figures 15.1 and 15.2) shows that at age 15, normal height can vary from 156 to 182 cm (5 ft 1 in. to 5 ft 11.5 in.) in males and from 151 to 174 cm (4 ft 11 in. to 5 ft 8.5 in.) in females. Likewise, weight is considered normal between 43 and 79 kg (95 and 174 lb) for males and 40 and 78 kg (88 and 172 lb) for females. This variability in growth is accompanied by similar variability in skill development, which one must take into account not only when evaluating exercise responses, but also when grouping children for activities. The risk for injury is increased in group activities in which size and skill play a role.

The Pulmonary System

Development of lung alveoli is virtually complete before the age of 6 years. Between age 6 and adulthood, increases in lung and alveolar size

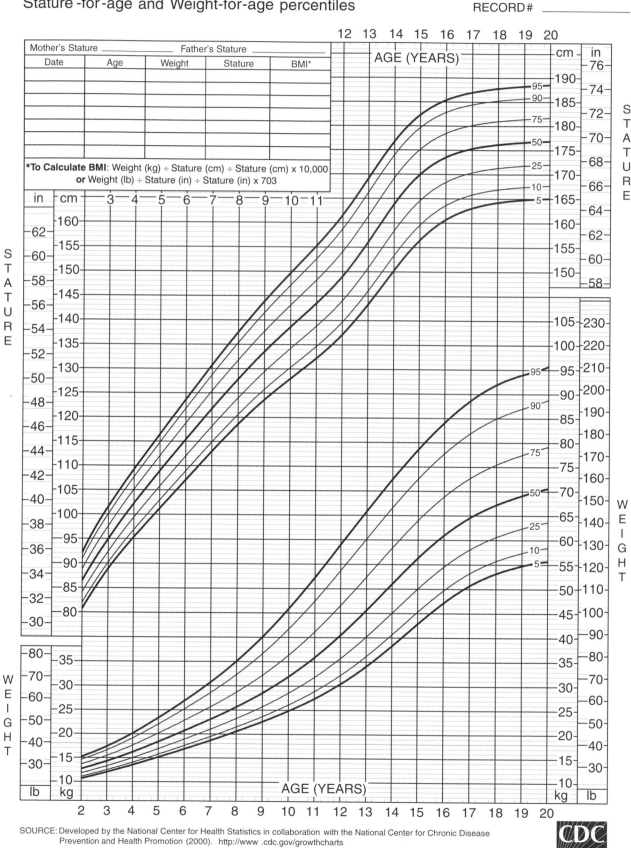

2 to 20 years: Boys
Stature-for-age and Weight-for-age percentiles

NAME _____

RECORD# _____

SOURCE: Developed by the National Center for Health Statistics in collaboration with the National Center for Chronic Disease Prevention and Health Promotion (2000). http://www .cdc.gov/growthcharts

Figure 15.1   Standard growth charts for males from ages 2 to 20.

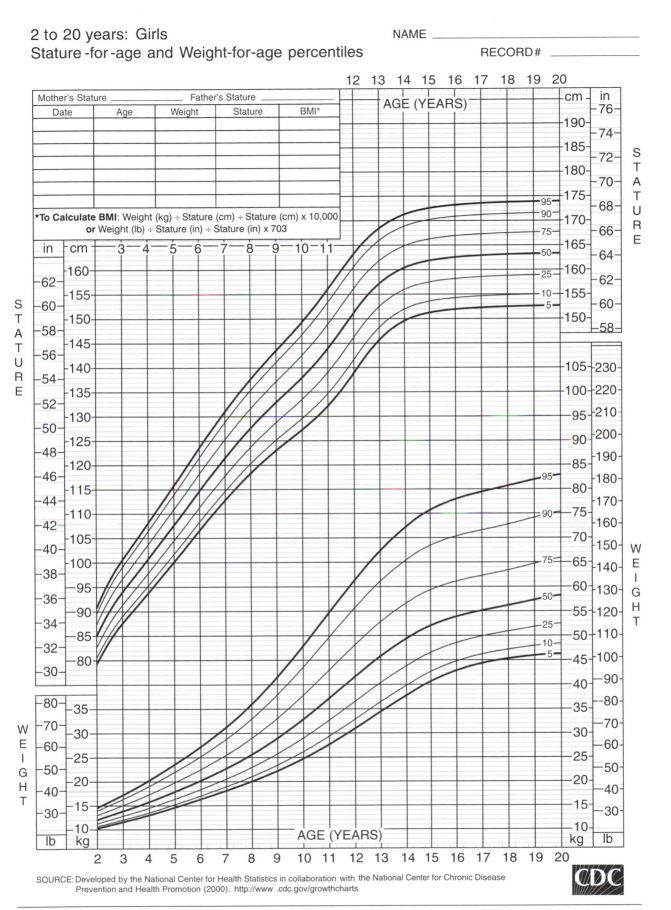

2 to 20 years: Girls
Stature-for-age and Weight-for-age percentiles

NAME _____

RECORD# _____

Mother's Stature		Father's Stature		
Date	Age	Weight	Stature	BMI*

***To Calculate BMI:** Weight (kg) ÷ Stature (cm) ÷ Stature (cm) x 10,000
or Weight (lb) ÷ Stature (in) ÷ Stature (in) x 703

AGE (YEARS)

STATURE

STATURE

WEIGHT

WEIGHT

Figure 15.2   Standard growth charts for females from ages 2 to 20.

parallel changes in the size of the thorax. The surface area available for gas exchange parallels increases in body size such that arterial blood gas values in children are the same as adult values.

The absolute compliance, or elasticity, of the respiratory system remains stable after the age of 2 to 3 years. However, compliance, normalized for height, increases during development, showing that growth-related changes in lung structure occur at a faster rate than changes in height. Airway resistance, normalized for height, decreases during development, showing that airway size also increases at a faster rate than height. Respiratory system compliance and airway resistance are primary determinants of tidal volume (Vt) and breathing frequency (Fb) (chapter 12). Resting Vt relative to body mass (Vt/kg) and Fb decrease from age 6 to 18, suggesting that breathing economy should be greater at age 18 than at age 6.

The surface area available for gas exchange parallels increases in body size such that arterial blood gas values in children are the same as adult values.

The growth-related changes in pulmonary system compliance and airway resistance are reflected in the ventilatory responses to exercise (figure 15.3). In absolute terms, children have a lower volume of air exhaled per minute ($\dot{V}_E$) at the same load, simply reflecting their smaller body mass relative to that of adults. When corrected for body size ($\dot{V}_E$/kg), however, total minute ventilation at the same exercise load is higher in children. This higher $\dot{V}_E$/kg results in a relative hyperventilation (reflected in lower estimated arterial CO_2 levels) and in a lower breathing economy (reflected in a higher $\dot{V}_E/\dot{V}O_2$) in children as compared to adults at the same exercise level (figure 15.4). Relative to adults, then, the work of breathing during exercise is higher in children.

Relative to adults, the work of breathing during exercise is higher in children.

The Cardiovascular System

The size of the resting heart, whether measured as mass or volume, increases from age 6 to adulthood. The volume relative to body size, however,

is higher in children, showing that during this period, increases in body size are greater than increases in heart size. Resting heart rate also is higher in children than in adults. The combination of higher relative heart volume and heart rate translates into a higher relative resting cardiac output in children compared to adults. Measures of cardiac contractility do not change over the age range of 6 to 18.

Submaximal as well as maximal exercise absolute stroke volume is lower in children compared to adults. Heart rate, on the other hand, is higher at any given exercise level in children compared to adults (figure 15.5). The higher heart rate, however, is insufficient to completely compensate for the lower stroke volume, so cardiac output, at any given exercise level, is lower in children than adults (figure 15.5). By itself, the higher relative heart size should give the child an exercise advantage in terms of the ability to pump blood to the periphery, but the age-related differences in heart rate and stroke volume result in a lower cardiac output in the child relative to the adult at the same exercise level. To maintain the same $\dot{V}O_2$ with a lower O_2 delivery, the arterial-to-venous difference (a-v O_2, Δ) is greater in the child, reflecting a greater removal of the O_2 from the blood. The child's increased capability to extract O_2 reflects a greater relative oxidative capacity in the muscles of children compared to adults.

Age-related differences in heart rate and stroke volume result in a lower cardiac output in children relative to adults at the same exercise level.

Case Study

Your friend, a local physical education teacher, has been doing some reading on aerobic fitness and wants to apply some of the principles she has read about in her classes. Just out of curiosity, she asks whether the heart rate guidelines for adults are applicable to children. What is your response?

The general recommendation, that people perform aerobic conditioning in a range of 60% to 85% of maximum capacity, would be appropriate for children. However, target heart rate recommendations based on adult values would be inappropriate. For instance, the target heart rates for exercise in the 20- to

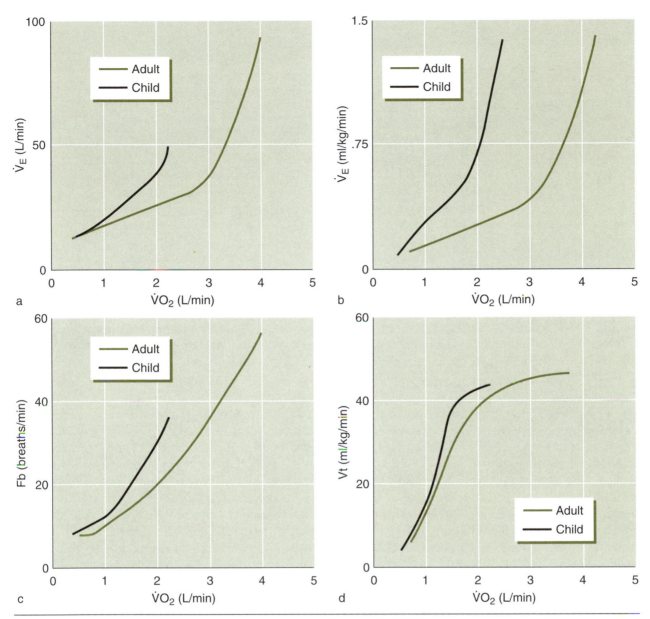

Figure 15.3   The ventilatory response to exercise in a typical 6-year-old child compared to a typical 18-year-old adult. (*a*) Minute ventilation in absolute terms ($\dot{V}_E$ in L/min); (*b*) minute ventilation relative to body weight ($\dot{V}_E$ in ml/kg/min); (*c*) breathing frequency (breaths/min); (*d*) tidal volume relative to body weight (Vt in ml/kg/min). Adapted from O. Bar-Or 1983.

30-year-old population are between 125 and 168 beats/min. The range for adolescent children would be on the order of 154 to 190 beats/min. The reason for the difference is that resting heart rate is somewhat higher in children (75 to 85 beats/min), as is maximal heart rate (195 to 215 beats/min). Calculations of target heart rate based on the Karvonen formula (see chapter 18) would give you the rate your friend should use.

➤ The Karvonen formula is a formula for calculating target heart rates where target heart rate = ([maximal heart rate – resting heart rate] × desired % maximal heart rate) + resting heart rate.

The Muscle/Metabolic System

Changes in the muscles with growth can greatly affect exercise performance. Some of these changes

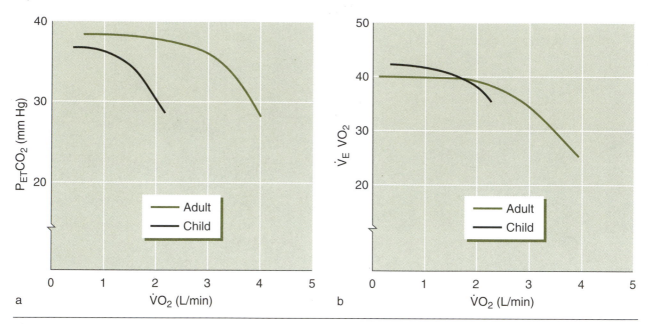

Figure 15.4   (*a*) End-tidal CO_2 ($P_{ET}CO_2$), a reflection of arterial CO_2, in a typical 6-year-old child and a typical adult at increasing O_2 consumption ($\dot{V}O_2$); (*b*) $\dot{V}_E/\dot{V}O_2$, a reflection of ventilatory economy, in the same two people. Adapted from O. Bar-Or 1983.

are related to the metabolic capability of the developing muscles. As with all aspects of pediatric exercise, one must recognize great variability in the developmental pattern.

The muscle metabolic profile in the preadolescent shows higher oxidative enzyme activities (succinate dehydrogenase, citrate synthetase) compared to those for adults (table 15.1). On the other hand, the number and size of mitochondria in the muscle are not different, suggesting that although a higher oxidative capacity is available, in terms of enzyme activities, that capacity cannot be fully utilized.

Growth and Development of Exercise Capacities

Physical capacities such as endurance, power, and strength are dependent on body or muscle mass. To determine whether age-related changes in these capacities are due simply to an increase in mass or to other maturational factors, these variables are normalized in some manner. For instance, $\dot{V}O_2$max and power output might be expressed per kilogram of body weight (ml/kg/min and W/kg, respectively); strength might be expressed per lean muscle mass or cross-sectional area of the involved muscles. In the follow-

ing discussion, most variables are expressed in both absolute and relative, or normalized, terms so that the simple effects of changing mass can be differentiated from the effects of other growth factors. In addition to growth-related influences on physical capacities, sex-related influences begin to become apparent during adolescence. The ability to discriminate among the effects of these various influences is important to the understanding of pediatric exercise responses.

Aerobic Capacity

Maximum O_2 consumption ($\dot{V}O_2$max) increases from age 6 to adolescence in both genders (figure 15.6). At puberty, $\dot{V}O_2$max continues to increase in males but tends to plateau in females as they progress through adolescence. Normalization of $\dot{V}O_2$max for body mass reduces some of the age-related increase (figure 15.6), indicating that most of the increase is due simply to the increase in body size. The linear relationship between changes in lean leg volume and $\dot{V}O_2$max (figure 15.6) further indicates that most of the size-related increase in $\dot{V}O_2$max is attributable to increases in lean muscle mass. An examination of children of the same mass, over a range of ages, however, still indicates a lower **aerobic capacity** in young children—suggesting an effect of maturation on $\dot{V}O_2$max that

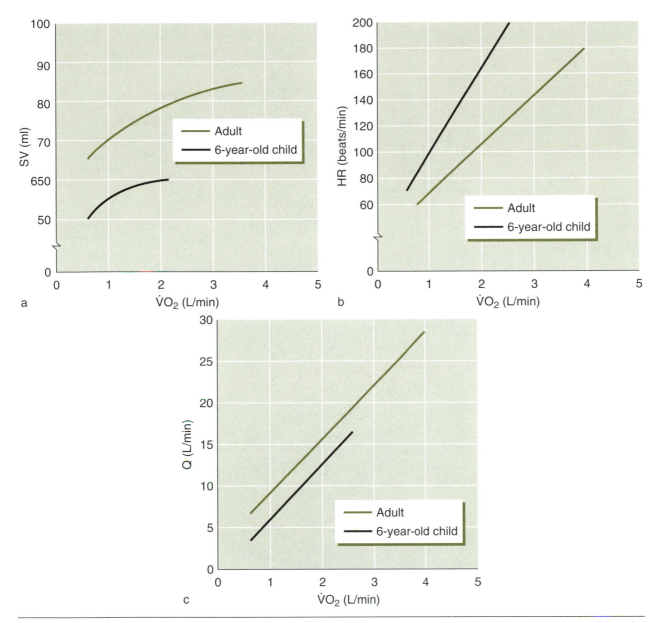

Figure 15.5   The relationship between cardiac function and O_2 consumption in a typical 6-year-old child and a typical 18-year-old adult. (*a*) Stroke volume (SV); (*b*) heart rate (HR); (*c*) cardiac output (Q).
Adapted from O. Bar-Or 1983.

is independent of body mass or body fat. Other age-related factors that could contribute to the change in $\dot{V}O_2$max with growth include alterations in myocardial contractility, blood volume, and muscle and metabolic variables.

> Some age-related factors that could contribute to the change in $\dot{V}O_2$max with growth include alterations in myocardial contractility, blood volume, and muscle and metabolic variables.

Normalization of $\dot{V}O_2$max with total body mass (ml/kg/min) or lean body mass (ml/kg lean body mass) or with a scaling factor ($\dot{V}O_2$max = kMb, where k is a constant, M is body mass, and b is a scaling exponent) reduces, but does not eliminate, sex differences (figure 15.6). It is clear that much of the sex difference in absolute $\dot{V}O_2$max (L/min) is due to the increase in body fat in females at puberty. Factors contributing to sex differences in $\dot{V}O_2$max include blood hemoglobin (Hb) levels (which become lower in females after

Table 15.1　Muscle Metabolic Characteristics of Children Compared to Adults

	Muscle levels	Utilization during exercise
ATP	No difference	No difference
CP	Lower	No difference or lower
SDH activity	Higher	
CS activity	Higher	
Mitochondria	No difference	
Glycogen	Lower	Lower
PFK activity	Lower	

ATP = adenosine triphosphate; CP = creatine phosphate; PFK = phosphofructose kinase; SDH = succinate dehydrogenase; CS = citrate synthetase.
Adapted from O. Bar-Or 1983.

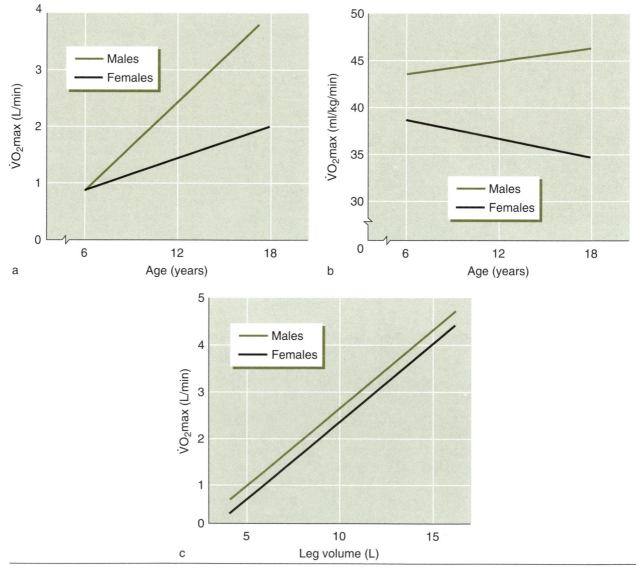

Figure 15.6　Changes in aerobic capacity from age 6 to adulthood. (*a*) The relationship between age and absolute $\dot{V}O_2$max; (*b*) the relationship between age and $\dot{V}O_2$max normalized for body mass; (*c*) the relationship between lean leg volume and $\dot{V}O_2$max. Adapted from O. Bar-Or 1983.

puberty), cultural influences that have discouraged activity (leading to reduced fitness levels) in young females, and other variables that are yet to be identified.

Factors contributing to sex differences in $\dot{V}O_2$max include blood hemoglobin (Hb) levels (which become lower in females after puberty), cultural influences that have discouraged activity (leading to reduced fitness levels) in young females, and other variables that are yet to be identified.

Aerobic "Reserve"

The energy requirement for performing movements such as walking or running, at the same velocities, is higher in young children than in adults. This lower movement economy has been attributed to greater extraneous movements, as reflected in greater relative vertical displacements, in children as compared to adults. Age-related differences in stride length also contribute to the greater energy cost in children moving at the same speed as compared to adults. Although movement economy can be improved with training, this does not eliminate differences between children and adults. One implication of lower movement economy in children is that at any given movement speed, children will be working closer to their maximum capacity, thereby reducing their "reserve"—the difference between en-

ergy requirement at submaximal work levels and their maximal capacity. One should take into account this lower aerobic reserve during walking and running tasks when prescribing activities for children based on adult references for predicted energy expenditures.

One implication of lower movement economy in children is that at any given movement speed, children will be working closer to their maximum capacity, thereby reducing their "reserve."

Non-Aerobic Capacity

The non-oxidative, or anaerobic, exercise capacity increases with age (figure 15.7). The lower slope in figure 15.7 b indicates that part of the age-related increase in **non-aerobic capacity** (the capacity of the non-oxidative or glycolytic metabolic pathway to produce energy) can be accounted for by the increase in body mass.

Case Study

Your teacher friend has accepted your recommendation regarding heart rate in children. You suddenly realize that you also need to explain something else that will affect how she approaches exercise in her students—reserve. How must the concept of reserve be taken into account during exercise in children?

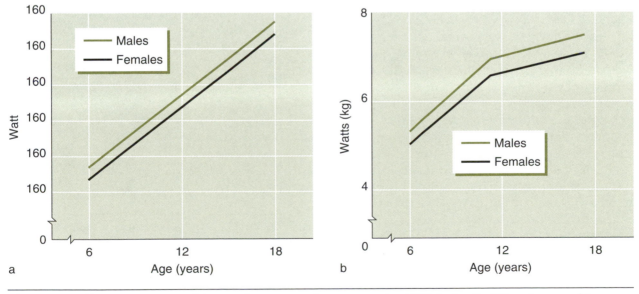

Figure 15.7 Changes in non-aerobic capacity, measured as power output (W) during the Wingate anaerobic power test, and age in males and females. (*a*) The relationship between absolute power output (W) and age; (*b*) the relationship between relative power output (W/kg) and age. Adapted from O. Bar-Or 1983.

Since young children doing the same exercise as an adult are going to be somewhat closer to their maximum capacity, they will have less reserve than an adult. Less economical movements increase the cost of any exercise. Those children who are less economical will be working proportionately closer to their maximum. Your friend should be aware of this and should consider scaling back on the target heart rate recommendation in these children by 10-12 beats/min.

Activity and Growth

It is not clear whether activity or inactivity during development actually affects growth. The inability to identify an activity-growth relationship is due to an inability to control for other factors that influence growth. It is impossible to differentiate growth patterns attributable to the normal growth of a child from patterns that may result from increased physical activity.

One approach to determining the effect of activity on development is to examine the effects of physical training on characteristics such as aerobic or non-oxidative capacity and strength. It appears that a training program of sufficient intensity, duration, and frequency to increase an adult's $\dot{V}O_2$max also will increase a child's $\dot{V}O_2$max. For prepubertal children, however, the changes appear to be less than those for postpubertal children. At this time there is no explanation for the apparent resistance to change in $\dot{V}O_2$max in younger children. Specific anaerobic training programs elicit improvements in anaerobic power in young children but no change in peak lactate levels, which would indicate a metabolic change in anaerobic capacity. Again, this apparent contradiction has not been explained. Finally, strength-training programs for children have elicited significant improvements in various measures of strength. It should be noted that most of these studies have used low-weight, higher-repetition (10-15 reps) programs to avoid excessive stress on the musculoskeletal systems of young children. In prepubertal children, the improvements are not related to increases in muscle mass, suggesting that strength gains are due primarily to neural adaptations.

Case Study

Your friend who teaches physical education was impressed by your knowledge of children and activity, so she approaches you with another question. What are the current recommendations regarding weight lifting for children?

You need to explain that excessive weight lifting for young children carries potential risks, including the separation of the growth plate from the bone. To avoid these risks but still promote physical activity in children, your friend should teach the children proper lifting techniques using light enough weights that allow 15 repetitions for the first set and then another two sets, for a total of three sets (figure 15.8). When these three sets can be done with 15 repetitions, the weight should be increased by 10%.

Figure 15.8 Prepubescent boy doing resistance training. With good teaching about proper technique and supervision from adults who understand the unique limitations of children, youngsters can benefit significantly from weight training.

What You Need to Know From Chapter 15

Key Terms

aerobic capacity

non-aerobic capacity

reserve

Key Concepts

1. You should be aware of the variability in growth and development among children.

2. You should be able to describe and explain differences between the adult responses to exercise and the responses among children of different ages. You should be able to do this for the pulmonary, cardiovascular, and muscle/metabolic systems.

Review Questions

1. What are some growth factors that could explain differences in exercise capacities—aerobic, non-aerobic, and strength—between young children and adults?

2. What are some sex-related factors that could explain differences in exercise capacities—aerobic, non-aerobic, and strength—between growing boys and girls?

3. Likewise, what are some age-related factors that can explain such differences between young people and adults?

4. The lower cardiac output but increased O_2 extraction seems to provide adequate O_2 for exercise in children. Under what circumstances might the already increased extraction become a liability?

Bibliography

Armstrong N, Welsman JR, and Kirby BJ. Peak oxygen uptake and maturation in 12-yr olds. *Med Sci Sports Exerc* 1998, 30:165–169.

Bar-Or O. *Pediatric sports medicine for the practitioner.* New York: Springer-Verlag, 1983.

Bar-Or O. The Wingate anaerobic test. An update on methodology, reliability and validity. *Sports Med* 1987, 4:381–394.

Cooper DM, Weiler-Ravell D, Whipp BJ, and Wasserman K. Aerobic parameters of exercise as a function of body size during growth in children. *J Appl Physiol* 1984, 56:628–634.

Cunningham DA, Patterson DH, Blimkie CJR, and Donner AP. Development of cardiorespiratory function in circumpubertal boys: A longitudinal study. *J Appl Physiol* 1984, 56:302–307.

Eriksson BO, Gollnick P, and Saltin B. The effect of physical training on muscle enzyme activities and fiber composition in 11-year old boys. *Acta Paediatr Belg* 1974, 28 (suppl):245–253.

Gaul CA, Docherty D, and Cicchini R. Differences in aerobic performance between boys and men. *Int J Sports Med* 1995, 16:451–455.

Goldberg B, ed. *Sports and exercise for children with chronic health conditions.* Champaign, IL: Human Kinetics, 1995.

Inbar O and Bar-Or O. Anaerobic characteristics in male children and adolescents. *Med Sci Sports Exerc* 1986, 18:264–269.

Maliszeewski AF and Freedson PS. Is running economy different between adults and children? *Ped Exerc Sci* 1996, 8:351–360.

Nevill AM, Holder RL, Baxter-Jones A, Round JM, and Jones DA. Modeling developmental changes in strength and aerobic power in children. *J Appl Physiol* 1998, 84:963–970.

Patterson DH, McLellean TM, Stella RS, and Cunningham DA. Longitudinal study of ventilation threshold and maximal O_2 uptake in athletic boys. *J Appl Physiol* 1987, 62:2051–2057.

Pfitzinger P and Freedson PS. Blood lactate responses to exercise in children: Part 1. Peak lactate concentration. *Ped Exerc Sci* 1997, 9:210–222.

Rowland T. *Exercise and children's health.* Champaign, IL: Human Kinetics, 1990.

Rowland T, Popowski B, and Ferrone L. Cardiac responses to maximal upright cycle exercise in healthy boys and men. *Med Sci Sports Exerc* 1997, 29:1146–1151.

Shephard RJ. *Physical activity and growth.* Chicago: Year Book Medical, 1982.

Zanconato S, Buchthal S, Barstow TJ, and Cooper DM. 31P-magnetic resonance spectroscopy of leg muscle metabolism during exercise in children and adults. *J Appl Physiol* 1993, 74:2214–2218.

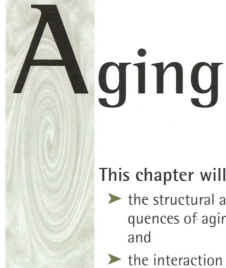Aging

This chapter will discuss

➤ the structural and physiological consequences of aging on various body systems and

➤ the interaction between physical activity and aging.

Despite abundant scientific research in recent years, the processes underlying aging (or **senescence**) are largely unknown. The most likely theory of aging involves an alteration in cellular genetic coding that results in abnormalities in synthesis of the cell's various constituents, accumulated over time. For example, any abnormality in the production or assembly of skeletal muscle contractile proteins would produce a deficit in force-producing capacity and could affect the person's ability to perform normal daily activities. Likewise, an alteration in fiber-type distribution within a skeletal muscle, likely related to a change in motor nerve innervation, produces a decline in muscle strength in aged people. Abnormalities in the synthesis of other proteins, such as those providing elastic properties of cartilage, tendons, and intramuscular connective tissue, result in decreased flexibility or loss of the ability to absorb and disperse energy. Mutation of other genes may result in inactivation of enzymes, which disrupts chemical reactions within cells necessary for maintaining homeostasis. Programmed cell death (**apoptosis**) and limited cell division (cells are limited in the number of times they can divide) are two additional factors that have been proposed to account for the aging process. Most

likely a combination of these processes causes senescence and gradually reduces a person's ability to cope with "stress" or external challenges. In many people, senescence advances until it compromises normal bodily function; independence is lost, and the person must depend on others to aid in the performance of simple activities of daily living.

➤ Mutation means a change in gene structure that results in altered cell constituents.

Primary Versus Secondary Aging

The study of mechanisms underlying the effects of lifelong physical activity on aging is complicated by the difficulty of separating **secondary aging** (the effects of disease, environmental influences, and injury on the decline in functional capacity) from **primary aging** (deterioration of physiological systems in the absence of these confounding factors). Someone who is 70 years of age chronologically may be relatively free of disease and physically active and have a "biologic age" closer to 50 years (figure 16.1). Conversely, a

Figure 16.1   Multi-generational biking. If a person exercises wisely and consistently, his biological age will tend to lag significantly behind his chronological age. Such a person may be far more healthy than a person in similar circumstances who is 30 years younger, but who fails to exercise.

50-year-old who has a debilitating chronic disease or who is sedentary or obese may have the functional capacity (or biologic age) of one who is 70. Problems inherent in separating primary from secondary aging have plagued investigators in the quest to uncover the aging mechanism.

> The study of mechanisms underlying the effects of lifelong physical activity on aging is complicated by the difficulty of separating secondary aging from primary aging.

Problems in Collecting Data

The effect of physical activity on the aging process has also been clouded by the difficulty of collecting data over the entire life span. Most studies have been cross-sectional in design, and this may bias results. For example, a large physically active group of 20-year-olds contains people genetically predetermined to have shorter lives.

The pool of people at 70 years of age is smaller and is genetically biased. Thus, it is difficult to determine the effect of lifelong physical activity on longevity using a cross section of 20-year-olds compared with a sample group of 70-year-olds. A better way to answer the question would be to perform a longitudinal study, following people throughout their lifetimes. The limitations of this type of study, however, are obvious because it is difficult to obtain data from a single individual over a long time span. Even if a longitudinal study were possible, the effects of dietary factors, environmental factors, and debilitating injury may affect longevity. Undetected diseases such as hypertension or coronary artery disease may also affect compliance or response to an exercise program.

➤ A cross-sectional study is an examination of a population at one particular time. A longitudinal study is observation of a group of people over an extended period, usually years.

Problems in Interpreting Data

Finally, misinterpretation (or misrepresentation) of available literature contributes to the difficulty in judging the benefits of exercise in delaying the aging process and/or prolonging life. For example, the decline in $\dot{V}O_2$max is widely used as a measure of decline in functional capacity related to senescence. In an attempt to calculate the average decline in $\dot{V}O_2$max with age, three different studies examined the decrease in $\dot{V}O_2$max in male subjects over time (figure 16.2) (taken from a study by Buskirk and Hodgson). One reported an average decline of 1.06 ml/kg/min per year, the second a decline of 0.45 ml/kg/min per year, and the third a rate of decline of zero! If you and two classmates had each read a different one of these studies, you would have widely different views on the decline (or lack of decline) in functional capacity with aging. Let's examine these studies more closely.

• **Study 1.** If the results of study 1 were accurate and could be applied to the general population, then an approximate decrease of 1 ml/kg/min per year, beginning at age 30, would leave an average male who had a $\dot{V}O_2$max of 45 ml/kg/min with a value of zero by the age of 75!—clearly a functional deficit that would be debilitating. A closer examination of the study shows that at initial testing these men were well-trained elite athletes; then when tested 27 years later, all either were not training or had dramatically reduced training—a confounding factor that obviously affected the results.

➤ A confounding factor is a factor that, if not separated from the measured variable, influences the outcome of the study.

• **Study 3.** Study 3 showed a reduction of zero in $\dot{V}O_2$max in physically active males over 10 years, strongly suggesting that physical activity prevents the age-related decline in cardiopulmonary function. This is hard to believe, in particular because the decline in maximum heart rate (HRmax) is about 5% per decade. You would expect the decline in $\dot{V}O_2$max to be at least that much. A closer look at the study shows that half of the group of exercisers had an increase in $\dot{V}O_2$max over the 10 years of the investigation, indicating an increase in activity levels. In addition, average body weight in the group decreased, which by itself affects $\dot{V}O_2$max when expressed relative to body mass (you increase your $\dot{V}O_2$max just by losing weight). Again, it is obvious that this study misrepresents the real rate of decline in $\dot{V}O_2$max with age.

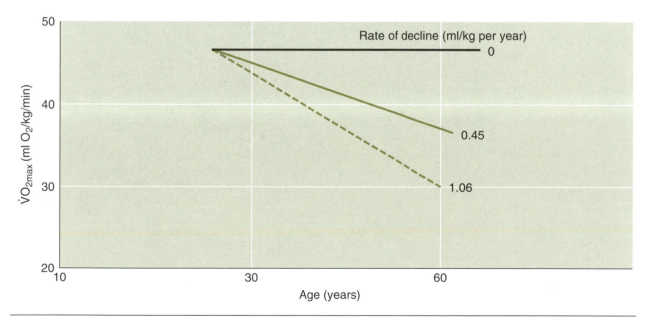

Figure 16.2   Results of three different studies that examined age-related rate of decline of $\dot{V}O_2$max. This example illustrates the importance of the ability to critically examine published articles (see text for explanation).

• **Study 2.** As it turns out, the generally accepted, broadly applied rate of decline in $\dot{V}O_2max$ is 0.45 ml/kg/min in males (found in study 2) and 0.35 ml/kg/min in females, representing an approximate 10% decline per decade beginning at around age 30. This study eliminated confounding variables and controlled or accounted for changes in activity levels and in body mass or composition.

It is clear that only one of these three articles would give you an accurate view of decline in cardiopulmonary function with aging. The critical analysis we just performed accentuates the problems inherent in correct interpretation of available information in the study of the aging process. (The danger of misinterpretation applies not only to this topic: it is universal in the scientific literature. You should not simply accept everything at face value, but must be able to read and analyze published articles critically.)

Getting back to our average decline in $\dot{V}O_2max$ with aging, we suggest that instead of an average decline that applies to everyone, realistically a family of curves affected by disease, injury, genetic endowment, diet, lifestyle, changes in body composition, level of physical activity, and environmental influences would better represent the decline in the general population (figure 16.3). Moreover, in any given person, the rate of decline is probably not linear, in relation to alterations at different stages of life, for any of the factors listed just mentioned (figure 16.3). Note that well-trained people with a high initial $\dot{V}O_2max$ have a greater rate of decline if they cease or reduce training than those who never trained in the first place. Those who maintain a high to moderate level of training into their later years and remain relatively disease-free tend to have the slowest rate of decline.

> Well-trained people with a high initial $\dot{V}O_2max$ have a greater rate of decline if they cease or reduce training than those who never trained in the first place.

Decline in Functional Capacity

The decline in $\dot{V}O_2max$ discussed in the previous section is an example of age-related decline in **functional reserve capacity** (maximum minus basal function) of the cardiovascular system. Deterioration of other systems, such as the neu-

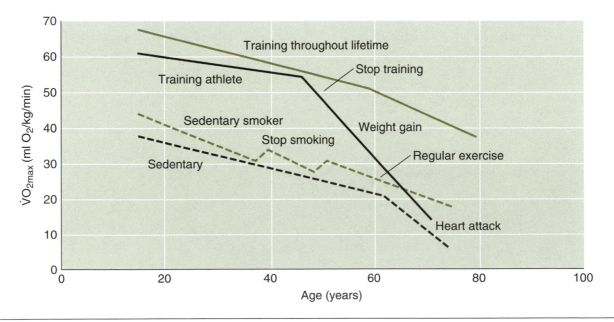

Figure 16.3 Age-related decline in $\dot{V}O_2max$. The age-related decline in $\dot{V}O_2max$ is a good indicator of decrease in functional reserve of the cardiopulmonary system. Rather than a single linear curve, it is likely that a family of curves, affected by changes in lifestyle and various pathologies throughout the life span, better represents decline in function. A person who remains relatively healthy and physically active has the slowest rate of decline of $\dot{V}O_2max$.

ral, musculoskeletal, metabolic, and pulmonary systems, also has a profound effect—initially on exercise performance and ability to participate in recreational activities or sport, but eventually on the performance of normal daily activities. Overwhelming evidence exists that intervention with aerobic and resistance-training programs, combined with sound nutrition, can preserve functional capacity, improve quality of life, and maximize independence of older adults.

The physiological effects of aging are summarized in table 16.1.

Overwhelming evidence exists that intervention with aerobic and resistance-training programs, combined with sound nutrition, can preserve functional capacity, improve quality of life, and maximize independence in older adults.

Cardiovascular

The benefits of physical activity in preventing the cardiovascular disease often associated with aging are clear to health care providers, but the majority of the population has not embraced this

Table 16.1 Physiological Effects of Aging

Physiological change	Deleterious effect
Cardiovascular system	
Decrease in muscle mass and heart volume	Decreased stroke volume and cardiac output
Decreased elasticity of blood vessels	Increased TPR, afterload
Decreased sympathetic nervous system stimulation of sinoatrial node	Decreased HR max
Pulmonary system	
Decreased elasticity	Increased work of breathing
Decreased number of pulmonary capillaries	Decreased ventilation/perfusion
Muscles and joints	
Decreased muscle mass	Loss of strength
Selective loss of innervation (type IIB)	Loss of strength/power
Increased unsteadiness during dynamic contractions	Higher incidence of falling
Decreased size and number of mitochondria	Decreased oxidative capacity
Decreased activity of oxidative enzymes	Decreased oxidative capacity
Increased muscle stiffness	Decreased flexibility
Increased joint stiffness	Decreased mobility and stability
Skeleton	
Loss of bone mass (osteoporosis)	Decreased tensile strength

practice as an integral part of their lifestyle. Whether longevity can be affected by physical activity per se is controversial, but certainly physical work capacity can be enhanced by regular exercise.

Cardiac

Body systems vary in their rate of decline in functional capacity, but clearly, the decline in heart function is one of the major factors contributing to decreased physical work capacity. Consider the decline in HRmax—about 5% per decade after age 30. Maximum heart rate can be estimated by 220 – age, so a 20-year-old has an approximate HRmax of 200, whereas a 60-year-old has a value of 160.

A Question on Heart Rate and Aging

Can you think of the mechanism that is responsible for the decline in maximum heart rate with aging? How would this decline in maximum heart rate affect $\dot{V}O_2$max? To check your answer, see page 264.

Even though resting heart rate is not different between older and younger individuals (and may even decrease with age), the heart rate reserve (maximum – resting heart rate) of the 60-year-old is much less than that of the youngster.

Two prevalent cardiovascular diseases, **hypertension (HTN)** and **coronary artery disease**, are associated with a decline in cardiovascular function as we age. As many as 50% of our population over 65 are hypertensive, and about 60% have coronary artery stenosis greater than 75%. Advances in cardiac imaging technology have enabled earlier detection of CAD, and more and more people are becoming aware of the need for early detection of HTN; but difficulties persist in identifying those who have either or both diseases. The result is confusion in separating the true effects that aging has on the decline in cardiac function from the effects attributable to these two diseases.

Figure 16.4 illustrates the differences in age-related decline in resting and maximum cardiac output between people who were free from HTN and CAD (from the Baltimore Longitudinal Study as reported by Hagberg) and those who were not

screened for the disease. Two differences between the groups are evident: (a) maximum cardiac output declined at a greater rate in the unscreened group than in the screened group; and (b) whereas resting cardiac output declined in the unscreened people, there was no statistically significant change in the screened group. Although not mentioned in this particular study, it is possible that the two groups had lifestyle differences (physical activity, diet, etc.) that may have affected the differences in rate of decline of cardiac output.

> As many as 50% of our population over 65 are hypertensive, and about 60% have coronary artery stenosis greater than 75%.

Interestingly, the rate of decline in maximum cardiac output in the screened group was about 5% per decade—the same as the decline in HRmax. This indicates that maximum stroke volume was unchanged, suggesting that heart function per se was not compromised in these people. Indeed, according to Lakkata, left ventricular (LV) end-diastolic volume (EDV, adjusted to body mass) is not reduced in healthy older men and women, either at rest or during exercise. This observation is a bit surprising given that the ventricular wall becomes stiffer with age. How then can EDV not become smaller? The answer lies in an amazing adaptation with aging that compensates for the increased ventricular stiffness—an increase in the size, and therefore filling, of the left atrium. As reported by Lakkata, a slight hypertrophy of the LV observed in older people also helps preserve stroke volume by maintaining contractility. These results show, then, that cardiac function can largely be preserved in healthy people who are free of HTN and CAD and who are physically active.

Of course, all bets are off in people who have HTN. If the condition is untreated, the constantly high LV afterload produces large changes in LV mass, increased stiffness, and infiltration of connective tissue and fat. This produces a very large, stiff LV with a very small chamber. End-diastolic volume is drastically reduced and contractile function is compromised, limiting stroke volume and reducing cardiac output. Cardiac function is also compromised by CAD, which limits O_2 and substrate supply to the heart muscle. This leads us to conclude that as the incidence of HTN and CAD increases with aging, cardiac output is

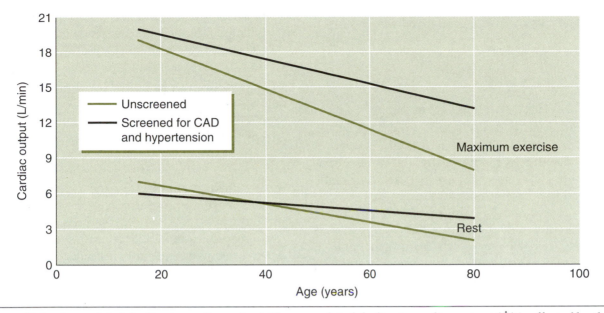

Figure 16.4   Age-related decline in cardiac output. The age-related decline in cardiac output ($\dot{Q}$) is affected by the two most prominent cardiovascular diseases, hypertension and coronary artery disease (CAD). People free of hypertension and CAD have no significant decrease in resting $\dot{Q}$ and about a 5% per decade decline in maximum $\dot{Q}$. Subjects not screened for these diseases have a noticeable decrease in resting $\dot{Q}$ and about twice the rate of decline in maximum $\dot{Q}$. Decreased maximum $\dot{Q}$ in the unscreened group is attributable to increased stiffness and decreased contractile function of heart.
Adapted from Gersenblith et al. 1987.

affected more than the 5% per decade attributable to the decrease in HRmax. Also, remember that cardiovascular function, and thus physical work capacity, are also affected indirectly by other diseases common in persons who are elderly (type II diabetes and osteoarthritis, for example), which may cause the person to decrease physical activity.

Cardiac function can largely be preserved in healthy people who are free of HTN and CAD and who are physically active.

As the incidence of HTN and CAD increases with aging, cardiac output is affected more than the 5% per decade attributable to the decrease in HRmax.

Vascular

Progressive increases in arterial wall thickness, as well as a loss of blood vessel elasticity with aging, results in increased peripheral resistance. Peripheral resistance is also affected by increased vascular smooth muscle tone. A higher total peripheral resistance (TPR) increases LV afterload, causing increased work of the heart and hypertrophy, a phenomenon that is a result of aging per se. People with essential HTN suffer detrimental effects to the heart over and above these changes. The loss of arterial elasticity also reduces propulsion of the blood through the vascular tree by elastic recoil, which may affect blood supply during exercise. Peripheral blood vessels are less responsive to vasodilators, which further compromises blood supply to the working muscles. The ability to vasoconstrict in the periphery is also reduced, affecting the redistribution of blood during exercise. Recall that the ideal situation during aerobic exercise is to reduce blood flow to all nonessential organs so that working muscles can be maximally supplied. Taken together, these alterations in the vascular system with aging compromise the ability to supply working muscle with O_2 and substrates, thus reducing physical work capacity and $\dot{V}O_2$max. Also affecting $\dot{V}O_2$max are the reductions in muscle mass and capillary density (which affects delivery of O_2), as well as in oxidative capacity (which reduces the ability to extract and utilize O_2).

Progressive increases in arterial wall thickness, as well as a loss of blood vessel elasticity with aging, result in increased peripheral resistance.

Question on Heart Rate and Aging Answered!

Heart rate is determined by the sinoatrial node and is regulated by the parasympathetic nervous system, which decreases rate, and the sympathetic nervous system, which increases rate. Maximum heart rate is under the control of the sympathetic nervous system; and as we age, this input declines—about 5% per decade. This is one consequence of aging that unfortunately is not affected by exercise training, so we're stuck with this limitation as we age. How would this affect $\dot{V}O_2$max? Because $\dot{V}O_2$max is a function of maximum cardiac output (delivery) and arterial-venous O_2 difference (extraction), you would expect $\dot{V}O_2$max to decline at least by 5% per decade unless you had a 5% increase in stroke volume (cardiac output = heart rate × stroke volume).

Skeletal Muscle

One of the most profound functional deficits with aging, especially after age 70, is the progressive decline in muscle strength and endurance. Coupled with the loss of muscle function, an age-associated increase in stiffness of intramuscular connective tissue (related to increased formation of collagen cross-bridges) affects range of motion and contributes to a gradual decline in frequency and intensity of daily physical activities. Loss of muscle function is often exacerbated by deterioration of joint structures, which reduces mobility and may be accompanied by pain or discomfort, further contributing to hypokinesis. Three of the most significant age-related changes that affect the ability to function are the decline in muscle strength, change in fiber-type distribution, and an increasing unsteadiness in eccentric muscle contractions.

➤ Hypokinesis is diminished or slow movement.

Three of the most significant age-related changes that affect the ability to function are the decline

in muscle strength, change in fiber-type distribution, and an increasing unsteadiness in eccentric muscle contractions.

• **Muscle strength.** The decline in muscle strength is attributable to both a decrease in mass and a decrease in intrinsic contractile capability. According to Frontera et al., the average annual decline in strength is estimated to be 1.5% to 2.5% after age 65 and is greater in muscles of the lower extremities compared with upper extremities. This latter observation has implications for strength-training programs for older adults. Resistance-training programs should contain some exercises specific to lower-extremity muscle groups to prevent decline in, or improve performance of, simple activities such as stair climbing, sit-to-stand, and ambulation (figure 16.5).

• **Fiber-type distribution.** There is also a change in fiber-type distribution, specifically a decrease in percentage of type II fibers. This is likely related to the reorganization of neural innervation. Enoka postulates that loss of fibers is related to motor neuron death and incomplete reinnervation of denervated muscle fibers by surviving neurons. This phenomenon may explain the age-related decline in percentage of type II fibers and increase in percentage of type I. Thus, older people have a decreased number of motor units but have more fibers per unit. More force per motor unit is possible, but the ability to produce small gradations in force is compromised.

• **Increasing unsteadiness.** Graves et al. have demonstrated that compared with young adults, older adults have a diminished capacity to exert steady submaximal forces during dynamic contractions. The differences are greater at lighter loads, and the unsteadiness is more pronounced with eccentric versus concentric contractions. Enoka has noted that the higher degree of unsteadiness in older adults during eccentric contractions may be associated with the higher incidence of falling during stair descent (eccentric contractions) compared with ascent (concentric contractions). This difference in steadiness is attributed to differences in muscle activation between young and old persons and is likely related to the neural reorganization discussed earlier. However, Enoka showed that when older subjects underwent a 12-week strength-training program, the magnitude of unsteadiness during contractions decreased. Accordingly, a resistance-training program for older people that specifically increases eccentric strength in muscle groups such as the quadriceps, hip extensors, and foot plantar flexors is recommended to help diminish the incidence of stair falling. Of course, reduced joint mobility or pain, or other neuromuscular deficits, may negate the effects of strength training alone.

Case Study

Your patient is a 72-year-old sedentary female with muscle weakness and unsteadiness, especially when descending stairs. She has reported one incident of falling while going down stairs. In terms of muscle function, what is the probable cause of your patient's muscle weakness and unsteadiness? What is the remedy for these problems?

The muscle weakness is attributable to a decrease in muscle mass and decline in type II fiber population that accompany aging. You can probably assume, since she is sedentary, that her activity levels were low to start with and declined further as she aged. The unsteadiness is likely due to a difference in muscle activation (fewer motor units), which causes older adults to lose fine gradation in force development. This unsteadiness is more pronounced in eccentric versus concentric contractions—hence the problems descending rather than ascending stairs.

The strength-training program should begin at low levels and increase volume gradually. A good place to start would be assisted bench stepping (up and down) and/or partially unloaded squats (e.g., pulling down on a pulley system to partially support the body weight, then performing a squat). At some point, when the patient has already experienced significant strength gains, you should introduce more vigorous, closely monitored, specific eccentric training of the quadriceps, hip extensor, and foot plantar flexor muscle groups. Keep in mind that older adults are more susceptible to exercise-induced muscle injury and take longer to heal.

Figure 16.5   Older adult doing resistance training. Older adults can significantly prolong their ability to remain physically independent if they engage in properly supervised lower body strength training.

Pulmonary

Lower maximal inspiratory and expiratory pressures in older people relate to a decline in strength of respiratory muscles and deterioration of lung elasticity. The decrease in expiratory pressure, in particular, increases the work of breathing, which may contribute to earlier fatigue during exercise and limit maximum work capacity. Decreased elasticity and stiffening of the chest wall cause a gradual decline in total lung capacity, as well as small increases in both residual volume and functional residual capacity. A change in the shape of alveoli (they become more rounded) reduces the surface area available for gas exchange, and, in combination with a decrease in alveolar capillary density, leads to a reduction in diffusion capacity. This and other alterations in lung architecture or function (or both) produce a mismatch in ventilation-to-perfusion ratio, causing a gradual decline in arterial partial pressure of O_2 (from approximately 95 mm Hg at age 20 to approximately 75 mm Hg at age 80). Many of these age-related changes in pulmonary function are exacerbated by pathologies such as emphysema and other chronic obstructive pulmonary diseases discussed in chapter 12. This decline in lung function is responsible, in part, for the gradual decline in exercise capacity seen with aging. Johnson et al. have suggested that regular exercise may prevent some of this decline, but it is difficult to separate the effects of regular exercise from the fact that subjects who exercise may do so simply because they have greater lung function.

Lower maximal inspiratory and expiratory pressures in older people relate to a decline in strength of respiratory muscles and deterioration of elasticity of the lungs.

Bone

An inevitable consequence of aging, bone loss (or osteoporosis) begins in women as early as age 30 but much later in men. Postmenopausal women are especially vulnerable to accelerated bone loss. The major factor in the loss of bone density is a gradual decline in bone formation as the person ages while the rate of bone resorption remains relatively constant. Because 99% of the body's calcium is contained in bone, it is thought that an imbalance in calcium regulation is a causative factor in osteoporosis. Indeed, the higher rate of age-related bone loss in women versus men, according to Kalu, may relate to a documented lower dietary calcium intake in women and the tendency of women to ingest less calcium with aging.

➤ Resorption is the reabsorption of osseous tissue.

Predicting Osteoporosis

Peak bone mass at maturity is a good predictor of the likelihood of onset of osteoporosis. The higher the peak bone mass, the less the chance of fractures related to osteoporosis later in life. It is interesting to note that, as reported by Kalu, dietary calcium intake and physical activity levels during maturation are related to peak bone mass at maturity.

After maturation, bone density is enhanced by regular physical activity. It is well established that bone undergoes remodeling in response to mechanical stress and becomes more dense. The opposite is also true—bed rest or "unweighting" causes a drastic reduction in bone density. Because people who return to weight-bearing activities after a period of non-weight-bearing experience a return of skeletal muscle mass and strength much sooner than return of bone density and tensile strength, an imbalance between muscle and bone strength may result. It is necessary to take appropriate precautionary measures to avoid bone fracture as a result of this imbalance, especially when one is prescribing resistance exercises.

Bone density is enhanced by regular physical activity.

Body Composition

A progressive increase in body fat, accompanied by a decrease in lean body mass, is commonly observed after age 30; according to Rogers and Evans, the most rapid rate of decline in men is between ages 40 and 60, but in women it is not until after age 60. These changes in body composition account for the observed age-related decreases in basal metabolic rate and in $\dot{V}O_2$max. The tendency for people to become more sedentary with age contributes to the problem of altered body composition. Any intervention, such as a resistance-exercise program that helps preserve muscle mass, or one that increases muscle

Figure 16.6 Putting the shot at the Senior Olympics. If you can instill in your patients a commitment to lifelong activity, they will experience significant gains in both length and quality of life.

mass in an aging, predominantly sedentary population, is beneficial.

A progressive increase in body fat, accompanied by a decrease in lean body mass, is commonly observed after age 30.

Physical Activity and Aging

The aging process cannot be prevented by physical activity, but adopting exercise as a lifelong commitment can enable a person to function at a higher work capacity at each chronological age (figures 16.6 and 16.7). Indeed, many characteristics of the decline in functional capacity with aging resemble changes observed with deconditioning or physical inactivity, undoubtedly related to our tendency to become less and less active as we age. In addition, lifelong physical activity reduces risk for cardiovascular disease, diabetes, cancer, and obesity, helping to prolong life and increase quality of living. As the average life span of our population increases (estimated to be around 90 years by 2040!) and the proportion of the population over age 65 continues to rise, it becomes increasingly important to promote regular physical activity, not just as a therapeutic modality for rehabilitation, but also as a preventive measure that improves both the quality and the "quantity" of life.

Because of the diminished functional reserve capacity of organ systems in older adults, a vigorous training program is inadvisable and may be hazardous. Special attention should be given to the intensity, duration, and frequency of any

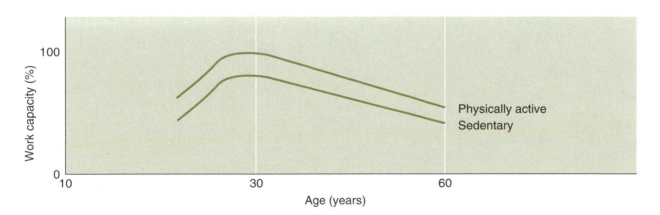

Figure 16.7 Relative work capacity for a person who has adopted physical activity as part of lifestyle and one who has not.

exercise program. The exercise program should begin at very low levels, and overload should increase gradually over a period of months. An initial training level of 40% of $\dot{V}O_2$max may be sufficient to elicit a measurable increase in $\dot{V}O_2$max. Also, recall that patients over 60 are more susceptible to exercise-induced muscle damage, and that repair of damaged muscle takes longer and may be incomplete. The higher risk for muscle damage particularly applies to sedentary or deconditioned older adults. Those who have adopted physical activity as an integral part of their lifestyle are less susceptible to exercise-induced injury.

> Because of the diminished functional reserve capacity of organ systems in older adults, a vigorous training program is inadvisable and may be hazardous.

Skeletal Muscle

Significant strength gains have been observed in older adults after a resistance-training program, but the mechanisms that underlie these strength gains are still controversial. Some investigators found that although increases in strength in older adults were similar to those in younger adults, the increase in mass of the muscle was disproportionately small; these findings suggest that increases in strength are related more to neural mechanisms than to muscle mass. Others have observed similar muscle hypertrophy in young and old subjects in response to resistance training, suggesting that increases in strength are mediated both by neural mechanisms and by an increase in mass. These findings support the benefits of resistance training for preserving or increasing muscle mass in older people, helping to reverse or delay the detrimental effects of age-related changes in body composition discussed previously. Specific resistance-training programs assist the older person in performing simple daily activities like ambulation, stair climbing, and household chores, helping maximize independence.

Resources for Health Professionals Working with Older Adults

Books

- Benyo R. *Running Past 50*. Champaign, IL: Human Kinetics, 1998.
- Chmiel D and Morris K. *Golf Past 50*. Champaign, IL: Human Kinetics, 2000.
- Cotton RT, Ekeroth CJ and Yancy H, eds. *Exercise for Older Adults: ACE's Guide for Fitness Professionals*. Champaign, IL: Human Kinetics, 1998.
- Francina S. *The New Yoga for People Over 50: A Comprehensive Guide for Midlife and Older Beginners*. Deerfield Beach, FL: Health Communications, 1997.
- Friel J. *Cycling Past 50*. Champaign, IL: Human Kinetics, 1998.
- Goldstein M and Tanner D. *Swimming Past 50*. Champaign, IL: Human Kinetics, 1999.
- Nelson ME. *Strong Women Stay Young*, revised edition. Champaign, IL: Human Kinetics, 2000.
- Osness W, ed. *Exercise and the Older Adult*. Dubuque, Iowa: Kendall Hunt Publishing Company, 1998.
- Pu CT and Nelson ME. "Aging, Function, and Exercise." In Frontera WR, Dawsin DA, and Slovik DM, eds. *Exercise in Rehabilitation Medicine*. Champaign, IL: Human Kinetics, 1999.
- Westcott WL and Baechle TR. *Strength Training for Seniors*. Champaign, IL: Human Kinetics, 1999.
- Westcott WL and Baechle TR. *Strength Training Past 50*. Champaign, IL: Human Kinetics, 1998.

Web Sites

- American Senior Fitness Association (SFA) **www.seniorfitness.net**
- Fifty Plus Fitness Association **www.50plus.org**

Journals

- Journal of Aging and Physical Activity. Champaign, IL: Human Kinetics.

Journal Articles

- William EJ. Exercise training guidelines for the elderly. *Med Sci Sports Exerc* 1999, 31:1, 12-17.

Cardiovascular, $\dot{V}O_2$max

Physically active older adults have a lower percent body fat and higher lean body mass than their sedentary counterparts. These differences in body composition alone help preserve $\dot{V}O_2$max (in ml/kg/min) with aging. In older men and women, endurance training elicits peripheral adaptations (increased capillarization, increased oxidative capacity) of approximately the same magnitude as in younger people. These improvements in delivery and extraction of O_2 help increase both physical work capacity and $\dot{V}O_2$max. Central adaptations to endurance training in younger adults include increased LV compliance and mass and increased contractility, which in turn increase stroke volume. These adaptations occur when older adults are trained, but the magnitude of the changes is less than in younger people. Note that the age-related decline in HRmax is not affected by training. Thus, endurance training in older adults produces

- changes in body composition;
- peripheral adaptations that are similar in magnitude to those in younger adults; and
- limited central adaptations.

All these changes preserve cardiovascular functional capacity.

Physically active older adults have a lower percent body fat and higher lean body mass than their sedentary counterparts.

Endurance training in older adults produces changes in body composition, peripheral adaptations that are similar in magnitude to those in younger adults, and limited central adaptations.

Case Study

Your patient is a 68-year-old male with moderate hypertension (HTN), high low-density lipoprotein/high-density lipoprotein (LDL/HDL) ratio (6:1), and CAD, although the heart muscle itself is undamaged (no previous myocardial infarction). His physician has prescribed a drug that lowers LDL and a beta blocker to help reduce blood pressure; she has advised an exercise program to raise HDLs and to improve cardiovascular fitness. The patient comes to you for advice about the exercise training program. Describe the limitations in EDV and end-systolic volume (ESV) in this patient as related to heart function during aerobic exercise. Compare these parameters to those of a healthy, physically active 68-year-old male who is free of HTN and CAD. What type of program would you recommend for this patient, and what adaptations would you expect, both in the heart and in the periphery (skeletal muscle)?

A healthy, physically active 68-year-old has an EDV that is unchanged from young adulthood, primarily due to increased volume of the atria. A slight hypertrophy of the LV attributable to increased peripheral resistance (related to decreased elasticity of blood vessels) helps maintain cardiac contractility, but stroke volume might be somewhat reduced because of the increased afterload (decreased ability to vasodilate blood vessels). So, ESV is likely to be somewhat higher (less ejection) than it would have been at an earlier age. In comparison, your patient has a large increase in afterload (increased stiffness and narrowing of all blood vessels), which has caused a large increase in mass of the LV and subsequent decrease in the size of the LV chamber. The heart muscle has become infiltrated with connective tissue and fat, which decrease the contractility and increase stiffness. Intrinsic changes in muscle fiber contractility also contribute to the decline in force-generating capacity. The small chamber size and increased stiffness lead to a greatly reduced EDV. The reduced contractility limits ejection capabilities, so ESV would be high. The combined low EDV and high ESV lead to a very small stroke volume. You would prescribe a low-level aerobic training program for your patient, with a gradual increase in frequency, intensity, and duration. You would monitor the patient for high diastolic or low systolic blood pressure, chest pain, dizziness, and undue fatigue. The HRmax attainable will be reduced because of the beta blocker, so the target heart rate should be adjusted accordingly.

The patient will improve his endurance through peripheral adaptations (increased blood flow and oxidative capacity), as well as central adaptations. Perhaps the most significant change to expect in this patient is an increase in the compliance of the heart and increased volume of the LV. This type of

training may also reduce the mass of the LV, which will allow greater filling. This will increase EDV and enhance cardiac output. The patient will increase his physical work capacity and also his sense of well-being.

What You Need to Know From Chapter 16

Key Terms

apoptosis

coronary artery disease (CAD)

functional reserve capacity

hypertension (HTN)

osteoporosis

primary aging

secondary aging

senescence

Key Concepts

1. You should be able to outline the problems separating primary from secondary aging and their effects on decline in functional capacity.

2. You should be able to describe the decline in functional capacity associated with the major body systems.

3. You should be able to describe the benefits of regular physical activity for improving the quality of life in older adults.

Review Questions

1. Compare and contrast the merits of cross-sectional versus longitudinal study designs for examining the effect of physical activity on aging.

2. Describe how various confounding variables (e.g., diet, environment, disease) would affect the outcome of longitudinal studies that examine the decline in functional capacity with aging.

3. Describe the age-related alterations in pulmonary function that limit work capacity.

4. Why are aging women more susceptible to osteoporosis than men?

5. What are the effects of bed rest and of weight-bearing exercise on bone density?

6. How do gradual changes in body composition that are observed with aging affect resting and maximum O_2 uptake?

7. Your 62-year-old patient has been recently diagnosed with borderline hypertension, and her physician has advised an exercise and weight-reduction program. You are examining her 12-lead electrocardiogram and comparing it with her records from 20 years ago. You notice that in the current recording, the R wave in the V6 lead has higher amplitude than that in V5, which is the opposite of what the older recording shows. What is the probable explanation?

Bibliography

Beissner KL, Collins JE, and Holmes H. Muscle force and range of motion as predictors of function in older adults. *Phys Ther* 2000, 80:556–563.

Buskirk ER and Hodgson JL. Age and aerobic power: The rate of change in men and women. *Fed Proc* 1987, 46:1824–1829.

Ehsani AA. Cardiovascular adaptations to exercise training in the elderly. *Fed Proc* 1987, 46:1840–1843.

Enoka RM. Neural Strategies in the control of muscle force. *Muscle Nerve* 1997, (suppl 5):S66–S69.

Frontera WR, Hughes VA, Fielding RA, Fiatarone MA, Evans WJ, and Roubenoff R. Aging of skeletal muscle: A 12-yr longitudinal study. *J Appl Physiol* 2000, 88:1321–1326.

Gerstenblith G, Renlund DG, and Lakatta EG. Cardiovascular response to exercise in younger and older men. *Fed Proc* 1987, 46:1834–1839.

Gibson GJ, Pride NB, O'Cain C, and Quagliato R. Sex and age differences in pulmonary mechanics in normal nonsmoking subjects. *J Appl Physiol* 1976, 41:20–25.

Graves AE, Kornatz KW, and Enoka RM. Older adults use a unique strategy to lift inertial loads with elbow flexor muscles. *J Neurophysiol* 2000, 83:2030–2039.

Hagberg JM. Effect of training on the decline of VO$_2$max with aging. *Fed Proc* 1987, 46:1830–1833.

Johnson BD, Reddan WG, Seow KC, and Dempsey J. Mechanical constraints on exercise hyperpnea in a fit aging population. *Am Rev Respir Dis* 1991, 143:968–977.

Kalu DB. Bone. In *Handbook of physiology. Section 11: Aging,* ed. EJ Masoro, pp. 395–412. New York: Oxford University Press, 1995.

Lakatta EG. Cardiovascular system. In *Handbook of physiology. Section 11: Aging,* ed. EJ Masoro, pp. 413–474. New York: Oxford University Press, 1995.

Lewis CB. Musculoskeletal changes with age: Clinical implications. In *Aging: The health care challenge,* 3rd ed., pp. 174–186. Philadelphia: Davis, 1996.

Rogers MA and Evans WJ. Changes in skeletal muscle with aging: Effects of exercise training. *Exerc and Sport Sci Rev* 1993, 21:65–102.

Sparrow D and Weiss ST. Respiratory system. In *Handbook of physiology. Section 11: Aging,* ed. EJ Masoro, pp. 474–484. New York: Oxford University Press, 1995.

Startzell JK, Owens DA, Mulfinger LM, and Cavanagh PR. Stair negotiation in older people: A review. *J Am Geriatr Soc* 2000, 48:567–580.

Zadai CC and Irwin SC. Cardiopulmonary rehabilitation of the geriatric patient. In *Aging: The health care challenge,* 3rd ed., pp. 196–226. Philadelphia: Davis, 1996.

Exercise in Various Environments

This chapter will discuss

➤ regulation of body temperature,

➤ compensations for decreased ambient O_2 pressure, and

➤ responses to exercise in the heat and at altitude.

To this point we have examined exercise responses and adaptations in physical environments that were neutral and did not place unusual stresses on the physiological systems. Physical activity performed in unfavorable environments, such as those in high heat and humidity or at moderate to high altitude, requires special preparation and knowledge of strategies for avoiding environmentally related illness.

Exercise in the Heat

Physiological stresses and adaptations associated with exercise in the heat are much more commonly experienced than those associated with cold environments. Thus, although we will note various facts regarding the body's response to cold, we will focus on how the body deals with the challenges it encounters during exercise in the heat.

Regulation of Body Temperature

Humans use a series of feedback mechanisms to maintain a core body temperature (at or near 37° C). Why do we need to regulate body temperature? We can turn to evolution for the answer.

Over the millennia, biochemical reactions in humans—which underlie all physiological processes—have evolved to function optimally at or near our regulated body temperature. This adaptation allows us to remain active and to maintain optimal conditions internally over a wide range of environmental temperatures. The efficiency with which we maintain a relatively stable core temperature is illustrated in figure 17.1. In this figure, you can see that humans are able to dissipate body heat at high environmental temperatures and conserve heat when temperatures are low.

Heat–Regulating Mechanisms

In order for life to be sustained, the simple equation heat production = heat loss must, ultimately, be balanced. Various regulatory mechanisms help us maintain a stable **core temperature** (T_c) by working toward preserving this balance between heat production and heat loss. Peripheral and central temperature-sensitive sensors are continually providing feedback about the external and internal environments to integration centers in the hypothalamus and spinal cord. Body temperature varies around a set point that works in much the same way as the setting on a thermostat in your home.

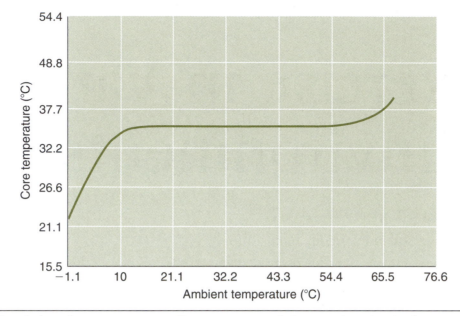

Figure 17.1　Regulation of core temperature. Core temperature in humans is maintained at or near the set point over a wide range of ambient temperatures, demonstrating the effectiveness of heat-conserving and -dissipating mechanisms.
Adapted from Guyton 1991.

When T_c falls, various heat-conserving mechanisms are activated in an effort to retain heat. Conversely, when body temperature increases, heat-dissipating mechanisms are turned on.

Temperature-Regulating Feedback

This schematic illustrates how temperature-regulating feedback mechanisms work.

Sensors　$\longrightarrow$　Integrators　$\longrightarrow$　Effectors
　↓　　　　　　↓　　　　　　↓
Skin, core　　Hypothalamus,　Raise/
　　　　　　spinal cord　　lower T_c

Peripheral and central temperature-sensitive sensors are continually providing feedback about the external and internal environments to integration centers in the hypothalamus and spinal cord.

Sweat glands, skin blood vessels, and skeletal muscle are the major effectors through which heat exchange is accomplished. When the environment is hot, the body responds with strategies that maximize heat dissipation. When cold, the body automatically minimizes heat loss. We will first examine the physical principles of heat dissipation and then turn to regulation of these processes in response to changes in conditions both outside the body (i.e., related to the weather) and inside the body (i.e., related to how hard a person is exercising).

The Physics of Heat Dissipation

Heat is lost from the body through four basic processes:

• **Conduction** occurs through the transfer of heat energy between molecules adjacent to one another. Put your hand on top of your desk. At first, the desktop feels cold; then, as heat is exchanged between your hand and the desk, the temperature differences disappear. Now, think of heat exchange through conduction between exposed skin and the air—not much heat is exchanged between adjacent molecules. Conduction is an ineffective method of losing heat into the air; but when you are submerged in water, which has a higher thermal conductivity, you lose heat much more rapidly. Onset of hypothermia is accelerated when the body is submerged in water even when water temperature is judged to be "comfortable."

• **Convection** occurs when air or water next to the body is heated by conduction and then moved away. Heat loss is a function of the velocity of the movement of air or water. Convection is much more effective in cooling the body than conduction by itself. In an uncomfortably hot working environment, simply adding fans to circulate the air and increase heat dissipation can increase productivity. The **windchill index** combines wind speed with ambient temperature (T_a) to provide a hypothetical temperature at which there is no wind movement. For example, a wind speed of 40 km/hr (25 mph) at a T_a of 30° F would produce a windchill of 0° F.

• **Radiation** is energy in the form of electromagnetic waves. Our bodies and all objects in our environment emit radiation. The rate of heat gain or loss depends on the temperature difference between your body and the average for all the objects in your environment. If your environment is warmer than you, your body will absorb energy in the form of radiant heat. If your environment is cooler than you, you will lose energy in the form of radiant heat. The sun is the greatest radiant heat source, and the amount of heat we absorb depends to a large degree on the area of skin exposed to the sun and the angle at which the sun's rays hit the body. Obviously, lying on the beach in a swimsuit at noontime provides the optimum condition for absorbing radiant heat. When working or exercising in the sun it is advisable to wear light-colored clothing, which absorbs radiant heat less readily than dark clothing.

• **Evaporation** uses heat energy to transform water from a liquid to a gas and is the most efficient form of cooling available to the body. Even in the absence of sweating, evaporation is occurring wherever body surfaces (i.e., the respiratory tract and skin surface) are exposed to the external environment.

Exercise Responses to Heat Stress

Compared with exercise in **thermoneutral** conditions, exercising in a hot environment

- stresses the cardiovascular system,
- produces higher perceived exertion, and
- results in an early onset of fatigue.

The added stress of exercise in the heat comes from the need to supply both the working muscle and the skin with blood. Evaporation of sweat is the major heat-dissipating mechanism activated during exercise in the heat. Profuse sweating increases the potential for dehydration and the onset of heat-related illness such as heat cramps, heat exhaustion, and heatstroke. In a person who is not well conditioned and/or not acclimatized, the dangers of **heat stress** are magnified.

Compared with exercise in thermoneutral conditions, exercising in a hot environment stresses the cardiovascular system, produces higher perceived exertion, and results in an early onset of fatigue.

Profuse sweating increases the potential for dehydration and the onset of heat-related illness such as heat cramps, heat exhaustion, and heatstroke.

Cardiovascular System

Exercise in the heat presents a unique challenge to the body systems, in particular the cardiovascular system. Let's first examine internal production of heat by contracting skeletal muscle.

Internal Heat Production

The efficiency with which skeletal muscle uses energy to produce work is surprisingly low. The highest theoretical efficiency in converting chemical to mechanical energy to contract muscle is 35%. That is, 35% of the energy provided by adenosine triphosphate (ATP) hydrolysis is used to perform work, and 65% is given off as heat. But, because of various constraints imposed on skeletal muscle by the arrangement of bony levers, differences in muscle length, force-velocity relationships, and so on, actual efficiency during most movements is even lower, maybe as low as 5% to 15%. When you consider that skeletal muscle composes approximately 40% of body mass, coupled with the fact that most energy during muscle contraction is given off as heat, the problem of coping with a large internal production of heat during exercise becomes apparent. Of course, as exercise intensity increases, the amount of heat produced increases as well; and unless dissipated, it is stored within the body and results in an increase in T_c. Now, if we add environmental heat stress to this scenario, the problems associated with heat regulation are exacerbated. During heavy exercise in hot environments, core temperatures of 40° C and muscle temperatures of 41° to 42° C are not uncommon. Tissue temperatures this high present a danger because protein breakdown, or denaturation, occurs at approximately 43° C (not a large margin of safety here). Because people are often exercising or working dangerously close to body temperatures at which permanent damage may occur, it is critical to be able to recognize characteristic signs of heat illness, and above all to act quickly to intervene and lower body temperature. Health care professionals may encounter scenarios such as these at athletic competitions or practices, work-site rehabilitation clinics (e.g., in factories), and orthopedic and sports medicine clinics.

Because people are often exercising or working dangerously close to body temperatures at which permanent damage may occur, it is critical to be able to recognize characteristic signs of heat illness, and above all to act quickly to intervene and lower body temperature.

Counteracting Internal Heat Production

Let's examine the impact of changes in skin blood flow on exercise performance. During most normal daily activities, a stable T_c is achieved mainly through increases or decreases in blood flow to the periphery, where the blood can transfer heat to the skin. From the skin, the heat can be lost through conduction, convection, or radiation.

Skin blood flow is regulated by the sympathetic nervous system, which is a powerful vasoconstrictor when T_c declines. Under cold conditions, skin blood vessels constrict to keep the blood in the central portion or core of the body, and heat is conserved. When T_c increases, vasoconstriction of the vessels in the skin is gradually removed, and skin blood flow increases a small amount. The ability to increase blood flow by removing sympathetic nervous system vasoconstriction is limited though, so skin blood vessels must be actively dilated to achieve the large increases in blood flow necessary to dissipate heat. Recall that active vasodilation in skeletal muscle is accomplished by various substances (adenosine, potassium, etc.), released from contracting

muscle fibers, that act directly on arterioles. There are no such substances in the skin, so another mechanism must be in place here. In fact, sympathetic vasodilator nerve fibers innervate skin arterioles and are responsible for most of the increase in blood flow during heat stress. So, the removal of vasoconstriction actually contributes very little to the overall increase in skin blood flow when body temperature begins to rise. Active vasodilation, controlled by sympathetic vasodilator nerve fibers, is largely responsible for the tremendous increase in flow to the skin.

Active vasodilation, controlled by sympathetic vasodilator nerve fibers, is largely responsible for the tremendous increase in flow to the skin.

An easy way to recall how skin blood flow is affected in cool or hot environments is to think of a large "inner shell" and small "outer shell" when blood flow is diverted to the center of the body, and a small inner shell and large outer shell when skin blood flow increases to dissipate heat (figure 17.2). Directing blood to the periphery in an attempt to stabilize T_c has a tremendous effect on exercise performance, largely because of the decrease in central blood volume and a concomitant reduction in ventricular preload.

Directing blood to the periphery in an attempt to stabilize T_c has a tremendous effect on exercise performance, largely because of the decrease in central blood volume and a concomitant reduction in ventricular preload.

Competition for Blood Flow

In chapter 13 we examined the principles of preload and afterload and how they affect the function of the heart. During exercise under conditions of heat stress, blood vessels in contracting skeletal muscle and those in the skin are competing for blood flow, a situation that affects both pre- and afterload. Remember, to perform prolonged exercise efficiently, it is necessary to have a closed loop between the heart and working muscle so that the optimum amounts of O_2 and substrates can be delivered in the blood and metabolic waste products can be removed. Recall that up to 90% of cardiac output in a conditioned athlete exercising at maximum levels (under thermoneutral conditions) is delivered to the con-

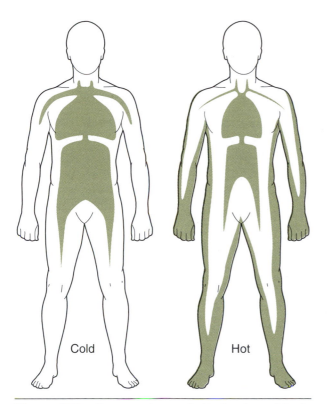

Cold   Hot

Figure 17.2   Conservation and dissipation of body heat. Skin blood vessels vasoconstrict under cold environmental conditions to keep blood in the core of the body and conserve heat. Under hot conditions, skin blood vessels dilate in an attempt to dissipate body heat. This decrease in central blood volume decreases preload, which lowers stroke volume and causes an increased heart rate. At any level of submaximal exercise in the heat, the heart is working at a higher percentage of maximum than under thermoneutral conditions.
Adapted from Rowell 1986.

tracting muscles. Blood vessels in all other nonessential organs are constricted to redistribute blood. It is very rare, though, that this ideal condition exists; so during most exercise bouts, we have competing responses—**thermoregulation** and the need to supply working muscle. Figure 17.3 illustrates the different responses of skin blood flow at rest and during exercise when T_c is increasing. Notice that resting skin blood flow is approximately 7 ml/100 g/min at a T_c of 37° C, but much lower at the same temperature during exercise. This means that when heat regulation is not a problem during exercise, skin blood vessels are constricted and flow is very low. As exercise progresses, skin blood flow rises at a very slow rate, reflecting the competition for flow between the skin blood vessels and the vessels in contract-

ing muscle. However, when body temperature increases under resting conditions, skin blood flow rises more rapidly, resulting in a more effective heat transfer from the core to the periphery. Thus, at the beginning of exercise, when no thermal stress is present, skin blood flow initially decreases. As exercise progresses and T_c begins to rise, skin blood flow increases, but at a much slower rate than at rest. Cutaneous circulation can accommodate up to 30% of the total cardiac output, so any significant increase in skin blood flow reduces central blood volume, which in turn has an adverse effect on stroke volume.

> Cutaneous circulation can accommodate up to 30% of the total cardiac output, so any significant increase in skin blood flow reduces central blood volume, which in turn has an adverse effect on stroke volume.

Remember that we are counting on blood delivery of O_2 and energy substrates to muscle to perform prolonged exercise. Look at figure 17.2 again and recall that any decrease in central blood volume

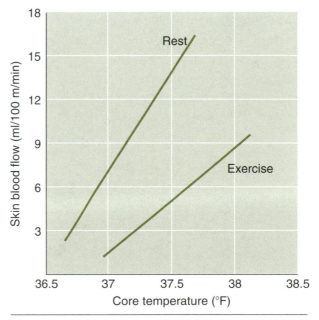

Figure 17.3   Changes in skin blood flow: skin blood flow under resting and exercise conditions as core temperature (T_c) increases. Skin blood flow at normal T_c (around 37° C) is approximately 7 ml/kg/min and falls when exercise begins. Also, skin blood flow rises at a slower rate during exercise as compared to resting conditions.
Adapted from Rowell 1986.

lowers ventricular filling, resulting in a smaller stroke volume. Heart rate has to increase to compensate for the lower stroke volume, causing the exerciser to work at a higher percentage of his or her heart rate reserve. A further complication occurs if the person is beginning the exercise in a hot environment and the body is already thermoregulating. In the sequence of events in this scenario,

1. stroke volume is likely to be low already;
2. exercise is added, causing more of a heat load through internal heat production;
3. more blood is diverted to the periphery, further decreasing stroke volume; and
4. the cardiovascular system may be forced to work at or near its limits at submaximal O_2 consumption.

If exercise continues and heat stress increases, more and more blood is sent to the skin in an attempt to dissipate heat. As a result, peripheral resistance is lowered and blood pressure drops. This exerciser may experience dizziness and/or disorientation—hallmarks of thermal imbalance.

Stress of the cardiovascular system is increased if a person begins to exercise in a hot environment and their body is already thermoregulating.

The reverse is true during exercise in a cool environment. Internal heat production is easily handled, and central blood volume is maintained at a higher level than during exercise in hot conditions. Ventricular preload is higher, leading to a large stroke volume. Heart rate is lower, the exerciser has plenty of heart rate reserve left, and perceived exertion is much lower. Obviously prolonged exercise in a cool environment is much less "stressful." Prolonged steady-state exercise under thermoneutral environmental conditions does, however, produce a phenomenon called cardiovascular drift.

Cardiovascular Drift

Cardiovascular drift occurs in the absence of an external heat stress and is characterized by a slow decline in stroke volume, which is mirrored by an equivalent increase in heart rate to maintain cardiac output (figure 17.4). Internal heat production from contracting muscles causes a slow, steady rise in T_c, accompanied by an increase in skin blood flow. Diversion of blood to the periphery causes a decline in left ventricle filling pressure (pulmonary arterial pressure), which is responsible for the decreased stroke volume. Other characteristics of cardiovascular drift include a decline in systemic arterial pressure and an increase in $\dot{V}O_2$. The decrease in arterial blood pres-

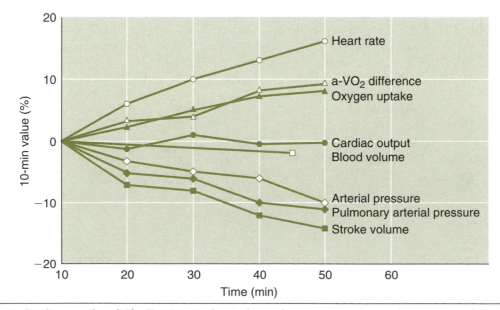

Figure 17.4 Cardiovascular drift. During prolonged steady-state exercise in thermoneutral conditions, core temperature rises slowly, resulting in elevated skin blood flow. Central venous pressure begins to drop; this lowers ventricular preload, which causes a lower stroke volume. Heart rate has to increase to maintain cardiac output.
Adapted from Rowell 1986.

sure is expected and easily explained, but the increase in $\dot{V}O_2$ is not so easily understood. What causes this increase in $\dot{V}O_2$? Remember, the exercise is steady state, so factors other than increased demand by contracting skeletal muscle must account for the increased $\dot{V}O_2$. O_2 uptake is a function of delivery (cardiac output) and extraction (a-v O_2 difference [arterial-to-venous difference]). Examine figure 17.4 and see if you can determine which of the two is increased, or whether both are. It is obvious that the approximate 8% increase in $\dot{V}O_2$ after 50 min of prolonged steady-state exercise is entirely accounted for by increased extraction. Does this, then, answer our initial question, how to account for the increased $\dot{V}O_2$? Only partially: this response is more complicated than it first appears.

Cardiovascular drift occurs in the absence of an external heat stress and is characterized by a slow decline in stroke volume, which is mirrored by an equivalent increase in heart rate to maintain cardiac output.

Let's look at a seeming paradox. Skin blood flow is obviously increasing as exercise progresses, while the O_2 demands for the contracting skeletal muscle are unchanged during this steady-state exercise but cardiac output remains the same. If skin blood flow is higher and cardiac output is unchanged, the implication is that muscle blood flow is lower. But blood flow likely remains unchanged, because blood is diverted from other vascular beds—splanchnic, renal, and so on. Recall from chapter 13 that blood flow redistribution during exercise is very precisely regulated. But our question is still not answered. What are the possible sources of increased $\dot{V}O_2$? We can suggest four possibilities:

1. Heart rate is increased, which increases myocardial O_2 consumption.

2. Increased temperature is a stimulus for increased ventilation, which increases activity of respiratory muscles.

3. Increased tissue temperature also speeds up cellular processes that use ATP (ATP must be replenished by oxidative or non-oxidative sources).

4. Lastly, a phenomenon called uncoupling of oxidative phosphorylation, which accompanies increased tissue temperature, may be contribut-

ing to the increased $\dot{V}O_2$. Basically, the production of ATP in the electron transport chain is uncoupled at high temperatures, and less than three ATPs are produced for each O_2 molecule consumed. Thus, more O_2 is consumed to produce the same amount of ATP.

Whether all or some of these factors contribute to the increased $\dot{V}O_2$ that accompanies cardiovascular drift is unknown. At the very least, observations such as these stimulate our thinking about the interplay among different body responses during exercise and hopefully contribute to our understanding of these responses. It is important to remember that even during exercise in thermoneutral environments, internal heat production may reduce the efficiency of exercise by increasing the energy cost.

Even during exercise in thermoneutral environments, internal heat production may reduce the efficiency of exercise by increasing the energy cost.

As mentioned previously, temperature regulation during exercise is largely a function of changes in skin blood flow and evaporation of sweat. Let's turn our attention now to our last line of defense—the sweating mechanism.

The Sweating Mechanism

During exercise in hot environments, or during moderate to intense exercise in thermoneutral conditions, conduction, convection, and radiation are inadequate to balance heat loss with heat production. Evaporation of sweat must also occur to prevent rapid and dangerous increases in T_c. When stimulated by sympathetic nerve fibers, sweat glands (figure 17.5) produce a "precursor" secretion in the coiled portion of the gland located in the dermis. Sweat contains varying amounts of sodium chloride, potassium, urea, and lactic acid. The duct portion of the gland transports the sweat to the surface of the skin; and when the sweat moves slowly, most of the sodium chloride and water have time to be reabsorbed. But when the glands are strongly stimulated, sweat is transported rapidly to the surface and not much sodium chloride is reabsorbed, so it forms a large proportion of the sweat. Thus, a person who is sweating profusely, especially some-

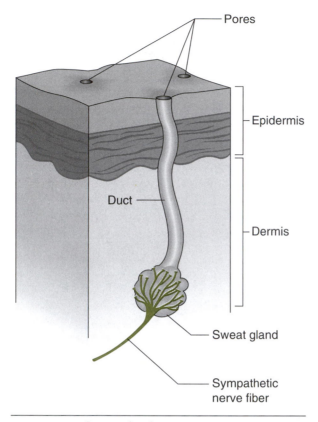

Pores

Epidermis

Duct

Dermis

Sweat gland

Sympathetic
nerve fiber

Figure 17.5   Sweat gland.
Adapted from Wilmore and Costill 1999.

one who is **unacclimatized**, loses a lot of water and sodium, and replacement of both becomes a priority. A person who is heat **acclimatized** sweats more than the unacclimatized person, but the sweat contains less sodium chloride. One of the major benefits of acclimatization, then, is the ability to cool the body more effectively by sweating more while retaining the important electrolyte, sodium.

One of the major benefits of acclimatization is the ability to cool the body more effectively by sweating more while retaining the important electrolyte, sodium.

The total amount of sweat evaporated depends on the surface area of the body exposed, wind speed, and, most importantly, the relative humidity (RH) in the surrounding environment. Relative humidity is the measure of water vapor pressure in the air and is affected by T_a—warmer air holds more water vapor. As RH increases, less and less sweat is evaporated: remember, the sweat needs to evaporate to cool the body. At very high RHs, sweat does not evaporate at all but merely

rolls off the body. If our last line of defense is ineffective in cooling the body in a hot, humid environment, physical activity performed under such conditions can become very dangerous, very quickly. Even when T_a is relatively low but RH is high, physical activity may result in dangerous levels of dehydration and in heat-related illness.

When one is planning or participating in an outdoor activity, a good practice is to figure out the value of T_a and RH added together. If the value is below 140 (approximately 90 when you are using degrees centigrade), heat stress is low, and activity can be considered safe. Values between 140 and 160 (approximately 105 with degrees centigrade) produce heat-stress conditions that are considered moderate to high, and preventive measures against dehydration and overheating should be undertaken. Vigorous long-term activity should be discouraged, and participants should hydrate at regular intervals. If the combined value of T_a and RH is greater than 160, heat stress is high to extreme, and one should consider canceling or changing the time of day of the activity, especially if children are involved. If the activity proceeds as planned, participants should take frequent breaks, rest in shaded areas, drink fluids at regular intervals, and be monitored closely for signs of heat-related illness.

The total amount of sweat evaporated depends on the surface area of the body exposed, wind speed, and, most importantly, the relative humidity in the surrounding environment.

Heat-Related Illness

Dangers associated with exercising in the heat are dehydration, heat cramps, heat exhaustion, and heatstroke. Every year, several deaths from heatstroke are reported among high school and college athletes. The most widely publicized incidents recently were the deaths of three college wrestlers—healthy athletes who were trying to "make weight" by performing intense, prolonged exercise while wearing rubber suits. In this situation the person sweats profusely, but the sweat does not cool the body because it does not evaporate. These wrestlers became severely dehydrated and overheated. Under these conditions the temperature-regulating mechanisms of the body begin to fail, and unless body temperature is reduced immediately, body organs begin to cease functioning.

Heat exhaustion and heatstroke are not limited to the sport of wrestling. Any activity that involves prolonged exposure to heat and humidity—especially those that require heavy equipment or wearing of helmets (e.g., football, lacrosse)—puts participants at risk. Let's examine very briefly some of the characteristics of heat-related illness.

• **Dehydration.** Profuse sweating in a heat-acclimatized person can result in the loss of body water at a rate up to 2 to 3 L/hr (4-6 pints/hr). Non-acclimatized people can produce about half that amount. If this water is not replaced, the person can become severely dehydrated in a short time. Dehydration during physical activity results in reduced production of sweat and a higher T_c. In addition, blood volume and cardiac output decrease, leading to a reduction in O_2-delivery capabilities. A complicating factor in the induction of heat illness is that the thirst mechanism lags behind the loss of water, so we cannot rely on thirst to trigger intake of fluids. This is especially true in children, who need to be instructed to drink fluids at regular intervals and who do not drink enough even when thirsty. Also, a child's T_c will rise at a greater rate than an adult's at any given level of hypohydration.

➤ Hypohydration means low levels of body water.

The thirst mechanism lags behind the loss of water, so we cannot rely on thirst to trigger intake of fluids.

• **Heat cramps.** Profuse sweating during exercise is often accompanied by cramping in the active skeletal muscles. This involuntary cramping of muscle has been attributed to an imbalance of electrolytes, in particular potassium and sodium, but the exact cause is unclear. An imbalance in calcium homeostasis, perhaps an impaired sarcoplasmic calcium uptake mechanism, may be a contributor. In past years, salt tablets were given to athletes prone to heat cramps, the theory being that cramping was due to excessive loss of sodium chloride in the sweat. Can you determine why this is inadvisable (see case study on page 282)?

• **Heat exhaustion.** Heat exhaustion occurs during exercise in the heat when a person becomes dehydrated and blood flow cannot meet the combined demands of the working muscles and increased cutaneous circulation. Under this condition, the rate of heat loss from the body does not keep up with heat gain. The symptoms are profuse sweating, rapid heart rate, and hypotension. The person is light-headed or dizzy and may become disoriented. Activity should be stopped, and the person should lie in a cool place and replace fluids with an electrolyte drink.

• **Heatstroke.** A major medical emergency, heatstroke is second behind head injuries among exercise- or sport-related fatalities. Heatstroke is the complete failure of the heat-regulating control centers in the brain, which causes body temperature to rise uncontrollably. Symptoms include hot, dry, red skin; extreme confusion or disorientation; and possibly convulsions and unconsciousness. Core temperature must be lowered through immediate intervention with ice packs on the groin, axillae, and neck or with immersion in slightly warm water. Because protein is breaking down and cell membrane permeability is altered, large amounts of protein circulating in the blood may cause permanent kidney or liver damage. Those who survive heatstroke are more susceptible than others to future heat-related illness, probably because of malfunction of sweat glands leading to a reduced ability to cool the body through sweating.

These heat illnesses are most commonly observed on athletic playing fields, especially during practices held under hot and humid conditions and, in particular, during activities involving high school or younger athletes.

Heatstroke is the complete failure of the heat-regulating control centers in the brain, which causes body temperature to rise uncontrollably.

Exercising in Heat

The major problems associated with exercising in the heat are as follows:

• There is competition for blood flow between the skin (for cooling) and the working muscle (maximum capacity of the cardiovascular system is easily reached during exercise in the heat).

• Heat is produced from the working muscles, adding to the heat load the body must dissipate.

• The body loses fluids and electrolytes rapidly through profuse sweating, leading to dehydration and an electrolyte imbalance. The exerciser should drink small amounts of fluid at frequent intervals to avoid dehydration.

• Large amounts of sodium chloride and smaller amounts of potassium are lost through sweat, and this may lead to muscle cramps or affect transmission of electrical signals. The ingested fluid should contain small amounts of sodium and potassium.

If a person appears fatigued or becomes dizzy or disoriented during exercise in the heat, he or she should cease activity immediately, should be moved to a shady, cool area, and should be given fluids (figure 17.6).

Figure 17.6 Receiving care for heat-related illness. Those who work with athletes should be acutely aware of the danger of heat-related illness, and should cultivate a similar awareness in the people with whom they work. Proper attention to hydration and to early warning signals of heat-related illness will go a long way toward preventing dangerous situations.

To help avoid dehydration during exercise in the heat it is advisable to prehydrate—that is, to take in fluids right before physical activity—and to drink small amounts at frequent and regular intervals during the exercise. Drinking large amounts of water after an exercise bout is not the best way to **rehydrate** because it results in hypo-osmolarity of the plasma, which stimulates urine output, resulting in inadequate restoration of fluid levels. Profuse sweating results in loss of large amounts of sodium and smaller amounts of potassium, in addition to the large water loss. Replacement of these electrolytes after an exercise bout is essential for effective rehydration. The recommendation is to drink fluids that contain moderately high levels of sodium (60-90 mmol/L) and smaller amounts of potassium (5-10 mmol/L). Small amounts of glucose (7%), such as those in sport drinks, are not necessary for rehydration but may improve intestinal uptake of sodium while increasing the palatability of the drink. More volume than was lost through sweating should be taken in, because water lost through urine production must also be replaced. A 7% carbohydrate (glucose) solution (1-1.5 L [2-3 pints]) ingested within the first 90 min of the conclusion of exercise enhances fluid uptake from the gut and helps return muscle glycogen levels to pre-exercise levels.

➤ Hypo-osmolarity is low concentration of solutes.

The recommendation is to drink fluids that contain moderately high levels of sodium (60-90 mmol/L) and smaller amounts of potassium (5-10 mmol/L).

Case Study

Football practice at Nocommonsense High, U.S., is being held between 3:30 and 5:30 p.m. on a sunny day in early fall when T_a is 77° F (25° C) and the RH is 82%. Wind is from the south at about 9 km/hr (about 6 mph). Full gear is required, and athletes must keep their helmets on at all times. Fluid breaks are infrequent. Some of the players experience muscle cramping in the lower extremities, while others complain they are light-headed or dizzy. The coach issues salt tablets to those suffering from heat cramps, theorizing that the cramping is due to loss of sodium chloride. Those who are dizzy are urged to drink large volumes of plain water to replace lost fluids. Do you agree with these remedies? Why or why not? What are some recommendations you would make to avoid these problems?

Ingesting salt tablets to replace sodium chloride lost through sweating is not a good idea because salt is rapidly absorbed into the blood and increases osmotic pressure. To balance the hyperconcentration of blood, water is drawn out of the interstitial space and from cells, exacerbating the tissue dehydration already present.

Drinking large quantities of pure water is not an effective way to rehydrate, because blood becomes diluted. As a result, urine production is stimulated, and body fluid levels remain in deficit. The best remedy is to drink a solution with moderately high levels of sodium and smaller levels of potassium to replace both water and electrolytes lost through sweating.

Acclimatization to Heat

Exposure to heat brings about alterations in heat-dissipating mechanisms over 7 to 10 days; when accompanied by exercise training, the acclimatization is much greater. Figure 17.7 illustrates

changes in sweat rate and the resulting benefits on T_c and heart rate during a standard exercise bout in the heat. Increase in the sweat rate is accompanied by a lowered threshold for the onset of sweating; both of these changes reduce the need to increase cutaneous circulation as a means of cooling the body. This preserves central blood volume, negating the drop in stroke volume and the concurrent increase in heart rate. Better distribution of sweat over the surface of the body also aids in heat dissipation and reduces skin blood flow. Acclimatization brings about increased secretion of **antidiuretic hormone** (ADH) and **aldosterone**. The ADH increases body fluid levels through decreased urine production, facilitating increased sweat production and an increased plasma volume. Aldosterone reduces sodium chloride loss, in spite of the higher sweat rate. This greater reabsorption of sodium chloride occurs in the duct portion of the sweat gland and is essential in helping to maintain tissue electrolyte balance during exercise in the heat.

As the body acclimatizes to hot conditions, an increase in the sweat rate is accompanied by a lowered threshold for the onset of sweating; both of these changes reduce the need to increase cutaneous circulation as a means of cooling the body.

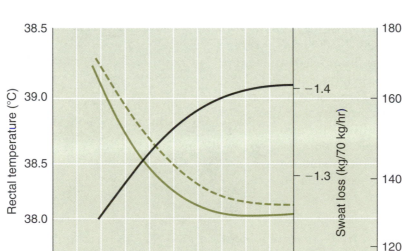

Figure 17.7   Acclimatization. Changes in sweat rate (black line), heart rate (green line), and core temperature (dashed green line) as a result of exercising for 100 min daily in the heat for nine days.
Adapted from Lind and Bass 1963.

Thus, a heat-acclimatized person has a lower heart rate and higher stroke volume as compared to an unacclimatized person at any level of submaximal exercise in the heat. This lowers stress on the cardiovascular system and produces less fatigue and lower perceived exertion.

Exercise at Altitude

Exercise at altitude creates stress on muscle and cardiovascular function because O_2 pressure in the inspired air is low. The fraction of O_2 in ambient air at altitude is not different from that at sea level (it remains about 20.9%), but the partial pressure of O_2 (PO_2) falls in tandem with the decline in barometric pressure as elevation increases. PO_2 of ambient air at sea level is approximately 159 mm Hg, but it falls to about 107 mm Hg at around 3000 m (10,000 ft) and is a mere 48 mm Hg at the summit of Mount Everest. The relationship between inspired and theoretical alveolar PO_2 at altitude is shown in figure 17.8. $\dot{V}O_2$max decreases about 3% per 30.5 m (100 ft) of elevation, beginning at about 1500 m (5000 ft). For example, a person with a $\dot{V}O_2$max of 50 ml/kg/min at sea level would have an approximate $\dot{V}O_2$max of 15 ml/kg/min at the summit of Mount Everest! No wonder these climbers have to rest after each step on the way up!

Much interest in understanding physiological responses to high-altitude training of athletes arose before and after the 1968 Mexico Olympics (altitude approximately 2300 m [7500 ft]), where long-distance events were dominated by athletes who lived and/or trained at high altitude. Today, high-altitude training is commonly used in an attempt to improve performance at lower altitudes. The popularity of high-altitude climbing has increased dramatically recently, also spurring interest in understanding limitations to physical work capacity at high elevations.

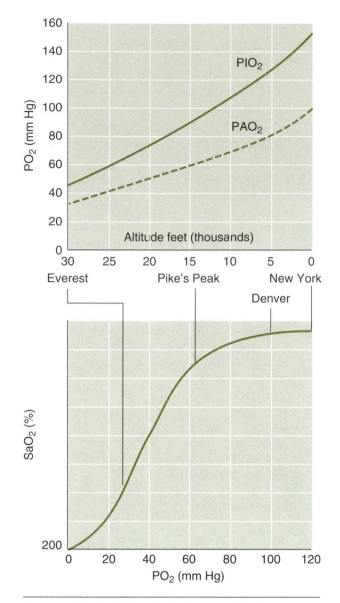

Figure 17.8   Relationship between altitude and oxygen pressure. The top panel shows the relationship between altitude and inspired (PIO_2) and alveolar (PAO_2) O_2 pressure. The lower panel shows the O_2 dissociation curve and its relationship to the effects of altitude.

Acute Exposure

Exposure to altitude results in an immediate increase in ventilation. The lower O_2 levels in the blood stimulate the carotid receptors, which signal the respiratory control center to increase ventilation. At moderate altitude the increase in ventilation can offset the decreased inspired O_2 levels, but at higher altitude even this mechanism is insufficient to maintain PAO_2 (figure 17.8). The hypoxia-induced increase in ventilation results in hyperventilation; this causes a more alkaline blood pH, reducing the stimulus to breathe. The decrease in alveolar O_2 pressure results in a lower pressure gradient for O_2 from the lung to the capillary. Thus diffusion is decreased at altitude, with the greater effects at higher altitudes.

Exposure to altitude results in an immediate increase in ventilation.

Heart rate is increased and stroke volume slightly decreased during exercise at altitude. The increase in heart rate is sufficient to balance the decrease in stroke volume at moderate altitudes. As a result of the respiratory and cardiovascular responses to acute exposure at altitude, submaximal exercise performance is preserved. These cardiorespiratory adjustments are not sufficient to meet the O_2 requirements of high-intensity exercise such as that performed during training or competition above about 1200 m [4000 ft] altitude. The effect of acute altitude exposure on $\dot{V}O_2$max is shown in figure 17.9. It is clear that even mild to moderate altitude exposure results in decrements in athletic performance. While altitude training has become popular as a mechanism for improving athletic performance, acute decreases in training intensity may negate this as a positive strategy in the short term.

Heart rate is increased and stroke volume slightly decreased during exercise at altitude.

Chronic Exposure

Continued exposure to altitude elicits a series of respiratory adjustments that enhance the gas

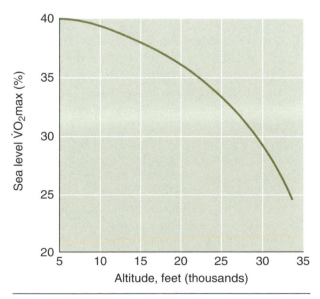

Figure 17.9 The change in maximum O_2 consumption with increasing altitude.
Adapted from West et al. 1983.

transfer and transport capacity of the body. These include

- bicarbonate excretion,
- increased red blood cell production, and
- changes in lung structure, in extreme cases.

Through excretion of **bicarbonate** (HCO_3^-), blood pH returns to normal, thus restoring the normal central drive to breathe. This increased central drive, along with the continued increase in carotid body drive, results in a level of ventilation that increases PAO_2. Of course the rise in PAO_2 also increases driving pressure, thereby enhancing diffusion of gas from the lung into the capillary.

Bicarbonate excretion, increased red blood cell production, and changes in lung structure, in extreme cases, are respiratory adjustments that enhance the gas transfer and transport capacity of the body.

Chronic exposure to altitude stimulates the production of **red blood cells** and their hemoglobin (Hb)-O_2-carrying capacity. The increase in O_2-carrying capacity has several positive effects, including a reduced ventilatory requirement to exchange O_2 and a reduction in the cardiac output needed to deliver the O_2 to the working muscles. As a consequence of these adaptations to chronic altitude exposure, most people are able to maintain activities of daily living at moderate altitude. For various reasons, the upper limit of continued altitude habitation appears to be slightly over 4250 m (14,000 ft). Populations in both the Andes and the Himalayans have made successful adaptations to exist in this harsh environment. At least part of their adaptation appears to be changes in lung structure to enhance gas exchange, an adaptation that has taken place over many years and possibly even many generations.

The chronic increase in Hb- and O_2-carrying capacity has led to the use of altitude training in an attempt to improve sea-level performance. This practice apparently has been successful in many athletes, but because of the reduction in training intensity has not succeeded in others. With the idea of counteracting the negative effects of altitude while retaining the positive effects, the practice of "living high and training low" has been investigated. According to Levine and Stray-Gundersen, living in a hypoxic environment at night and during much of the non-training time of the day, but training at or near sea level, stimulates increases in Hb with no decrement in training intensity. Several national teams for a variety of countries now promote this practice.

Acute Mountain Sickness

Millions of people travel to recreational paradises at high altitude only to suffer from **acute mountain sickness (AMS)** shortly after arriving. Acute mountain sickness occurs as a result of the sudden exposure to lowered inspired O_2 pressure (PIO_2) by lowlanders when traveling (usually in an airplane) to destinations at high elevations. It is best to acclimatize to altitude through gradual exposure to successively higher elevations. People vary in their response to acute altitude exposure, but many people experience the symptoms of mountain sickness—headache, nausea, insomnia, loss of appetite—at elevations greater than 2450 m (8000 ft). Severe AMS occurs in about 2 of every 10,000 visitors to altitude.

Acute mountain sickness can affect multiple body systems, but the symptoms are generally associated with the pulmonary and central nervous systems. The sudden exposure to hypoxia stimulates a general vasoconstriction of the pulmonary vasculature, resulting in high pulmonary arterial pressures that can lead to pulmonary edema. The pulmonary edema can be subclinical (not measurable by standard tests) or can lead to severe pulmonary distress. Subclinical pulmonary edema may be expressed as a continuous cough and shortness of breath. Severe pulmonary edema can be identified by crackles heard with chest wall auscultation and even the coughing up of blood. Likewise, the exposure to hypoxia may stimulate swelling around the brain. This cerebral edema is expressed as mild symptoms of dizziness and headache in many, but can lead to nausea or vomiting and even coma in severe cases.

▶ Auscultation means listening to sounds of internal organs.

The sudden exposure to hypoxia stimulates a general vasoconstriction of the pulmonary vasculature, resulting in high pulmonary arterial pressures that can lead to pulmonary edema.

Initial treatment of AMS entails reducing or eliminating the hypoxia. Immediate treatment may include administration of O_2 along with the objective of moving the victim to a lower altitude as soon as possible. Diuretics can reduce the pulmonary edema. The incidence of AMS can be reduced through a slower ascent to altitude and / or by the administration of acetazolamide before ascent.

▶ Acetazolamide is a drug that inhibits the action of carbonic anhydrase in the kidney, causing an increase in urinary excretion of sodium, potassium, and bicarbonate and a decreased excretion of ammonium. This leads to greater production of urine (loss of body water) and lowered blood pH.

Case Study

You have planned a hard-earned break from your job in New York City (elevation approximately 9 m [30 ft]) and fly to Aspen, Colorado (elevation 2900 m [9500 ft]), for a glorious two-week vacation. You arrive in the evening, and the next morning you hit the slopes to go downhill skiing. After a few runs you experience difficulty breathing and extreme muscle fatigue. You pack in the skiing, return to the lodge feeling nauseous, and develop a headache. The lunch buffet doesn't look all that appealing. Your glorious vacation is not much fun anymore. What are the likely causes of the breathlessness and fatigue, and what could (should) you have done to avoid these problems?

The most obvious cause of shortness of breath is, of course, the hypoxia. The lower O_2 levels stimulate the carotid receptors to increase breathing. Of course this process starts as soon as you are exposed to altitude. Exercise exacerbates this stimulus through several mechanisms. First, exercise itself stimulates breathing, and this increased level of ventilation on top of already higher resting ventilation leads to enhanced sensations of shortness of breath. Second, skiing is an activity involving forceful static muscle contractions, leading to periods when blood flow to the muscle is reduced or eliminated. This limitation enhances the need for non-oxidative energy production and subsequent lactic acid formation. The lactic acid further stimulates breathing, leading to increased ventilation and the sensation of breathlessness. The increased reliance on non-oxidative metabolism and lactic acid production also leads to premature fatigue.

By planning to stay at a lower altitude (1500-1800 m [5000-6000 ft]) for one to two days before going to Aspen, or by spending time to acclimatize upon arrival at Aspen without skiing or any other exercise, you would have minimized much of the breathlessness and fatigue. Within 24 to 48 hr of exposure to altitude, the body begins to adjust blood acid/base balance in a direction to reduce sensations of breathlessness. Small changes in blood O_2-carrying capacity and distribution also occur in this period, and both of these can reduce reliance on non-oxidative metabolism. For more severe symptoms, another option is to have your physician prescribe acetazolamide, which is a diuretic and also lowers blood pH.

What You Need to Know From Chapter 17

Key Terms

acclimatized
acute mountain sickness (AMS)
aldosterone
antidiuretic hormone
bicarbonate
conduction
convection
core temperature
dehydration
evaporation
heat cramps
heat exhaustion
heat stress
heatstroke
radiation
red blood cells
rehydrate
thermoneutral
thermoregulation
unacclimatized
windchill index

Key Concepts

1. You should be able to describe the basic mechanisms by which temperature regulation is achieved.

2. You should be able to state the significance of breakdowns in the body's temperature regulating systems.

3. You should be able to describe cardiovascular responses and limitations to exercise in the heat.

4. You should be able to describe how the body acclimatizes to chronic heat exposure.

5. You should be able to describe the effects of both acute and chronic altitude exposure in terms of the cardiopulmonary systems and how altitude affects exercise performance.

Review Questions

1. Describe how body temperature is regulated around the set point.

2. Which three of the four heat-dissipating mechanisms are most important in regulating body temperature under thermoneutral conditions?

3. How is skin blood flow regulated?

4. Compare the skin blood flow response to an increase in core temperature under resting conditions and during exercise.

5. Compare changes in central blood volume, ventricular preload, and stroke volume between exercising in a hot environment and exercising in a cold environment.

6. Describe the cardiovascular responses to prolonged steady-state exercise in a

(continued)

thermoneutral environment. What accounts for the increase in O_2 consumption despite the fact that the exercise is steady state?

7. Describe adaptations to acute and chronic exposure to altitude.

8. Compare the changes in plasma volume that accompany heat acclimatization versus acute exposure to high altitude. What are the benefits of these changes?

9. For vacation this year, you want to experience some "fun in the sun" instead of taking your normal ski trip to the mountains. You have been running regularly for six months prior to the trip because you plan to enter the annual 15K "Snowbird" road race at the end of your seven-day stay. You leave your home in Flint, Michigan, on a cold December morning and arrive in Miami Beach that afternoon (T_a 77° F or 25° C; RH 60%; wind from the south at 6 mph, or 10 km/hr). You decide to go for a 40 min run. You have a choice between running against the wind on the way out, then turning and running with the wind to get back to your starting point, and doing the opposite (running with and then against the wind). Wind speed is about the same as the speed you run. Your heart rate and ventilation will be higher than normal during your run. What are the physiological bases for these problems, and what is the major heat-dissipating mechanism? What precautions will you take to avoid heat-related illness, and which option will you choose regarding the wind? If you run each day before the race, do you expect to be back to your normal performance levels by the time of the 15K race? Why or why not?

Bibliography

American College of Sports Medicine. Prevention of thermal injuries during distance running. *Med Sci Sports Exerc* 1987, 19:529–533.

Bar-Or O. Invited review: Climate and the exercising child. *Int J Sports Med* 1980, 1:53–65.

Bar-Or O. Effects of age and gender on sweating pattern during exercise. *Int J Sports Med* 1998, 19 (suppl 2):S103–S107.

Bebout DE, Story D, Roca J, Hogan MC, Poole DC, Gonzalez-Camarena R, Ueno O, Haab P, and Wagner PD. Effects of altitude acclimatization on pulmonary gas exchange during exercise. *J Appl Physiol* 1989, 67:2286–2295.

Brooks GA. Temperature, skeletal muscle mitochondrial function and O_2 debt. *Am J Physiol* 1971, 220:1013–1019.

Ekelund LG. Circulatory and respiratory adaptations during prolonged exercise. *Acta Phsyiol Scand* 1967, 70 (suppl 292):1–38.

Green HJ, Sutton J, Young P, Cymerman A, and Houston CS. Operation Everest II: Muscle energetics during maximal exhaustive exercise. *J Appl Physiol* 1989, 66:142–150.

Johnson JM. Exercise and the cutaneous circulation. *Exerc and Sport Sci Rev* 1992, 20:59–98.

Levine BD and Stray-Gundersen J. Living high-training low: Effect of moderate-altitude acclimatization with low-altitude training on performance. *J Appl Physiol* 1997, 83:102–112.

Lind AR and Bass DE. Optimal exposure time for development of acclimatization to heat. *Fed Proc* 1963, 22:704.

Maughan RJ. Rehydration and recovery after exercise. *Sports Sci Exch: Exercise and Environment* 1996, 9 (3).

Meyer F, Bar-Or O, MacDougall D, and Heigenheiser G. Sweat electrolyte loss during exercise in the heat: Effects of gender and maturation. *Med Sci Sports Exerc* 1992, 24:776–781.

Nadel ER. Limits imposed on exercise in a hot environment. *Sports Sci. Exch: Exercise and Environment* 1990, 3 (27).

Nadel ER, Fortney SM, and Wenger CB. Effect of hydration state on circulatory and thermal regulations. *J Appl Physiol* 1980, 49:715–721.

Rowell LB. Human cardiovascular adjustments to exercise and thermal stress. *Physiol Rev* 1974, 54:75–159.

Rowell LB, Marx HJ, Bruce RA, Conn RD, and Kusumi F. Reductions in cardiac output, central blood volume and stroke volume with thermal stress in normal men during exercise. *J Clin Invest* 1966, 45:1801–1816.

Sawka MN and EF Coyle. Influence of body water and blood volume on thermoregulation and exercise performance in the heat. *Exerc and Sport Sci Rev* 1999, 27:167–218.

Sutton JR, Reeves JT, Wagner PD, Groves BM, Cymerman A, Malconian MK, Rock PB, Young PM, Walater SD, and Houston CS. Operation Everest II: Oxygen transport during exercise at extreme altitude. *J Appl Physiol* 1988, 64:1309–1321.

Welfel EE, Groves BM, Brooks GA, Butterfield GE, Mazzeo RS, Moore LG, Sutton JR, Bender PR, Dahms TE, McCullough RE, McCullough RG, Huang S-Y, Sun S-F, Grover RF, Hultgren HN, and Reeves JT. Oxygen transport during steady-state submaximal exercise in chronic hypoxia. *J Appl Physiol* 1991, 70:1129–1136.

West JB, Boyer SJ, Graber DJ, Hackett PH, Maret KH, Milledge JS, Peters RM, Pizzo CJ, Samaja M, Sarnquist FH, Schoene RB, and Winslow RM. Maximal exercise at extreme altitudes on Mount Everest. *J Appl Physiol* 1983, 55:688–698.

West JB, Hackett, PH, Maret, KH, Milledge, JS, Peters, RM, Pizzo, CJ, and Winslow, RM. Pulmonary gas exchange on the summit of Mt. Everest. *J Appl Physiol* 1983, 55:678-687.

Exercise Testing and Prescription

This chapter will discuss

➤ the reasons for exercise testing, the types of tests, and safety considerations for these tests; and

➤ basic concepts of exercise prescription in healthy and patient populations.

We have learned that exercise places stress on most of the body's systems. The amount of stress—that is, the exercise intensity and duration—and the health of the body's systems determine the physiologic responses to the exercise. A functional deficit in the system that results from a disease process may not affect a person at rest; but when the demands of exercise are imposed, the deficit may be reflected in an abnormal response to exercise and/or a reduced exercise capacity. The capacity of exercise to expose physiologic and functional deficits makes exercise testing useful in clinical practice. This chapter presents information on available exercise tests and the use of these tests to develop exercise programs for both healthy populations and patients.

Reasons for Exercise Testing

An exercise test is rarely needed in a healthy person. However, a $\dot{V}O_2\text{max}$ test is frequently performed for athletes. Knowledge of $\dot{V}O_2\text{max}$ provides feedback about an athlete's aerobic potential, or capacity to perform endurance exercise. Repeated $\dot{V}O_2\text{max}$ tests also provide feedback regarding an athlete's conditioning program.

An exercise test is rarely needed in a healthy person; however, a $\dot{V}O_2\text{max}$ test is frequently performed for athletes.

Case Study

A friend of yours has reached the point where he wants to run a 10K (6.2 mile) race. He has always shown good athletic capacity and has been running recreationally for over five years. He is quite proud of this capacity and announces that his goal is to run the distance in under 35 min. On the basis of some of his times at shorter distances, you are reasonably sure that he has the ability to run at this speed; but it strikes you that this is a running speed that would require a pretty high **aerobic capacity** ($\dot{V}O_2\text{max}$). You aren't sure that your friend has ever done anything to suggest that he has the aerobic capacity to accomplish this goal. How would you determine this?

The first step is to calculate the O_2 cost of running at this speed. Using the formulas presented in chapter 1 (page 3), you calculate the cost of running at this speed:

$$10 \text{ km} = 10{,}000 \text{ m}$$

$$10{,}000 \text{ m} / 35 \text{ min} = 286 \text{ m}/\text{min}$$

$$\dot{V}O_2 = 286 \times 0.2 \text{ ml/kg/min} + 3.5 =$$
$$61 \text{ ml } O_2/\text{kg/min}$$

The next step is to determine your friend's $\dot{V}O_2$max. You perform one of the tests described later in this chapter and measure his $\dot{V}O_2$max as 57 ml O_2/kg/min. You now have evidence that your friend's goals are unrealistic.

Since conditioning can increase $\dot{V}O_2$max, you offer to help your friend by developing a conditioning program designed to optimize his $\dot{V}O_2$max. To monitor his progress, you schedule him for repeated $\dot{V}O_2$max tests every four weeks. The results, in $\dot{V}O_2$max (ml O_2/kg/min), are:

Week 1	57
Week 4	59
Week 8	61
Week 12	60

On the basis of your conditioning program and the results of the $\dot{V}O_2$max testing, you are happy to tell your friend that he is very close to having the aerobic capacity to reach his target time. You recognize that his aerobic capacity with his enhanced non-oxidative capacity probably gives him the energy systems he needs to complete the 10K in 35 min or better.

An exercise test also may be recommended to measure a person's maximum work capacity. This maximum might be expressed in terms of $\dot{V}O_2$max but is generally expressed as **peak work capacity**, or functional capacity, since this information is more valuable than true $\dot{V}O_2$max. Knowledge of a person's true $\dot{V}O_2$max is desirable in many athletic situations but is not required in other exercise-testing applications. In a clinical situation, an exercise test commonly is performed to determine

- **functional work capacity** and
- **symptom-limited work capacity** (the exercise level at which an abnormal response is measured).

For these determinations it is not necessary to measure O_2 consumption. In these cases the test can be used to establish the starting point of therapy, project a therapeutic goal, and document any changes in the exercise response as a result of therapy. A functional capacity test would be performed to answer the question "What is the maximum work capacity of this person?" This might be important to know for a disability or return-to-work question. A symptom-limited test might be performed to determine the level at which angina might limit a cardiac patient or at which a pulmonary patient feels dyspneic. Likewise, an exercise test might be performed to determine the exercise level at which an electrocardiogram (ECG) abnormality is observed in a cardiac patient or hemoglobin desaturation is observed in a pulmonary patient.

Knowledge of a person's true $\dot{V}O_2$max is desirable in many athletic situations but is not required in other exercise-testing applications.

➤ Dyspnea means shortness of breath.

Reasons for Exercise Testing

- Measurement of $\dot{V}O_2$ max
- Examination of complaints of shortness of breath or dyspnea on exertion
- Evaluation of physiological effects of acute and chronic disease (functional capacity)
- Documentation of exercise-induced asthma
- Determination of exercise capacity for prescription of therapy
- Assessment of the effects of therapy
- Establishment of safe limits for activity
- Determination of insurance claims or disability

An exercise test is frequently performed to document the presence of exercise-induced asthma or **exercise-induced bronchospasm** (EIB). Since 80% to 90% of patients with asthma respond to exercise, an exercise test, performed to induce EIB, also may be used to diagnose asthma.

Since 80% to 90% of patients with asthma respond to exercise, an exercise test, performed to induce EIB, also may be used to diagnose asthma.

Types of Exercise Tests

The selection of an exercise test depends on the information sought (see earlier section, "Reasons for Exercise Testing"). Standard testing protocols have been designed to measure $\dot{V}O_2$max in healthy athletes and non-athletes. Because of the clinical utility of exercise testing in cardiopulmonary rehabilitation, a range of standard tests have been developed to allow determination of exercise capacity and the level of exercise at which an abnormal response can be identified (figure 18.1).

Testing in the Lab

An exercise test can be performed on a variety of standard devices (figure 18.2), including treadmills and arm or leg cycle **ergometers** (an exercise machine capable of precise, calibrated work,

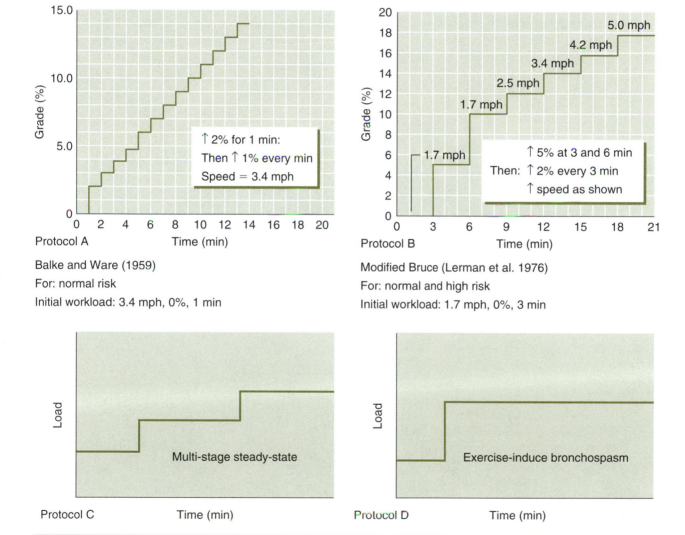

Figure 18.1   A variety of protocols available for exercise testing. Protocol A is a generalized graded, progressive, incremental test that can be adapted for many purposes. Protocol B has become a standard protocol in many clinics. Protocol C is used in cases in which maximum testing is not necessary. Protocol D is used to test for exercise-induced bronchospasm.

power, speed, or grade settings). In selected circumstances a device may be designed for a specific use, for example, a wheelchair ergometer or a rowing machine. Because exercise tests are used to measure a person's work capacity, the work variables on the selected device must be precisely calibrated. In the case of a treadmill, the speed and elevation must be controlled; on cycle-type ergometers the power output, in Watts, must be controlled. The treadmill has an advantage in that walking or running is a natural activity for most patients. The disadvantage of the treadmill is that monitoring of blood pressure and even ECG is sometimes difficult because of the movements associated with walking or running. The noise from the treadmill motor contributes to these monitoring difficulties. Monitoring is less of a problem with the cycle ergometer because body movements are minimized. Cycling, however, is not a natural movement, and for many patients cycling is not a normal activity. Muscle

fatigue due to such an unaccustomed activity may result in premature test termination.

The most common exercise test is the **graded exercise test (GXT)**. There are many accepted protocols for performing a GXT, but all use regular increments of work intensity. The most common incremental tests follow some modification of the **Balke protocol**, popularized by Dr. Bruno Balke. Balke designed both a treadmill and a cycle ergometer test that used increments in work intensity designed to increase the demand by about 1 MET for each increment (figure 18.1, protocol A). The choice of protocol, and even speed, grade, or Watt increments within a protocol, depends on the population being tested. The ideal GXT lasts between 12 and 16 min. To attain this goal with an athlete, for instance, the speed should be increased to 4 or 4.5 mph, whereas a patient with a reduced exercise capacity might be tested at 3 mph or even slower. The **Bruce protocol** (figure 18.1, protocol B) uses increments of both speed and grade to

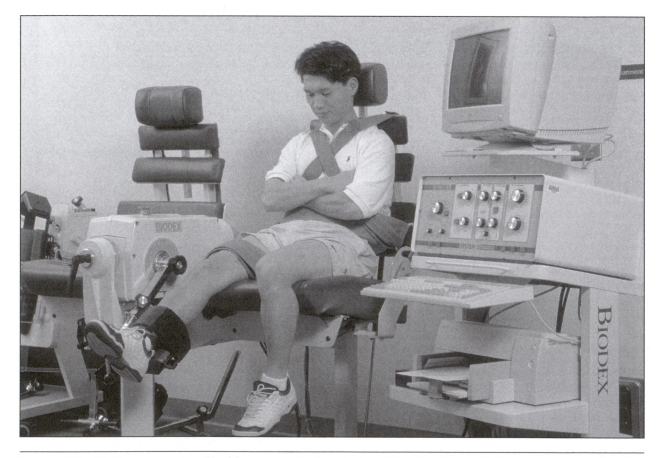

Figure 18.2 Testing ergometer. Health care professionals should be familiar with the major types of testing ergometers and with how to use them.

provide load increases. The widespread use of this protocol in most cardiac stress-testing laboratories allows easy comparisons across facilities and among patients. For instance, if a cardiologist describes the exercise capacity of a patient as Stage 3 Bruce, other laboratories know exactly what this means in terms of exercise intensity.

MacDougall, Wenger, and Green have described testing protocols for athletes that use the same progressive, graded principles as described for the standard GXT. Increments for athletes, however, must be greater than for a non-athletic population, or the test would last for more than the optimal 12 to 16 min.

The most common exercise test is the graded exercise test.

When the information sought does not require measurement of the peak response, a steady-state test may be performed (figure 18.1, protocol C). This test is useful for measuring a patient's response to one or more levels of exercise where the measurements require a steady state. You might use this test if you needed to measure arterial O_2 saturation or arterial blood gases at several exercise intensities to determine a patient's need for supplemental O_2. You might also use a steady-state test if you needed to document the effectiveness of therapy. For instance, if a cardiac patient had documented angina or a pulmonary patient had desaturation at a specific exercise level, you might have the patient exercise at that level on a regular basis to determine whether the angina or desaturation was reduced as a result of therapy.

Exercise-induced bronchospasm can be identified by means of a two-stage exercise test (figure 18.1, protocol D). The first stage is a short (1 or 2 min) stage to accommodate the patient to the exercise. After this accommodation period, the load is increased to a level predicted to elicit a level of ventilation that will cause sufficient airway cooling and/or water loss to induce asthma (page 198). This second stage is performed for 7 to 8 min during breathing of cool, dry air. In most cases, air breathed from a bag filled from a compressed-air cylinder is sufficiently dry to induce bronchospasm in patients with asthma. Lung functions are measured prior to the test and at selected intervals for approximately 30 min after the test. Decreases in flow rates (FEV_1, $FEF_{25-75\%}$) of 15% or greater are considered diagnostic for exercise-induced asthma.

When the information sought does not require measurement of the peak response, a steady-state test may be performed.

➤ FEV_1 is forced expired volume in one second; $FEF_{25-75\%}$ is forced expired flow between 25% and 75% of the vital capacity.

Testing in the Field

When an exercise laboratory is not available, or when a formal test is not required, field tests can be used to determine exercise capacity or to periodically monitor changes in a patient's response to exercise. As with ergometer tests, predicted normal values are available for many of these field tests. The three most commonly used tests are those based on

- time,
- distance, or
- incremental pace adjustments.

When an exercise laboratory is not available, or when a formal test is not required, field tests can be used to determine exercise capacity or to periodically monitor changes in a patient's response to exercise.

• **Time.** The simplest of the **field tests** are based on time (12 min test, 6 min test) or distance (mile run [1.6 km], 600 yd [550 m] run/walk). Most people are familiar with the 12 min test of aerobic power developed by Dr. Kenneth Cooper and known as the **Cooper protocol**. Cooper, and others, have been able to show that actual $\dot{V}O_2$max can be predicted from the distance covered over 12 min. Tests of shorter or longer duration were poorer predictors in the general population. The duration has been altered for specific populations: a 6 min test has been developed for children and patients with low exercise capacity, and a 15 min test has been used for athletes. The 6 min test is used in a variety of clinical situations, with the hall serving as a walkway (figure 18.3).

• **Distance.** Tests designed to determine how long it takes to cover a specified distance also provide an estimate of exercise capacity. The mile run and the 600 yd run are frequently used in health and physical education classes to document fitness levels and changes in fitness levels in school-age children.

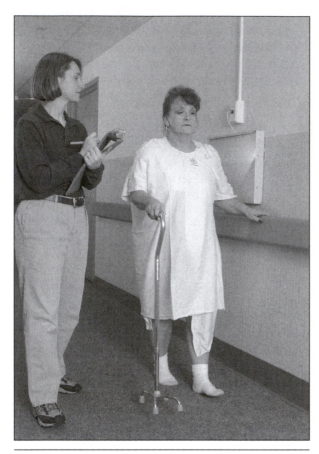

Figure 18.3 The 6 min test. This test is easily administered and is extremely useful in a wide variety of clinical situations. The handrail is used only to steady oneself and not for support.

• **Incremental pace adjustments.** Incremental shuttle tests have been developed to simulate a graded treadmill or cycle ergometer test in the field. Each variation requires participants to walk or run between two points that are a known distance apart. Speed of movement between the points is controlled by a metronome, starting at a slow pace and increasing in pace at specified intervals. Recording the pace or exercise intensity at which the test is terminated provides a means of monitoring peak work capacity; the measurement of heart rate at each level provides information regarding the response to exercise.

Field tests are most useful when one is testing large numbers of people and precision is not required. These tests also provide a rough estimate of peak work capacity that can be monitored over time to assess the effects of conditioning or therapy. Thus they may be useful at intervals

between more formal ergometer testing of patients. Monitoring physiological responses during these tests is difficult, but heart rate measured during or immediately after the test can provide information regarding conditioning status.

Field tests are most useful when one is testing large numbers of people and precision is not required.

Safety Considerations

Patient or client safety is enhanced when all aspects of the testing environment are carefully thought out (see General Recommendations for Exercise Tests). The laboratory and all equipment should be well maintained and should be inspected prior to each test. In most clinical situations a "crash cart" containing a source of O_2, drugs, a bag respirator, and a defibrillator is required. All personnel should be trained in cardiopulmonary resuscitation and should be familiar with emergency procedures. These procedures should be posted where personnel can consult them quickly and easily if an emergency should arise.

General Recommendations for Exercise Tests

- Have emergency procedures in place.
- Select appropriate exercise protocol.
- Perform pre-exercise clinical assessment and testing.
- Determine variables to be monitored.
- Perform postexercise evaluation and monitoring.

Patients should be told to refrain from eating large meals, smoking, or drinking caffeine for 2 hr before the test and should receive recommendations regarding proper dress for the test. They should receive instructions regarding suspension of drugs. Personnel should review the patient's history and, if appropriate, perform an exam. Pertinent history could include x-rays, ECG, blood gases, drug therapy, and pulmonary function tests. During the test, a minimum of two people should be present to monitor the equipment and the patient. The patient also should be observed and monitored for 5 to 10 min after termination of the test and should be allowed a cool-down period before leaving the monitoring area for a shower.

Variables to be monitored during the test depend on the patient and the information being sought. Clinically, personnel should monitor the person's color, level of exertion, and signs of shortness of breath and cyanosis. For people in whom cyanosis may be difficult to identify such as African-Americans, personnel must pay particular attention to the other signs of difficulty. Clinical monitoring can be enhanced through use of the Borg scale (page 303) to elicit patient feedback regarding exercise intensity. Most tests require the measurement of heart rate. For athletes, a heart rate monitor is sufficient; but in most patients, ECG and blood pressure should be monitored for signs of arrhythmia or pressure decompensation during and after the test. Arterial saturation, by oximetry, should be monitored in pulmonary patients. In more advanced laboratories, it may be possible to measure minute ventilation and its components (frequency and tidal volume) via open circuit spirometry or even cardiac output via one of the rebreathing techniques. Many laboratories can monitor these variables using automated, computerized equipment.

An exercise test can be terminated by the examiner or by the person being tested. People being tested should be told that they may terminate the test at any time. If personnel suspect malingering, they should give encouragement to continue. Personnel should terminate the test using the following criteria.

Termination of the Exercise Test

- Patient request
- Clinical observation of pallor, cyanosis, or extreme exertion
- Electrocardiogram arrhythmia
- Decrease in systolic or increase in diastolic pressure with increasing exercise intensity
- Decreases in O_2 saturation to <75-80%

An exercise test can be terminated by the examiner or by the person being tested. People being tested should be told that they may terminate the test at any time.

Exercise Prescription

Exercise programs are developed to enhance health, promote fitness, increase work capacity for employment, and rehabilitate. An effective exercise program provides appropriate **overload** through manipulation of exercise intensity, duration, and frequency. Appropriate overload places sufficient stress on the body to stimulate the desired adaptive response without doing harm. If acquisition of strength is the goal, the resistance used in the exercise program must be high enough to stimulate increases in contractile proteins and other cellular adaptations that will be reflected in greater strength. If increased cardiovascular capacity is desired, the duration and intensity of the exercise must be sufficient to stimulate changes in the cardiovascular system and the oxidative capacity of the muscles. In either case, the overload should be applied in progressively increasing intensities.

Overload should be targeted to the goals of the person's program, which are based on an assessment of each person's status and the objective to be met. At the cellular level, the adaptive responses are quite specific to the stimulus applied. If the objective of a program is to return a patient to work, the program must be designed to ensure that the specific requirements of the work task are mimicked in the rehabilitation program. This **specificity** requires some knowledge of the cardiovascular requirements of the work task, as well as the strength and particular movement requirements of the task.

Exercise programs are developed to enhance health, promote fitness, increase work capacity for employment, and rehabilitate.

Case Study
You are frequently asked to lecture to athletic groups regarding conditioning programs. You need to develop a means to explain the principle of specificity.

First you need to give the audience an appreciation for muscle fiber structure. Ask them to visualize the strands (fibers) of meat seen in a piece of beef, chicken, or turkey. Next they need to understand that each movement recruits only a few such fibers, and that the specific fibers recruited are movement dependent. Now they should realize that if they wish to improve strength for a specific movement, their strength exercises should involve only that movement: performing other movements with the strengthening exercises will likely recruit fibers other than those they wish to train. A good example is the bench press, which U.S. football players, rugby players, and others often emphasize. Although the bench press is a good general strengthening exercise, that particular movement is rarely performed on the field. The only time such players push out with both hands, with their back against a fixed object, is when they are at the bottom of a pileup! Suggest that a better strengthening exercise would be one that more closely simulated the on-field movements of each player.

It is beyond the scope of this text to describe specific exercise programs for the great variety of purposes and the many populations such programs are used for. In the following sections we describe some important elements common to all exercise programs. The references listed in Resources for Conditioning Programs provide additional information.

Resources for Conditioning Programs

- American Association of Cardiovascular and Pulmonary Rehabilitation. *Guidelines for pulmonary rehabilitation programs.* 2nd ed. Champaign, IL: Human Kinetics, 1993.

- American Association of Cardiovascular and Pulmonary Rehabilitation. *Guidelines for cardiac rehabilitation programs.* 2nd ed. Champaign, IL: Human Kinetics, 1995.

- American College of Sports Medicine. *ACSM's guidelines for exercise testing and prescription.* 5th ed. Baltimore: Williams & Wilkins, 1995.

- American College of Sports Medicine. *ACSM's resource manual for guidelines for exercise testing and prescription.* 3rd ed. Baltimore: Williams & Wilkins, 1998.

- Australian Sports Commission. *Physiological tests for elite athletes.* Champaign, IL: Human Kinetics, 2000.

- Baechle TR, ed. *Essentials of strength training and conditioning.* 2nd ed. Champaign, IL: Human Kinetics, 2000.

- MacDougall D, Wenger H, and Green H. *Physiological testing of the high performance athlete.* 2nd ed. Champaign, IL: Human Kinetics, 1991.

- Skinner J, ed. *Exercise testing and exercise prescription for special cases.* 2nd ed. Philadelphia: Lea & Febiger, 1993.

- Wilmore J and Costill DL. *Training for sport and activity.* Champaign, IL: Human Kinetics, 1988.

Healthy Populations

Development of an exercise program should consider the elements of flexibility, strength, and endurance.

• **Flexibility.** A general stretching program includes stretches for the major muscle groups of the upper and lower body—the hip ad-/abductors and flexors/extensors, the gastrocnemius, and the arm ad-/abductors. Stretches should be passive as much as possible and should be maintained statically for 20 to 30 s at a time. Active stretches whereby the muscles are loaded and unloaded quickly should be avoided. There has been little scientific examination of the role of and appropriate place for stretching in an exercise program. Evidence does suggest that most, if not all, exercisers can benefit from the increased flexibility stimulated by stretch. Consensus also supports the importance of stretching after exercise to reduce the common phenomenon of "muscle tightening" that occurs after unaccustomed exercise.

• **Strength.** Strengthening exercises can be categorized as general or specific. General strengthening exercises are those that emphasize large muscle groups in gross movements. General strengthening exercises should emphasize the same muscle groups as listed for the flexibility exercises. Where the objectives of the exercise program are clear, these general strengthening exercises should use the muscles involved in the targeted activities. Specific strengthening exercises are those that emphasize specific muscles by using finer movements. For most health and fitness programs, general exercises are appropriate. If, on the other hand, you are conditioning someone for a specific employment task, you need to consider using exercises that provide the necessary specificity for that task. Each strengthening exercise can be performed for a specific number of repetitions that are grouped into one or more sets. The weight at which you start depends on the existing level of conditioning. A good place to start is with a weight, or resistance, that allows the person to perform 10 repetitions. Initially the person might perform this exercise for one or two sets with 1 or 2 min between sets. Resistance should increase by about 10% when he or she can perform three sets of 10 repetitions easily.

• **Endurance.** Endurance is that quality that allows one to exercise for longer periods of time with a minimal amount of fatigue. We can speak of muscular endurance or cardiovascular endurance. Muscular endurance defines the ability of a specific muscle or specific groups of muscles to perform work over time. Muscle endurance can be developed using muscle-specific resistance exercises with low resistance but high repetition. For example, programs using weights that can be lifted 12 to 15 times develop muscle endurance, whereas programs using high resistance and low repetition are prescribed for strength conditioning. Muscle endurance also can be developed using task-specific exercises, or repetition of a specific work task, over a period similar to that in the work environment. In this way you can prepare a patient for return to work.

Development of an exercise program should consider the elements of flexibility, strength, and endurance.

Cardiovascular endurance is achieved by increasing the capacity of the cardiovascular system. It is difficult to differentiate changes in the cardiovascular system from the changes that take place within the oxidative energy-producing system, so cardiovascular endurance also is called aerobic capacity. As you might imagine based on your understanding of the oxidative metabolic system, any increase in the capacity to produce energy requires an increase in the delivery of O_2

(the cardiopulmonary system) and in the capacity to utilize this O_2 within the mitochondria. The former involves improvements in the myocardial capacity to pump the blood and in the vascular system to deliver it to the muscle, whereas the latter involves increases in the size and number of mitochondria. As also might be expected, then, cardiovascular capacity will be improved little with exercise that lasts 4 min or less. The optimal development of cardiovascular capacity involves exercise low enough in intensity that it can last more than 10 to 15 min. The usual recommendation is that the exercise last 30 min at an intensity of between 55% and 85% of the person's maximum capacity.

It is recommended that exercise last 30 min at an intensity of between 55% and 85% of the person's maximum capacity.

Case Study

Jason is a 17-year-old college freshman interested in improving his endurance capacity, but he has also expressed an interest in "bulking up" a bit to impress the campus female cohort. Unfortunately he is struggling a little in his classes, so he needs to accomplish his goals in the least amount of time possible. He thinks he can afford about 40 min three times a week. How would you design his program?

For endurance, Jason should exercise for periods greater than 10 to 15 min, and ideally around 30 min. For strength, you need to select resistance exercises that work large muscle groups and resistances at which he can perform 8 to 10 sets for three repetitions. Given his time constraint, you realize that you have to make a few compromises. If you compromise slightly on the weight program, you can have Jason complete his exercise routine in the allotted time using both resistance-exercise and endurance principles.

You should design a circuit training program in which Jason performs a series of resistance and more rhythmic, endurance, exercises at stations in a specified rotation with little rest between each station. For instance, you might select three lower-body resistance exercises (leg extension, leg flexion, and squat) and three upper-body resistance exercises (e.g., bench press, extended arm raises, and overhead press) to be interspersed with brief (2-3 min) periods on the treadmill, stepping device, and cycle. When a person performs these exercises with little break between each, the activity retains its endurance characteristics while at the same time accomplishing strength goals. Jason should perform the endurance exercise at an intensity that will raise the heart rate to 65% to 75% of his maximum. The compromise would be to decrease the resistance for the strength exercises so that he can perform 10 to 15 repetitions, making it easier to move to the next station without a long recovery period. A circuit training program has an additional advantage: it is easier to maintain interest in a program that has this variety than in one that involves the use of only one or two devices for longer periods of time (figure 18.4). The rotation might look like the following, repeated three times:

15 leg extensions	15-30 s
Treadmill	2 min
15 bench press	15-30 s
Cycle exercise	2 min
15 squats	15-30 s
Stepping exercise	2 min
15 extended arm raise	15-30 s
Treadmill	2 min
15 leg flexion	15-30 s
Cycle exercise	2 min
15 overhead press	15-30 s

Figure 18.4 Equipment for circuit training. Circuit training is an efficient way to combine strength and aerobic training. You should be skilled in creating such programs for patients or clients who want to exercise but who have little time to spend on it.

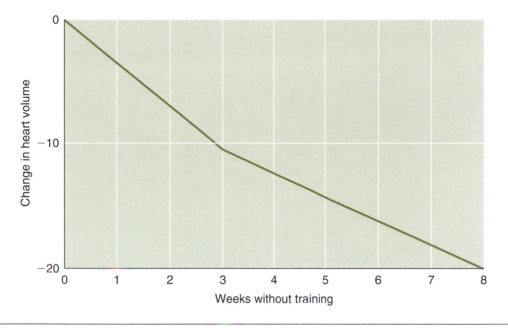

Figure 18.5 Changes in heart volume. Heart volume decreases dramatically if training is withheld. One must take this rapid detraining effect into account when prescribing activity after even short periods of inactivity. Adapted from Martin and Coyle 1986.

Patient Populations

Most patients are **detrained**. This detraining may be the result of a sedentary lifestyle or disease-related bed rest. Figure 18.5 shows that failing to train results in a dramatic decrease in heart volume. In any case, one must consider detraining when developing a conditioning program. To avoid making false assumptions regarding a patient's status, you should perform an exercise assessment as the first step in developing an exercise program. The principles of overload and specificity that we have discussed should form the basis of the program, and elements of flexibility, strength, and endurance should be part of the exercise prescription.

Exercise rehabilitation programs are typically divided into **phases I through IV**. These phases and the associated exercise capacities are depicted in figure 18.6. Although much of the information on rehabilitation programs was presented in chapter 14 with reference to cardiac rehabilitation, it is repeated here because it applies also to more general rehabilitation programs.

• **Phase I** defines the acute, usually inpatient, phase of exercise rehabilitation. This phase

may last for several days to two weeks. Once the acute, unstable stage of a disease exacerbation is past, patients should be encouraged to become active. Exercises during this phase may be confined to efforts to perform activities of daily living, for example walking to the washroom. Most patients can perform bedside activities such as leg extensions with or without weights or by pulling against an elastic band attached to the bed rail. Over time, the patient should be encouraged to increase walking distance and perhaps even to climb stairs. Phase I is an important time to prepare the patient for moving into the outpatient setting and eventually into a less structured exercise/activity program. It is during phase I that patients should receive education regarding diet, their disease, lifestyle factors that might contribute to the disease process, medications, and the rehabilitation program.

• **Phase II** of the rehabilitation program usually lasts 8 to 12 weeks and should be preceded by a GXT to establish safe limits of activity. Phase II entails close monitoring of the patient during exercise and the use of devices that can be precisely controlled in terms of intensity.

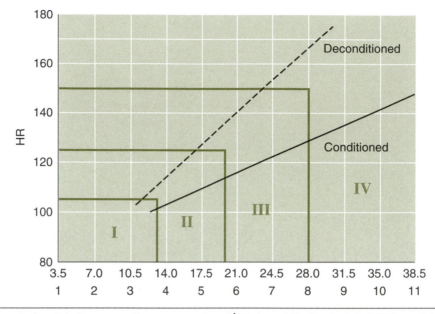

Figure 18.6 The relationship between O_2 consumption ($\dot{V}O_2$ in ml O_2/kg/min), METs, and heart rate (HR). In general, patients with peak exercise MET levels of <4 participate in phase I rehabilitation programs, while those with peak MET levels of 4 to <6, 6 to <8, and 8 to <10 participate in phase II, II, and IV programs, respectively. The exact response is dependent on age, gender, medication, and other factors.

Monitoring should include heart rate but may also include ECG, blood pressure, and arterial O_2 saturation. Patients should perform formal exercise three to four days a week, emphasizing endurance activities lasting more than 10 min. As the patient progresses through phase II, monitoring is reduced and more variety is incorporated into the exercise program. During the later stages of this phase, outside activity should be encouraged, and patients need to be counseled as to their choice of appropriate activities. It is important at this time for patients to identify barriers to exercise and to establish support structures, such as an exercise partner or group, to ease the transition to an active lifestyle. Progress is monitored using one of the field tests that have been described. Another GXT should be performed at the end of phase II.

• **Phase III** is also called the maintenance program. The purpose is to maintain the progress made in phases I and II and to prepare the patient for moving into a non-medically oriented environment. Patients in phase III still may require monitoring initially, but they must be weaned from the monitoring as the phase progresses. Any activity should be encouraged in this phase, but 30 min of exercise three to four days a week should be the minimum.

• **Phase IV** exercise is virtually indistinguishable from that performed by a healthy person. The patient should do another GXT before being passed into phase IV.

The Cardiovascular Exercise Prescription

Heart rate is the standard tool for monitoring exercise intensity. Once the exercise capacity has been defined, the target heart rate can be calculated (figure 18.7). The target heart rate is

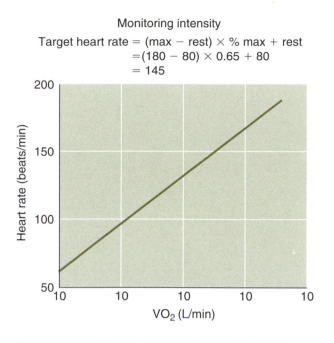

Monitoring intensity

Target heart rate = (max − rest) × % max + rest
=(180 − 80) × 0.65 + 80
= 145

Figure 18.7. Monitoring intensity. The linear relationship between O_2 consumption ($\dot{V}O_2$) and heart rate allows the use of heart rate to monitor exercise intensity. With the Karvonen formula one can calculate the heart rate at any given desired exercise intensity.

a RPE scale		b Modified CR10 scale	
6	No exertion at all		
7		0	
8	Extremely light	00.5	Very, very light
9	Very light	01.0	Very light
10		02.0	Light
11	Light	03.0	Moderate
12		04.0	Somewhat hard
13	Somewhat hard	05.0	Hard
14		06.0	
15	Hard (heavy)	07.0	Very hard
16		08.0	
17	Very hard	09.0	
18		10.0	Very, very hard
19	Extremely hard		
20	Maximal exertion		

Borg RPE scale
© Gunnar Borg, 1970, 1985, 1994, 1998

Borg CR10 scale
© Gunnar Borg, 1981, 1982, 1998

Figure 18.8 The Borg scales for assessing exercise intensity. (*a*) shows an RPE scale while (*b*) shows a scale that many find easier to use. For correct usage, see Borg, G. 1998. *Borg's Perceived Exertion and Pain Scales*, Champaign, IL: Human Kinetics.

calculated using the **Karvonen formula** so that exercise intensity is established at between 55% and 85% of the maximum. The lower target heart rate is appropriate for patients in the initial stages of rehabilitation and for healthy people who have been previously sedentary. Measuring heart rate during or immediately after exercise is difficult for some and inconvenient for many. In many cases, patients can more easily monitor exercise intensity using the **Borg rating of perceived exertion (RPE) scale** or the Borg CR10 scale (figure 18.8). While both scales can be applied to monitor exercise intensity, the 0- to 10-point category ratio scale is easier for most patients to understand than the 6- to 20-point scale. In addition, attaining a 6 on the 10-point scale roughly corresponds to exercising at 60%

of the maximum capacity—a useful relationship one can use to monitor relative exercise intensity.

Activities can be prescribed according to general intensity guidelines (figure 18.9 and chapter 4). A key to obtaining **compliance** with an exercise prescription is to engage the patient in activities of his or her liking. You must discover these preferences (or help the patient discover them, if he or she has none at the outset) so you can select appropriate activities based on the relative intensities shown in the figure. During phase III and early phase IV, select activities classified as low intensity. Activities that span the range from moderate to high-intensity exercise afford additional flexibility to the program.

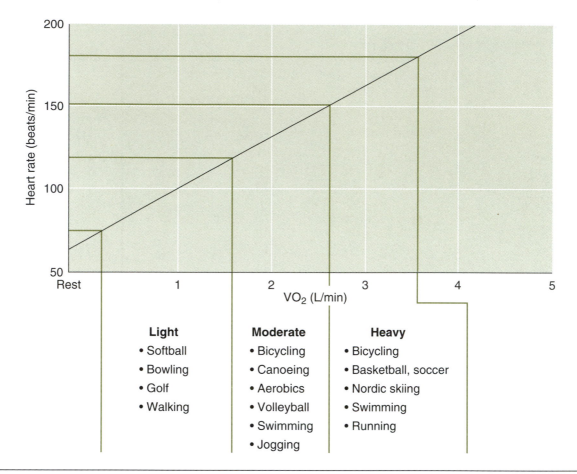

Figure 18.9 Heart rate and exercise prescription. Using the relationship between O_2 consumption and heart rate, one can assign a variety of exercises to relative exercise intensities. Activities that the client prefers can be chosen to attain the desired exercise prescription.

What You Need to Know From Chapter 18

Key Terms

aerobic capacity	flexibility
Balke protocol	functional capacity
Borg rating of perceived exertion (RPE) scale	graded exercise test (GXT)
Bruce protocol	Karvonen formula
compliance	overload
Cooper protocol	peak work capacity
detrained	phase I-IV rehabilitation
endurance	shuttle tests
ergometer	specificity
exercise-induced bronchospasm	strength
field tests	symptom-limited work capacity
	$\dot{V}O_2$max

Key Concepts

1. You should understand the differences between tests for $\dot{V}O_2$max, peak work capacity, functional capacity, and symptom-limited capacity, as well as the circumstances under which each of these tests is appropriate.

2. You should be familiar with common laboratory protocols and safety procedures.

3. You should be able to use exercise test information to design elementary exercise conditioning or rehabilitation programs.

Review Questions

Jane, 68 years old, has been referred for cardiac rehabilitation. She has been employed as a secretary for 31 years at the same company. Her job is sedentary, and she has never been very physically active during her free time. She has been married for 42 years and has one child and three grandchildren. She was unable to have more children because she has had diabetes since age 7. Prior to her myocardial infarction, her diabetes was under good control. Over the past several years, she has monitored her blood glucose three to four times a day with a portable glucose meter and has been able to keep her blood glucose levels between 80 and 125 mg/100 ml blood through insulin injections and dietary control. Before that, however, her blood glucose would change greatly throughout the day and she experienced many instances of both hypo- and hyperglycemia.

Jane was admitted to the hospital six weeks ago. She had been complaining of "chest tightness" for several days but had done nothing about it until it became unbearable. At the time of her admission her ECG showed an ST segment depression of 3 mm, indicating significant myocardial ischemia. Her blood level of creatine phosphokinase—a heart muscle enzyme that enters the blood when there is damage to the myocardium—was six times higher than normal, indicating the possibility of significant damage to the heart.

She was given medication to reduce the work of her heart and was stabilized within 48 hr after admission. Four days after admission her ECG showed sign of damage to the left anterior portion of the heart, but the medication seemed to have eliminated the myocardial ischemia since the ST segment was not depressed. On the fifth day after admission, Jane was encouraged to get up to go to the washroom with assistance. Her ECG monitor showed no sign of ischemia. On day 3, a therapist helped Jane walk slowly for 6 min in the hall of the ward. Again, the monitor showed no signs of ischemia. Jane progressed over the next three days so that she could walk for 10 min with no assistance and no further signs of ischemia. This progression is typical of phase I in cardiac rehabilitation.

At the time of discharge, 8 days after her admission, Jane performed a treadmill exercise test using the Bruce protocol. The first stage was at 4.7 km/hr at 0% grade for 3 min. The second stage was 4.7 km/hr at 5% grade, at which time she began to complain of mild angina and the ECG showed a 1 mm ST depression. Her heart rate at the report of angina was 115 beats/min. At the next stage, 4.7 km/hr and 10% grade, the ST depression was over 3 mm, and she had moderate angina. The test was terminated at this time. Her resting heart rate was 82, and the heart rates during the three exercise levels were 98, 116 and 127 beats/min, respectively. Her blood pressure was 132/89 at rest and 145/95, 157/88, and 169/100 at the three exercise levels, respectively.

Over the next four weeks Jane participated in a rehabilitation program in which her ECG was monitored continuously.

1. How would you have calculated the upper and lower heart rate limits for Jane's exercise program?

2. Jane exercised at this level for 10 min initially, then progressed to 25 min at the end of the three-week period. During this same period she attended education sessions. What topics would be important for these sessions?

3. Jane continued to monitor her blood glucose and found that she was having difficulty keeping it in the desirable range. She learned that her heart attack had probably made her diabetes worse and also that she needed to

(continued)

adjust her insulin replacement to account for the fact that exercise increases the sensitivity of the insulin receptors. In general, what is your advice for her insulin?

4. For the past two weeks Jane has been exercising with only intermittent monitoring, and her ECG shows no signs of ischemia if she stays within her target range. She is now ready to move into your phase III program. In your initial interview with Jane, you record her history and need to ask some questions that will help you formulate her exercise prescription. What information do you need to optimize your exercise prescription?

Bibliography

American Association of Cardiovascular and Pulmonary Rehabilitation. *Guidelines for pulmonary rehabilitation programs.* 2nd ed. Champaign, IL: Human Kinetics, 1993.

American Association of Cardiovascular and Pulmonary Rehabilitation. *Guidelines for cardiac rehabilitation programs.* 2nd ed. Champaign, IL: Human Kinetics, 1995.

American College of Sports Medicine. *ACSM's guidelines for exercise testing and prescription.* 5th ed. Baltimore: Williams & Wilkins, 1995.

American College of Sports Medicine. *ACSM's resource manual for guidelines for exercise testing and prescription.* 3rd ed. Baltimore: Williams & Wilkins, 1998.

Australian Sports Commission. *Physiological tests for elite athletes.* Champaign, IL: Human Kinetics, 2000.

Baechle TR, ed. *Essentials of strength training and conditioning.* 2nd ed. Champaign, IL: Human Kinetics, 2000.

Buchfurer MJ, Hansen JE, Robinson TE, Sue DY, Wasserman K, and Whipp BJ. Optimizing the exercise protocol for cardiopulmonary assessment. *J Appl Physiol* 1983, 55:1558–1564.

Council on Scientific Affairs. Indications and contraindications for exercise testing. *JAMA* 1981, 246:1015–1018.

Cropp GJA. The exercise bronchoprovocation test: Standardization of procedures and evaluation of response. *J Aller Clin Immunol* 1979, 64:627–633.

Froelicher VF, Thompson AJ, Noguera I, Davis G, Stewart AJ, and Triebwasser JH. Prediction of maximal oxygen consumption: Comparison of the Bruce and Balke treadmill protocols. *Chest* 1975, 68:331–336.

Greiwe JS, Kaminsky LA, Whaley MH, and Dwyer GB. Evaluation of the ACSM submaximal ergometer test for estimating VO$_2$max. *Med Sci Sports Exerc* 1995, 27:1315–1320.

Howley ET, Bassett DR, and Welch HG. Criteria for maximal oxygen uptake: review and commentary. *Med Sci Sports Exerc* 1995, 27:1292–1301.

Kline GM, Porcari JP, Hintermeister R, Freedson PS, Ward A, McCarron RF, Ross J, and Rippe JM. Estimation of VO$_2$ max from a one-mile track walk, gender, age and body weight. *Med Sci Sports Exerc* 1987, 19:253–259.

MacDougall D, Wenger H, and Green H. *Physiological testing of the high performance athlete.* 2nd ed. Champaign, IL: Human Kinetics, 1991.

McGavin CR. Twelve minute walking test for assessing disability in chronic bronchitis. *Br Med J* 1976, 1:822–823.

McGuire LB. The uses and limits of standard exercise test. *Arch Int Med* 1981, 141:2290232.

Rochmis P and Blackburn H. Exercise tests: A survey of procedures, safety, and litigation experience in approximately 170,000 tests. *JAMA* 1971, 217:1061–1066.

Sawka MN, Glaser RM, Laubach LL, Al-Samkari O, and Suryaprasad AG. Wheelchair exercise performance of the young, middle-aged, and elderly. *J Appl Physiol* 1981, 50:824–828.

Skinner J, ed. *Exercise testing and exercise prescription for special cases.* 2nd ed. Philadelphia: Lea & Febiger, 1993.

Wilmore J and Costill DL. *Training for sport and activity.* Champaign, IL: Human Kinetics, 1988.

Zhang Y, Johnson MC, Chow N, and Wasserman K. Effects of exercise testing protocol on parameters of aerobic function. *Med Sci Sports Exerc* 1991, 23:625–630.

Calculation of Oxygen Consumption ($\dot{V}O_2$)

In its simplest form, the calculation of O_2 consumption can be expressed as the amount (volume) of O_2 taken into the lungs minus the amount (volume) of O_2 exhaled from the lungs.

$$\dot{V}O_2 = \dot{V}_I O_2 - \dot{V}_E O_2$$

Of course, nothing can remain that simple! To calculate the elements on the right side of the equation, we need to know the volume of air inspired ($\dot{V}_I$) and exhaled ($\dot{V}_E$) per minute, or minute ventilation and the fraction of O_2 in the inspired ($F_I O_2$) and exhaled ($F_E O_2$) air. We already know that the $F_I O_2$ is 0.209 or 20.9% of the ambient air. In most laboratory situations, we actually measure the $F_E O_2$ and the $F_E CO_2$ along with the $\dot{V}_E$.

A typical situation is to have the subject breathe on a two-way valve so that the inspired and exhaled air go in two different directions, and to measure volume, O_2, and CO_2 only from the exhaled air. Once we know the volume of air exhaled in 60 s and the fraction of O_2 and CO_2 in that air, we can calculate the amount (volume) of O_2 exhaled in that 60 s. We know the $F_I O_2$ and have measured the $\dot{V}_E$ and the $F_E O_2$.

Before we go on, we need to consider the fact that any volume of air is dependent on the temperature and barometric pressure conditions under which it is measured. In the system described in the preceding paragraph, the volume of air is measured under ATPS conditions (i.e., ambient temperature and pressure and saturated water vapor). To eliminate the effects of varying temperature and pressure conditions on the expression of $\dot{V}O_2$, we convert the volume measurement to standard temperature and pressure and dry air (STPD) conditions.

$$\dot{V}_{STPD} = \dot{V}_{ATPS} \left(\frac{273}{273 + T^\circ C} \right) \left(\frac{P_B - P_{H_2O}}{760} \right)$$

The only part of the equation that is still unknown is the volume inspired. We cannot assume that the volume inspired is the same as the volume exhaled, because the amount of CO_2 produced from each breath is not the same as the amount of O_2 consumed. This inequality is based on the differences in O_2 consumed and CO_2 produced during the metabolism of fats and carbohydrates as fuels and is reflected in the respiratory exchange ratio, or respiratory quotient (see chapter 4). An estimate of the inequality can be made on the basis of differences between inspired and exhaled N_2. Since N_2 is neither consumed nor produced, any change in its concentration has to be due to changes in the other gases present, O_2 and CO_2. Because of this relationship between the concentrations of O_2 and CO_2 combined and N_2, we use our O_2 and CO_2 values to calculate changes in N_2 concentration, and by that estimate the difference between inspired and exhaled volume. The equation for calculating $\dot{V}_I$ from $\dot{V}_E$ is as follows:

$$\dot{V}_I = \dot{V}_E \left(\frac{1 - (F_E O_2 + F_E CO_2)}{F_I N_2} \right)$$

where F_IN_2 is 0.7904. Substituting this calculated value for $\dot{V}_I$ we can now calculate $\dot{V}O_2$.

In practice, this method of measuring $\dot{V}O_2$ is used only in laboratory environments. Clinical environments use automated, electronic equipment that provides minute-by-minute or even breath-by-breath values for $\dot{V}O_2$ and multiple other variables. An understanding of this laboratory measurement of $\dot{V}O_2$ is necessary, however, to properly use, troubleshoot, and interpret data from this automated equipment.

Energy Expenditure Calculations for Treadmill or Cycle Ergometer Exercise

Horizontal: Use these conversion factors to calculate the energy cost of horizontal movement during walking or running.

0.1 ml O_2/kg/min for every m/min walking (50 to 100 m/min)

0.2 ml O_2/kg/min for every m/min running (>134 m/min)

Vertical: Use these conversion factors to calculate the energy cost of vertical movement during running or walking.

1.8 ml O_2/kg/min for every m/min speed walking

0.9 ml O_2/kg/min for every m/min speed running

and multiply the product by % grade, as a fraction

Rest: Resting expenditure normalized for body mass = 3.5 ml O_2/kg/min.

Calculations:

1. What is the $\dot{V}O_2$ for a 60 kg person walking at a speed of 80 m/min, 0% grade?

   a. Calculate the horizontal component:

   $$\dot{V}O_2 = 0.1\ \frac{\text{ml } O_2 / \text{kg} / \text{min}}{\text{m} / \text{min}} \times 80\ \text{m} / \text{min} =$$
   $$8\ \text{ml } O_2 / \text{kg} / \text{min}$$

   b. Calculate the vertical component:

   $$0\% \times 1.8\ \text{ml } O_2 / \text{kg} / \text{min} \times 2\ \text{m} / \text{min} =$$
   $$0.0\ \text{ml } O_2 / \text{kg} / \text{min}$$

   c. Add the resting component:

   $$3.5\ \text{ml } O_2 / \text{kg} / \text{min}$$

Total = 11.5 ml O_2/kg/min

Of course, the whole-body $\dot{V}O_2$ would be 11.5 ml O_2/kg/min $\times$ body weight, in kg.

$$\text{Total } \dot{V}O_2 = 11.5\ \text{ml } O_2 / \text{kg} / \text{min} \times 60\ \text{kg} =$$
$$690\ \text{ml } O_2 / \text{min or } 0.690\ \text{L} / \text{min}$$

How many kcal are expended for this activity?

How many METs are represented?

2. The same person is now walking up a slope of 5% grade (5% is expressed as 0.05).

   a. Calculate the horizontal component:

   $$\dot{V}O_2 = 0.1\ \frac{\text{ml } O_2 / \text{kg} / \text{min}}{\text{m} / \text{min}} \times 80\ \text{m} / \text{min} =$$
   $$8.0\ \text{ml } O_2 / \text{kg} / \text{min}$$

   b. Calculate the vertical component:

   $$1.8\ \text{ml } O_2 / \text{kg} / \text{min} \times 80\ \text{m} / \text{min} \times 0.05 =$$
   $$7.2\ \text{ml } O_2 / \text{kg} / \text{min}$$

c. Add the resting component:

$$3.5 \text{ ml O}_2/\text{kg}/\text{min}$$

Total = 18.7 ml O_2/kg/min

The whole-body $\dot{V}O_2$ is 18.7 ml O_2/kg/min $\times$ body weight, in kg.

$$18.7 \text{ ml O}_2 / \text{kg} / \text{min} \times 60 \text{ kg} =$$
$$1122 \text{ ml O}_2 / \text{min or } 1.122 \text{ L O}_2 / \text{min}$$

How many kcal are expended for this activity?

How many METs are represented?

3. Calculate the $\dot{V}O_2$ for a 60 kg person running at 8 mph on a level surface.

   a. Convert 8 mph to m/min:

   $$8 \text{ mph} \times 26.8 \frac{\text{m} / \text{min}}{1 \text{ mph}} = 214 \text{ m} / \text{min}$$

   b. Calculate the horizontal component:

   $$\dot{V}O_2 = 0.2 \frac{\text{ml O}_2 / \text{kg} / \text{min}}{\text{m} / \text{min}} \times 214 \text{ m} / \text{min} =$$
   $$42.8 \text{ ml O}_2 / \text{kg} / \text{min}$$

   c. Calculate the vertical component:

   $$0\% \times 1.8 \text{ ml O}_2/\text{kg}/\text{min} \times 214 \text{ m}/\text{min} =$$
   $$0.0 \text{ ml O}_2/\text{kg}/\text{min}$$

   d. Add the resting component:

   $$3.5 \text{ ml O}_2/\text{kg}/\text{min}$$

Total = 46.3 ml O_2/kg/min

What is the whole-body $\dot{V}O_2$?

How many kcal are expended for this activity?

How many METs are represented?

4. This same person is now running up a 5% grade. Calculate the $\dot{V}O_2$.

   a. Calculate the horizontal component:

   $$\dot{V}O_2 = 0.2 \frac{\text{ml O}_2 / \text{kg} / \text{min}}{\text{m} / \text{min}} \times 214 \text{ m} / \text{min} =$$
   $$42.8 \text{ ml O}_2 / \text{kg} / \text{min}$$

b. Calculate the vertical component:

$$0.9 \text{ ml O}_2/\text{kg}/\text{min} \times 214 \text{ m}/\text{min} \times 5 \% =$$
$$9.6 \text{ ml O}_2/\text{kg}/\text{min}$$

c. Add the resting component:

$$3.5 \text{ ml O}_2/\text{kg}/\text{min}$$

Total = 55.9 ml O_2/kg/min

What is the whole-body $\dot{V}O_2$?

How many kcal are expended for this activity?

How many METs are represented?

In the case of a cycle ergometer there is no horizontal component, and the vertical component is based on the power output (kgm/min). Remember, expressions of energy expenditure during running and walking are normalized for body mass (ml O_2/kg/min), whereas energy expenditure for an activity that is not weight bearing, such as cycling, is not dependent on body mass so expressions of energy expenditure during these activities are expressed in absolute terms (L O_2/min).

Use the following factors for calculating energy expenditure during cycle ergometer exercise:

- **Horizontal:** No conversion factors.
- **Vertical:** Use the following conversion factors to calculate the energy cost of vertical movement during cycling.

2.0 ml O_2/min for each kgm/min

where kgm/min = kg resistance $\times$ meters $\times$ revolutions/min

(Note: The distance per revolution is 6 m for Monark ergometers and 3 m for Tunturi and Bodyguard ergometers.)

- **Rest:** Resting expenditure = 3.5 ml O_2/kg/min $\times$ body mass in kg

5. Calculate $\dot{V}O_2$ for a 50 kg person working on a Monark ergometer with a tension of 3 kg and a pedal rate of 60 rpm.

   a. Calculate the vertical component:

   $$\dot{V}O_2 = 2.0 \text{ ml O}_2/\text{min} \times (3 \text{ kg} \times 6 \text{ m} \times 60 \text{ rpm}) =$$
   $$1080 \text{ ml O}_2/\text{min}$$

b. Add the resting component:

3.5 ml O_2/kg/min $\times$ 50 kg =175 ml O_2/min

Total = 1255 ml O_2/min

What is the normalized O_2 consumption?

How many kcal are expended for this activity?

How many METs are represented?

Solutions to the metabolic equation questions

1. The first calculation determined that the relative $\dot{V}O_2$ for a 60 kg person walking at 80 m/min, 0% grade is 11.5 ml O_2/kg/min and that the total $\dot{V}O_2$ is 0.690 L/min.

The kcal expended for this activity would be 0.690 L/min $\times$ 5 kcal/L O_2 = 3.45 kcal/min.

The number of METs would be 11.5 ml O_2/kg/min $\div$ 3.5 = 3.3 METs.

2. For this same person walking up a 5% grade, the $\dot{V}O_2$ is 18.7 ml O_2/kg/min or 1.22 L O_2/min.

The kcal expended would be 1.22 L O_2/min $\times$ 5 kcal/L O_2 = 6.1 kcal/min.

The number of METs would be 18.7 ml O_2/kg/min $\div$ 3.5 = 5.34 METs.

3. The 60 kg person running at 8 mph would consume 46.3 ml O_2/kg/min.

The total-body $\dot{V}O_2$ would be 46.3 ml O_2/kg/min $\times$ 60 kg = 2778 ml O_2/min or 2.78 L O_2/min.

The total kcal would be 2.78 L O_2/min $\times$ 5 kcal/L O_2 = 13.9 kcal/min.

The total METs would be 46.3 ml O_2/kg/min $\div$ 3.5 = 13.2 METs.

4. This same person running up a 5% grade would consume 55.9 ml O_2/kg/min.

The total $\dot{V}O_2$ would be 55.9 ml O_2/kg/min $\times$ 60 kg = 3354 ml O_2/min or 3.354 L/min.

The kcal energy expenditure for this activity would be 3.354 L/min $\times$ 5 kcal/L O_2 = 16.8 kcal/min.

The number of METs would be 55.9 ml O_2/kg/min $\div$ 3.5 = 16 METs.

5. The $\dot{V}O_2$ for the 50 kg person exercising on the ergometer would be 1255 ml O_2/min.

The normalized $\dot{V}O_2$ would be 1255 ml O_2/min $\div$ 50 kg = 25.1 ml O_2/kg/min.

The kcal expended would be 1.255 L O_2/min $\times$ 5 kcal/L O_2 = 6.28 kcal/min.

The number of METs would be 25.1 ml O_2/kg/min $\div$ 3.5 = 7.2 METs.

Compendium of Physical Activities

METs	Activity category	Specific activity
8.5	Bicycling	Bicycling, BMX or mountain
4.0	Bicycling	Bicycling, <10 mph, general, leisure, to work or for pleasure
6.0	Bicycling	Bicycling, 10-11.9 mph, leisure, slow, light effort
8.0	Bicycling	Bicycling, 12-13.9 mph, leisure, moderate effort
10.0	Bicycling	Bicycling, 14-15.9 mph, racing or leisure, fast, vigorous effort
12.0	Bicycling	Bicycling, 16-19 mph, racing/not drafting or >19 mph drafting, very fast, racing, general
16.0	Bicycling	Bicycling, >20 mph, racing, not drafting
5.0	Bicycling	Unicycling
5.0	Conditioning exercise	Bicycling, stationary, general
3.0	Conditioning exercise	Bicycling, stationary, 50 W, very light effort
5.5	Conditioning exercise	Bicycling, stationary, 100 W, light effort
7.0	Conditioning exercise	Bicycling, stationary, 150 W, moderate effort
10.5	Conditioning exercise	Bicycling, stationary, 200 W, vigorous effort
12.5	Conditioning exercise	Bicycling, stationary, 250 W, very vigorous effort
8.0	Conditioning exercise	Calisthenics (e.g., push-ups, pull-ups, sit-ups), heavy, vigorous effort
4.5	Conditioning exercise	Calisthenics, home exercise, light or moderate effort, general (e.g., back exercises), going up and down from floor
8.0	Conditioning exercise	Circuit training, general
6.0	Conditioning exercise	Weightlifting (free weight, Nautilus or universal type), power lifting or body building, vigorous effort
5.5	Conditioning exercise	Health club exercise, general
6.0	Conditioning exercise	Stair-treadmill ergometer, general
9.5	Conditioning exercise	Rowing, stationary ergometer, general
3.5	Conditioning exercise	Rowing, stationary, 50 W, light effort

METs	Activity category	Specific activity
7.0	Conditioning exercise	Rowing, stationary, 100 W, moderate effort
8.5	Conditioning exercise	Rowing, stationary, 150 W, vigorous effort
12.0	Conditioning exercise	Rowing, stationary, 200 W, very vigorous effort
9.5	Conditioning exercise	Ski machine, general
6.0	Conditioning exercise	Slimnastics
4.0	Conditioning exercise	Stretching, hatha yoga
6.0	Conditioning exercise	Teaching aerobic exercise class
4.0	Conditioning exercise	Water aerobics, water calisthenics
3.0	Conditioning exercise	Weightlifting (free, Nautilus or universal type), light or moderate effort, light workout, general
1.0	Conditioning exercise	Whirlpool, sitting
6.0	Dancing	Aerobic, ballet or modem, twist
6.0	Dancing	Aerobic, general
5.0	Dancing	Aerobic, low impact
7.0	Dancing	Aerobic, high impact
4.5	Dancing	General
5.5	Dancing	Ballroom, fast (disco, folk, square)
3.0	Dancing	Ballroom, slow (e.g., waltz, fox-trot, slow dancing)
5.0	Fishing and hunting	Fishing, general
4.0	Fishing and hunting	Digging worms, with shovel
5.0	Fishing and hunting	Fishing from river bank and walking
2.5	Fishing and hunting	Fishing from boat, sitting
3.5	Fishing and hunting	Fishing from river bank, standing
6.0	Fishing and hunting	Fishing in stream, in waders
2.0	Fishing and hunting	Fishing, ice, sitting
2.5	Fishing and hunting	Hunting, bow and arrow or crossbow
6.0	Fishing and hunting	Hunting, deer, elk, large game
2.5	Fishing and hunting	Hunting, duck, wading
5.0	Fishing and hunting	Hunting, general
6.0	Fishing and hunting	Hunting, pheasants or grouse
5.0	Fishing and hunting	Hunting, rabbit, squirrel, prairie chicken, raccoon, small game
2.5	Fishing and hunting	Pistol shooting or trap shooting, standing
2.5	Home activities	Carpet sweeping, sweeping floors
4.5	Home activities	Cleaning, heavy or major (e.g., wash car, wash windows, mop, clean garage), vigorous effort
3.5	Home activities	Cleaning, house or cabin, general
2.5	Home activities	Cleaning, light (dusting, straightening up, vacuuming, changing linen, carrying out trash), moderate effort
2.3	Home activities	Washing dishes, standing or in general (not broken into stand / walk components)
2.3	Home activities	Washing dishes; clearing dishes from table, walking

METs	Activity category	Specific activity
2.5	Home activities	Cooking or food preparation, standing or sitting or in general (not broken into stand/walk components)
2.5	Home activities	Serving food, setting table, implied walking or standing
2.5	Home activities	Cooking or food preparation, walking
2.5	Home activities	Putting away groceries (e.g., carrying groceries, shopping without a grocery cart)
8.0	Home activities	Carrying groceries up stairs
3.5	Home activities	Food shopping, with grocery cart
2.0	Home activities	Standing—shopping (non-grocery shopping)
2.3	Home activities	Walking—shopping (non-grocery shopping)
2.3	Home activities	Ironing
1.5	Home activities	Sitting—knitting, sewing, light wrapping (presents)
2.0	Home activities	Weaving at loom, sitting
2.0	Home activities	Implied standing—laundry, folding or hanging clothes, putting clothes in washer or dryer, packing suitcase
2.3	Home activities	Implied walking—putting away clothes, gathering clothes to pack, putting away laundry
2.0	Home activities	Making bed
5.0	Home activities	Maple syruping/sugar bushing (including carrying buckets, carrying wood)
6.0	Home activities	Moving furniture, household
5.5	Home activities	Scrubbing floors, on hands and knees
4.0	Home activities	Sweeping garage, sidewalk, or outside of house
7.0	Home activities	Moving household items, carrying boxes
3.5	Home activities	Standing—packing/unpacking boxes, occasional lifting of household items, light-moderate effort
3.0	Home activities	Implied walking—putting away household items, moderate effort
9.0	Home activities	Moving household items up stairs, carrying boxes or furniture
2.5	Home activities	Standing—light (pump gas, change light bulb, etc.)
3.0	Home activities	Walking—light, noncleaning (get ready to leave, shut/lock doors, close windows, etc.)
2.5	Home activities	Sitting—playing with child(ren), light
2.8	Home activities	Standing—playing with child(ren), light
4.0	Home activities	Walk/run—playing with child(ren), moderate
5.0	Home activities	Walk/run—playing with child(ren), vigorous
3.0	Home activities	Child care: sitting/kneeling—dressing, bathing, grooming, feeding, occasional lifting of child, light effort
3.5	Home activities	Child care: standing—dressing, bathing, grooming, feeding, occasional lifting of child, light effort
3.0	Home repair	Airplane repair
4.5	Home repair	Automobile body work

METs	Activity category	Specific activity
3.0	Home repair	Automobile repair
3.0	Home repair	Carpentry, general, workshop
6.0	Home repair	Carpentry, outside house, installing rain gutters
4.5	Home repair	Carpentry, finishing or refinishing cabinets or furniture
7.5	Home repair	Carpentry, sawing hardwood
5.0	Home repair	Caulking, chinking log cabin
4.5	Home repair	Caulking, except log cabin
5.0	Home repair	Cleaning gutters
5.0	Home repair	Excavating garage
5.0	Home repair	Hanging storm windows
4.5	Home repair	Laying or removing carpet
4.5	Home repair	Laying tile or linoleum
5.0	Home repair	Painting, outside house
4.5	Home repair	Painting, papering, plastering, scraping, hanging Sheetrock, remodeling inside house
3.0	Home repair	Applying and remove tarp—sailboat
6.0	Home repair	Roofing
4.5	Home repair	Sanding floors with power sander
4.5	Home repair	Scraping and painting sailboat or power boat
5.0	Home repair	Spreading dirt with a shovel
4.5	Home repair	Washing and waxing hull of sailboat, car, power boat, airplane
4.5	Home repair	Washing fence
3.0	Home repair	Wiring, plumbing
0.9	Inactivity, quiet	Lying quietly, reclining (watching television), lying quietly in bed—awake
1.0	Inactivity, quiet	Sitting quietly (riding in a car, listening to a lecture or music, watching television or a movie)
0.9	Inactivity, quiet	Sleeping
1.0	Inactivity, light	Standing quietly (standing in a line)
1.0	Inactivity, light	Reclining—writing
1.0	Inactivity, light	Reclining—talking or talking on phone
1.0	Inactivity, light	Reclining—reading
5.0	Lawn and garden	Carrying—loading or stacking wood, loading/unloading or carrying lumber
6.0	Lawn and garden	Chopping wood, splitting logs
5.0	Lawn and garden	Clearing land, hauling branches
5.0	Lawn and garden	Digging sandbox
5.0	Lawn and garden	Digging, spading, filling garden
6.0	Lawn and garden	Gardening with heavy power tools, tilling a garden (see occupation, shoveling)
5.0	Lawn and garden	Laying crushed rock

METs	Activity category	Specific activity
5.0	Lawn and garden	Laying sod
5.5	Lawn and garden	Mowing lawn, general
2.5	Lawn and garden	Mowing lawn, riding mower
6.0	Lawn and garden	Mowing lawn, walk, hand mower
4.5	Lawn and garden	Mowing lawn, walk, power mower
4.5	Lawn and garden	Operating snow blower, walking
4.0	Lawn and garden	Planting seedlings, shrubs
4.5	Lawn and garden	Planting trees
4.0	Lawn and garden	Raking lawn
4.0	Lawn and garden	Raking roof with snow rake
3.0	Lawn and garden	Riding snow blower
4.0	Lawn and garden	Sacking grass, leaves
6.0	Lawn and garden	Shoveling, snow, by hand
4.5	Lawn and garden	Trimming shrubs or trees, manual cutter
3.5	Lawn and garden	Trimming shrubs or trees, power cutter
3.5	Lawn and garden	Walking, applying fertilizer or seeding a lawn
1.5	Lawn and garden	Watering lawn or garden, standing or walking
4.5	Lawn and garden	Weeding, cultivating garden
5.0	Lawn and garden	Gardening, general
3.0	Lawn and garden	Implied walking/standing—picking up yard, light
1.5	Miscellaneous	Sitting—card playing, playing board games
2.0	Miscellaneous	Standing—drawing (writing), casino gambling
1.3	Miscellaneous	Sitting—reading
1.8	Miscellaneous	Sitting—writing, desk work
1.8	Miscellaneous	Standing—talking or talking on phone
1.5	Miscellaneous	Sitting—talking or talking on phone
1.8	Miscellaneous	Sitting—studying, general, including reading and/or writing
1.8	Miscellaneous	Sitting—in class, general, including note-taking or class discussion
1.8	Miscellaneous	Standing—reading
1.8	Music playing	Accordion
2.0	Music playing	Cello
2.5	Music playing	Conducting
4.0	Music playing	Drums
2.0	Music playing	Flute (sitting)
2.0	Music playing	Horn
2.5	Music playing	Piano or organ
3.5	Music playing	Trombone
2.5	Music playing	Trumpet
2.5	Music playing	Violin
2.0	Music playing	Woodwind

METs	Activity category	Specific activity
2.0	Music playing	Guitar, classical, folk (sitting)
3.0	Music playing	Guitar, rock and roll band (standing)
4.0	Music playing	Marching band, playing an instrument, baton twirling (walking)
3.5	Music playing	Marching band, drum major (walking)
4.0	Occupation	Baking, general
2.3	Occupation	Bookbinding
6.0	Occupation	Building road (including hauling debris, driving heavy machinery)
2.0	Occupation	Building road, directing traffic (standing)
3.5	Occupation	Carpentry, general
8.0	Occupation	Carrying heavy loads, such as bricks
8.0	Occupation	Carrying moderate loads up stairs, moving boxes (16-40 lb)
2.5	Occupation	Chambermaid
6.5	Occupation	Coal mining, drilling coal, rock
6.5	Occupation	Coal mining, erecting supports
6.0	Occupation	Coal mining, general
7.0	Occupation	Coal mining, shoveling coal
5.5	Occupation	Construction, outside, remodeling
3.5	Occupation	Electrical work, plumbing
8.0	Occupation	Farming, baling hay, cleaning barn, poultry work
3.5	Occupation	Farming, chasing cattle, nonstrenuous
2.5	Occupation	Farming, driving harvester
2.5	Occupation	Farming, driving tractor
4.0	Occupation	Farming, feeding small animals
4.5	Occupation	Farming, feeding cattle
8.0	Occupation	Farming, forking straw bales
3.0	Occupation	Farming, milking by hand
1.5	Occupation	Farming, milking by machine
5.5	Occupation	Farming, shoveling grain
12.0	Occupation	Firefighter, general
11.0	Occupation	Firefighter, climbing ladder with full gear
8.0	Occupation	Firefighter, hauling hoses on ground
17.0	Occupation	Forestry, ax chopping, fast
5.0	Occupation	Forestry, ax chopping, slow
7.0	Occupation	Forestry, barking trees
11.0	Occupation	Forestry, carrying logs
8.0	Occupation	Forestry, felling trees
8.0	Occupation	Forestry, general
5.0	Occupation	Forestry, hoeing
6.0	Occupation	Forestry, planting by hand
7.0	Occupation	Forestry, sawing by hand

METs	Activity category	Specific activity
4.5	Occupation	Forestry, sawing, power
9.0	Occupation	Forestry, trimming trees
4.0	Occupation	Forestry, weeding
4.5	Occupation	Furriery
6.0	Occupation	Horse grooming
8.0	Occupation	Horse racing, galloping
6.5	Occupation	Horse racing, trotting
2.6	Occupation	Horse racing, walking
6.0	Occupation	Lifting 22 lb 1 m
8.0	Ocupation	Lifting 45 lb 1 m
11.0	Occupation	Lifting 65 lb 1 m
3.5	Occupation	Locksmith
2.5	Occupation	Machine tooling, machining, working sheet metal
3.0	Occupation	Machine tooling, operating lathe
5.0	Occupation	Machine tooling, operating punch press
4.0	Occupation	Machine tooling, tapping and drilling
3.0	Occupation	Machine tooling, welding
7.0	Occupation	Masonry, concrete
4.0	Occupation	Masseur, masseuse (standing)
7.0	Occupation	Moving, pushing heavy objects, 75 lb or more (desks, moving-van work)
2.5	Occupation	Operating heavy-duty equipment/automated, not driving
4.5	Occupation	Orange grove work
2.3	Occupation	Printing (standing)
1.5	Occupation	Police, directing traffic (standing)
2.0	Occupation	Police, driving a squad car (sitting)
1.3	Occupation	Police, riding in a squad car (sitting)
8.0	Occupation	Police, making an arrest (standing)
2.5	Occupation	Shoe repair, general
8.5	Occupation	Shoveling, digging ditches
9.0	Occupation	Shoveling, heavy (>16 lb · mile^{-1})
6.0	Occupation	Shoveling, light (<10 lb · mile^{-1})
7.0	Occupation	Shoveling, moderate (10-15 lb · mile^{-1})
1.5	Occupation	Sitting—light office work, in general (chemistry lab work, light use of hand tools, watch repair or microassembly, light assembly/repair)
1.5	Occupation	Sitting—meetings
2.5	Occupation	Sitting, moderate (heavy levers, riding mower/forklift, crane operation)
2.5	Occupation	Standing, light (bartending, store clerk, assembling, filing, photocopying, putting up Christmas tree)

METs	Activity category	Specific activity
3.0	Occupation	Standing, light/moderate (assemble/repair heavy parts, welding, stocking, auto repair, packing boxes for moving, etc.), patient care (as in nursing)
3.5	Occupation	Standing, moderate (assembling at fast rate, lifting 50 lb, hitch/twisting ropes)
4.0	Occupation	Standing, moderate/heavy (lifting >50 lb, masonry, painting, paper hanging)
5.0	Occupation	Steel mill, fettling
5.5	Occupation	Steel mill, forging
8.0	Occupation	Steel mill, hand rolling
8.0	Occupation	Steel mill, merchant mill rolling
11.0	Occupation	Steel mill, removing slag
7.5	Occupation	Steel mill, tending furnace
5.5	Occupation	Steel mill, tipping molds
8.0	Occupation	Steel mill, working in general
2.5	Occupation	Tailoring, cutting
2.5	Occupation	Tailoring, general
2.0	Occupation	Tailoring, hand sewing
2.5	Occupation	Tailoring, machine sewing
4.0	Occupation	Tailoring, pressing
6.5	Occupation	Truck driving, loading and unloading truck (standing)
1.5	Occupation	Typing, electric, manual, or computer
6.0	Occupation	Using heavy power tools (e.g., pneumatic tools, jackhammers, drills)
8.0	Occupation	Using heavy tools (not power) such as shovel, pick, tunnel bar, spade
2.0	Occupation	Walking on job, <2.0 mph (in office or lab area), very slow
3.5	Occupation	Walking on job, 3.0 mph, in office, moderate speed, not carrying anything
4.0	Occupation	Walking on job, 3.5 mph, in office, brisk speed, not carrying anything
3.0	Occupation	Walking, 2.5 mph, slowly and carrying light objects <25 lb
4.0	Occupation	Walking, 3.0 mph, moderately and carrying light objects <25 lb
4.5	Occupation	Walking, 3.5 mph, briskly and carrying objects <25 lb
5.0	Occupation	Walking or walking down stairs or standing, carrying objects about 25-49 lb
6.5	Occupation	Walking or walking down stairs or standing, carrying objects about 50-74 lb
7.5	Occupation	Walking or walking down stairs or standing, carrying objects about 75-99 lb
8.5	Occupation	Walking or walking down stairs or standing, carrying objects about 100 lb and over

METs	Activity category	Specific activity
3.0	Occupation	Working in scene shop, or as theater actor, backstage employee
6.0	Running	Jog/walk combination (jogging component of <10 min)
7.0	Running	Jogging, general
8.0	Running	Running, 5 mph (12 min · mile^{-1})
9.0	Running	Running, 5 mph (12 min · mile^{-1})
8.0	Running	Running, 5.2 mph (11.5 min · mile^{-1})
10.0	Running	Running, 6 mph (10 min · mile^{-1})
11.0	Running	Running, 6.7 mph (9 min · mile^{-1})
11.5	Running	Running, 7 mph (8.5 min · mile^{-1})
12.5	Running	Running, 7.5 mph (8 min · mile^{-1})
13.5	Running	Running, 8 mph (7.5 min · mile^{-1})
14.0	Running	Running, 8.6 mph (7 min · mile^{-1})
15.0	Running	Running, 9 mph (6.5 min · mile^{-1})
16.0	Running	Running, 10 mph (6 min · mile^{-1})
18.0	Running	Running, 10.9 mph (5.5 min · mile^{-1})
9.0	Running	Running, cross country
8.0	Running	Running, general
8.0	Running	Running, in place
15.0	Running	Running, stairs, up
10.0	Running	Running, on a track, team practice
8.0	Running	Running, training, pushing wheelchair, marathon wheeling
2.5	Self-care	Standing—getting ready for bed, in general
1.0	Self-care	Sitting on toilet
2.0	Self-care	Bathing (sitting)
2.5	Self-care	Dressing, undressing (standing or sitting)
1.5	Self-care	Eating (sitting)
2.0	Self-care	Talking and eating or eating only (standing)
2.5	Self-care	Sitting or standing—grooming (washing, shaving, brushing teeth, urinating, washing hands, putting on makeup)
4.0	Self-care	Showering, toweling off (standing)
1.5	Sexual activity	Active, vigorous effort
1.3	Sexual activity	General, moderate effort
1.0	Sexual activity	Passive, light effort, kissing, hugging
3.5	Sports	Archery (nonhunting)
7.0	Sports	Badminton, competitive
4.5	Sports	Badminton, social singles and doubles, general
8.0	Sports	Basketball, game
6.0	Sports	Basketball, nongame, general
7.0	Sports	Basketball, officiating
4.5	Sports	Basketball, shooting baskets

METs	Activity category	Specific activity
6.5	Sports	Basketball, wheelchair
2.5	Sports	Billiards
3.0	Sports	Bowling
12.0	Sports	Boxing, in ring, general
6.0	Sports	Boxing, punching bag
9.0	Sports	Boxing, sparring
7.0	Sports	Broomball
5.0	Sports	Children's games (hopscotch, 4-square, dodgeball, playground apparatus, T-ball, tetherball, marbles, jacks, arcade games)
4.0	Sports	Coaching: football, soccer, basketball, baseball, swimming, etc.
5.0	Sports	Cricket (batting, bowling)
2.5	Sports	Croquet
4.0	Sports	Curling
2.5	Sports	Darts, wall or lawn
6.0	Sports	Drag racing, pushing or driving a car
6.0	Sports	Fencing
9.0	Sports	Football, competitive
8.0	Sports	Football, touch, flag, general
2.5	Sports	Football or baseball, playing catch
3.0	Sports	Frisbee playing, general
3.5	Sports	Frisbee, ultimate
4.5	Sports	Golf, general
5.5	Sports	Golf, carrying clubs
3.0	Sports	Golf, miniature, driving range
5.0	Sports	Golf, pulling clubs
3.5	Sports	Golf, using power cart
4.0	Sports	Gymnastics, general
4.0	Sports	Hacky sack
12.0	Sports	Handball, general
8.0	Sports	Handball, team
3.5	Sports	Hang gliding
8.0	Sports	Hockey, field
8.0	Sports	Hockey, ice
4.0	Sports	Horseback riding, general
3.5	Sports	Horseback riding, saddling horse
6.5	Sports	Horseback riding, trotting
2.5	Sports	Horseback riding, walking
3.0	Sports	Horseshoe pitching, quoits
7.5	Sports	In-line skating, 10 mph
8.5	Sports	In-line skating, 11 mph

METs	Activity category	Specific activity
10.0	Sports	In-line skating, 12 mph
12.0	Sports	Jai alai
10.0	Sports	Judo, jujitsu, karate, kick boxing, tae kwon do
4.0	Sports	Juggling
7.0	Sports	Kickball
8.0	Sports	Lacrosse
4.0	Sports	Motocross
9.0	Sports	Orienteering
10.0	Sports	Paddleball, competitive
6.0	Sports	Paddleball, casual, general
8.0	Sports	Polo
10.0	Sports	Racketball, competitive
7.0	Sports	Racketball, casual, general
11.0	Sports	Rock climbing, ascending rock
8.0	Sports	Rock climbing, rappelling
8.0	Sports	Rollerskiing, 10 mph, no grade
10.0	Sports	Rollerskiing, 11 mph, no grade
11.0	Sports	Rollerskiing, 12 mph, no grade
12.0	Sports	Rollerskiing, 9 mph, 6% grade
12.0	Sports	Rope jumping, fast
10.0	Sports	Rope jumping, moderate, general
8.0	Sports	Rope jumping, slow
10.0	Sports	Rugby
3.0	Sports	Shuffleboard, lawn bowling
5.0	Sports	Skateboarding
7.0	Sports	Skating, roller
3.5	Sports	Sky diving
10.0	Sports	Soccer, competitive
7.0	Sports	Soccer, casual, general
5.0	Sports	Softball or baseball, fast or slow pitch, general
4.0	Sports	Softball, officiating
6.0	Sports	Softball, pitching
12.0	Sports	Squash
4.0	Sports	Table tennis, Ping-Pong
4.0	Sports	Tai chi
7.0	Sports	Tennis, general
6.0	Sports	Tennis, doubles
8.0	Sports	Tennis, singles
3.5	Sports	Trampoline
4.0	Sports	Volleyball, competitive, in gymnasium

METs	Activity category	Specific activity
3.0	Sports	Volleyball, noncompetitive; 6- to 9-member team, general
8.0	Sports	Volleyball, beach
6.0	Sports	Wrestling (one match = 5 min)
7.0	Sports	Wallyball, general
2.0	Transportation	Automobile or light truck (not a semi) driving
2.0	Transportation	Flying airplane
2.5	Transportation	Motor scooter, motor cycle
6.0	Transportation	Pushing plane in and out of hangar
3.0	Transportation	Driving heavy truck, tractor, bus
7.0	Walking	Backpacking, general
3.5	Walking	Carrying infant or 15 lb load (e.g., suitcase), level ground or down stairs
9.0	Walking	Carrying load up stairs, general
5.0	Walking	Carrying 1 to 15 lb load, up stairs
6.0	Walking	Carrying 16 to 24 lb load, up stairs
8.0	Walking	Carrying 25 to 49 lb load, up stairs
10.0	Walking	Carrying 50 to 74 lb load, up stairs
12.0	Walking	Carrying 74+ lb load, up stairs
7.0	Walking	Climbing hills with 0 to 9 lb load
7.5	Walking	Climbing hills with 10 to 20 lb load
8.0	Walking	Climbing hills with 21 to 42 lb load
9.0	Walking	Climbing hills with 42+ lb load
3.0	Walking	Down stairs
6.0	Walking	Hiking, cross country
8.0	Walking	Ice climbing
6.5	Walking	Marching, rapidly, military
2.5	Walking	Pushing or pulling stroller with child
6.5	Walking	Race walking
8.0	Walking	Rock or mountain climbing
8.0	Walking	Up stairs, using or climbing up ladder
4.0	Walking	Using crutches
2.0	Walking	Walking, <2.0 mph, level ground, strolling, household walking, very slow
2.5	Walking	Walking, 2.0 mph, level, slow pace, firm surface
3.0	Walking	Walking, 2.5 mph, firm surface
3.0	Walking	Walking, 2.5 mph, downhill
3.5	Walking	Walking, 3.0 mph, level, moderate pace, firm surface
4.0	Walking	Walking, 3.5 mph, level, brisk, firm surface
6.0	Walking	Walking, 3.5 mph, uphill
4.0	Walking	Walking, 4.0 mph, level, firm surface, very brisk pace
4.5	Walking	Walking, 4.5 mph, level, firm surface, very, very brisk

METs	Activity category	Specific activity
3.5	Walking	Walking, for pleasure, work break, walking the dog
5.0	Walking	Walking, grass track
4.0	Walking	Walking, to work or class
2.5	Water activities	Boating, power
4.0	Water activities	Canoeing, on camping trip
7.0	Water activities	Canoeing, portaging
3.0	Water activities	Canoeing, rowing, 2.0-3.9 mph, light effort
7.0	Water activities	Canoeing, rowing, 4.0-5.9 mph, moderate effort
12.0	Water activities	Canoeing, rowing, >6 mph, vigorous effort
3.5	Water activities	Canoeing, rowing, for pleasure, general
12.0	Water activities	Canoeing, rowing, in competition, or crew or sculling
3.0	Water activities	Diving, springboard or platform
5.0	Water activities	Kayaking
4.0	Water activities	Paddleboat
3.0	Water activities	Sailing, boat and board sailing, windsurfing, ice sailing, general
5.0	Water activities	Sailing, in competition
3.0	Water activities	Sailing, Sunfish/Laser/Hobby Cat, keel boats, ocean sailing, yachting
6.0	Water activities	Skiing, water
7.0	Water activities	Skimobiling
12.0	Water activities	Skin diving or scuba diving as frogman
16.0	Water activities	Skin diving, fast
12.5	Water activities	Skin diving, moderate
7.0	Water activities	Skin diving, scuba diving, general
5.0	Water activities	Snorkeling
3.0	Water activities	Surfing, body or board
10.0	Water activities	Swimming laps, freestyle, fast, vigorous effort
8.0	Water activities	Swimming laps, freestyle, slow, moderate or light effort
8.0	Water activities	Swimming, backstroke, general
10.0	Water activities	Swimming, breaststroke, general
11.0	Water activities	Swimming, butterfly, general
11.0	Water activities	Swimming, crawl, fast (75 yards $\cdot$ mile^{-1}), vigorous effort
8.0	Water activities	Swimming, crawl, slow (50 yards $\cdot$ mile^{-1}), moderate or light effort
6.0	Water activities	Swimming, lake, ocean, river
6.0	Water activities	Swimming, leisurely, not lap swimming, general
8.0	Water activities	Swimming, sidestroke, general
8.0	Water activities	Swimming, synchronized
10.0	Water activities	Swimming, treading water, fast vigorous effort
4.0	Water activities	Swimming, treading water, moderate effort, general
7.0	Water activities	Swimming, underwater, 1 mph

METs	Activity category	Specific activity
10.0	Water activities	Water polo
3.0	Water activities	Water volleyball
5.0	Water activities	Whitewater rafting, kayaking, or canoeing
6.0	Winter activities	Moving ice house (set up/drill holes, etc.)
9.0	Winter activities	Skating, figure skating
5.5	Winter activities	Skating, ice, 9 mph or less
7.0	Winter activities	Skating, ice, general
9.0	Winter activities	Skating, ice, rapidly, >9 mph
15.0	Winter activities	Skating, speed, competitive
7.0	Winter activities	Ski jumping (climb up carrying skis)
7.0	Winter activities	Skiing, general
14.0	Winter activities	Skiing, competitive, short periods
7.0	Winter activities	Skiing, cross-country, 2.5 mph, slow or light effort, ski walking
8.0	Winter activities	Skiing, cross-country, 4.0-4.9 mph, moderate speed and effort, general
9.0	Winter activities	Skiing, cross-country, 5.0-7.9 mph, brisk speed, vigorous effort
14.0	Winter activities	Skiing, cross-country, >8.0 mph, racing
16.5	Winter activities	Skiing, cross-country, hard snow, uphill, maximum
5.0	Winter activities	Skiing, downhill, light effort
6.0	Winter activities	Skiing, downhill, moderate effort, general
8.0	Winter activities	Skiing, downhill, vigorous effort, racing
7.0	Winter activities	Sledding, tobogganing, bobsledding, luge
8.0	Winter activities	Snow shoeing
3.5	Winter activities	Snowmobiling

Readers should convert measurements from English units to metric units as needed.

Reprinted from Ainsworth et al. 1993.

Body Fat Estimates

Body fat is commonly estimated using hydrostatic weighing, skinfold measurements, anthropometry, bioelectrical impedance devices, dual-energy x-ray absorbitometry (DEXA), infrared interactance, and body plethysmography (for example, BodPod, a body plethysmograph). We present the technique for hydrostatic weighing, since it is considered the gold-standard technique, and for skinfold measurement, since it is the most common field technique. In any case, the equations that are used were developed from specific population samples and may not apply to groups outside that population or even at the periphery of the chosen population. The descriptions that follow will give you a general idea of how to perform the measurements and include the equations used for each technique. More specific instructions are available in laboratory manuals designed for that purpose.

Hydrostatic Weighing

Most of us have heard the story of how Archimedes, wishing to ascertain the gold content of the king's crown, discovered that the mass of an object placed in water displaces an equivalent mass of water. Archimedes knew the mass of the crown in room air and was able to determine its weight in the water. The difference in weight, in kg, was equivalent to the volume, in cm^3, of the crown. Density of the crown could be calculated as D = mass/volume. Since Archimedes knew the density of pure gold and since the calculated density of the crown was less than that of gold, he was able to tell the king that the crown was not pure gold.

Now, if we wanted to apply these principles to measuring the density of the human body, we would measure the mass (kg) of the body in room air, say 70 kg, and measure the mass of the body completely submerged. The loss in weight in the water, say 65 kg, would be equivalent to the volume of the body, so that density would be 70/65 or 1.077 kg/cm^3. Since we know the relative densities of fat and fat-free tissues, we can estimate their proportions in this body based on the density estimate.

At a water temperature of 4°C, the volume of the water displaced is equal to the volume of the object in the water. Since we do hydrostatic weighing at much higher temperatures, we must correct for the effects of the warmer water on volume displaced. The volume of air in the lungs also affects the weight in water; therefore it is important to have the subject blow out as much air as possible under water and to estimate the residual volume (RV = the amount of air left in the lungs at the end of a maximal exhalation; see chapter 11) so that this can be taken into account. Residual volume is either directly measured or assumed to be 30% of the vital capacity, expressed at body temperature and pressure and saturated with water vapor (BTPS) conditions. The formula for taking into account RV and the water temperature correction is

$$Db = \frac{M_a}{\frac{M_a - M_w}{D_w} - RV}$$

where M_a = weight in air in g, M_w = weight in water in g, D_w = density of water at actual water temperature, and RV = residual volume, in ml, estimated from vital capacity.

A typical college age-student might weigh 80 kg in air, weigh only 2.2 kg in water, and have an RV of about 2.5 L. If we assume a water temperature of 77° F (25° C), the correction factor would be 0.997. Density of this person would be

$$Db = \frac{80000 \text{ g}}{\dfrac{80000 \text{ g} - 2200 \text{ g}}{0.997} - 2500 \text{ ml}} = 1.059$$

One of several equations can be used to estimate body fat from this density measurement; Brozek's is % body fat = (457/D) – 414.2, and Siri's is % body fat = (495/D) – 450. For our student, the % fat would be 17.4%.

Skinfold Measurements

Analysis of human cadaver specimens has shown that the amount of subcutaneous fat is reflective of the total body fat. Estimates of total body fat have been made using caliper measurements of single and multiple sites. The objective is to find the fewest number of appropriate sites that, in combination, will provide an acceptable estimate of body fat as measured with densiometric (hydrostatic) analyses. Because of the ease of measurement and the reasonableness of the prediction, we present the technique of body fat estimation from three sites. These equations, developed for the general adult population, show less predictability when used in extremely thin or fat people or in the pediatric population.

The accuracy of skinfold measurements to estimate body fat depends on adherence to standard protocols using acceptable instruments. These techniques should be practiced and monitored by trained personnel before implementation in the field. Standard skinfold calipers are manufactured to provide a known amount of pressure (10 g/mm²) over the full range of measurement (0-50 mm).

The subject should stand in a relaxed position. All sites should be measured on the right side of the body. The selected fold of skin and subcutaneous fat should be grasped between the left thumb

and forefinger and pulled gently away from the body. The caliper should be placed approximately 1 cm (0.4 in.) from the fingers. As the caliper is released, the pressure exerted by the fingers should decrease so that the caliper pressure on the fold is greater. When the reading is stable, record the number to the nearest 0.1 mm. Be sure to open the calipers before removing them from the fold. You should obtain three consistent readings from each site.

Although there are many equations for predicting body fat from skinfold measurements using a variety of combinations of sites, the simplest form uses three sites and is presented here. For females, the sites are the tricep, the suprailiac, and the thigh; for males they are the chest (pectoral), the umbilical, and the thigh.

The tricep skinfold is located midway between the acromion and olecranon processes on the back of the right upper arm. The suprailiac site is located immediately above the crest of the ilium. The thumb is placed over the crest, and the fold is taken medial to the midline at a slight angle following the normal slope of the crest. The chest site is located above and to the right of the right nipple. This fold is taken at a 45° angle to the horizontal following a line from the nipple to the anterior axillary line. The thigh measurement is taken midway between the patella and the inguinal fold of the right leg. The right foot is placed forward; full weight is placed on the left foot. The fold is lifted parallel to the axis of the leg.

To calculate body density from these skinfold measurements the following equations should be used.

For males:

Density = 1.109380 – 0.0008267(Σ 3 skinfolds) + 0.0000016(Σ 3 skinfolds)² – 0.0002574(age),

where Σ = sum.

For females:

Density = 1.0994921 – 0.0009929(Σ 3 skinfolds) + 0.0000023(Σ 3 skinfolds)² – 0.0001392(age),

where Σ = sum.

Once body density is determined, this number can be substituted in the Brozek or Siri equation for calculating % body fat. To determine a body-weight target for a desired body fat percentage, the total amount of existing body fat is calculated.

Fat weight = body weight in air − % fat

Then the lean body weight (LBW) is calculated.

LBW = body weight in air − fat weight

Finally the estimated "ideal" or desired weight is calculated assuming that the LBW isn't going to change.

Desired body weight =
LBW/(100% − desired % fat) × 100

In our example we have already determined the % body fat to be 17.4. The fat weight would be 80 kg × 0.174 = 13.9 kg, and the LBW would be 80 kg − 13.9 kg = 66.1 kg. If we assume that the desired % body fat might be 12%, the person's target weight would be 66.1/(100 − 12) × 100 = 75 kg. In this case you would expect the student to lose 5 kg of fat weight.

Bibliography

Brozek J, Grande F, Anderson JT, and Keys A. Densiometric analysis of body composition: Revision of some quantitative assumptions. *Ann NY Acad Sci* 1963, 110:113-140.

Heyward VN and Stolarczyk LM. *Applied body composition assessment.* Champaign, IL: Human Kinetics, 1996.

Jackson AS and Pollock ML. Generalized equations for predicting body density of men. *Br J Nutr* 1978, 61:497-504.

Jackson AS, Pollock ML, and Ward A. Generalized equations for predicting body density of women. *Med Sci Sports Exerc* 1980, 12:175-182.

Siri WE. The gross composition of the body. In *Advances in biological physics,* vol. 4, ed. CA Tobias and JH Lawrence, pp. 239-280. New York: Academic Press, 1956.

Questionnaires

Baecke Questionnaire

Questionnaire, Codes, and Method of Calculation of Scores on Habitual Physical Activity

1. What is your main occupation? (Low level [1] occupations include clerical work, driving, shopkeeping, teaching, studying, housework, medical practice, and all other occupations with a university education; middle level [3] occupations include factory work, plumbing, carpentry, and farming; high level [5] occupations include dock work, construction work, and sport.)

   1 3 5

2. At work I sit

   never seldom sometimes often always   1 2 3 4 5

3. At work I stand

   never seldom sometimes often always   1 2 3 4 5

4. At work I walk

   never seldom sometimes often always   1 2 3 4 5

5. At work I lift heavy loads

   never seldom sometimes often very often   1 2 3 4 5

6. After working I am tired

   very often often sometimes seldom never   5 4 3 2 1

7. At work I sweat

   very often often sometimes seldom never   5 4 3 2 1

8. In comparison with others of my own age I think my work is physically

   much heavier heavier as heavy lighter much lighter   5 4 3 2 1

9. Do you play sports?   yes   no

   If yes:

Which sport do you play most frequently?	Intensity	0.76 1.26 1.76	
How many hours a week? 1 1-2 2-3 3-4 4	Time	0.5 1.5 2.5 3.5 4.5	
How many months a year? 1 1-3 4-6 7-9 9	Proportion	0.04 0.17 0.42 0.67 0.92	

If you play a second sport:

Which sport?	Intensity 0.76 1.26 1.76
How many hours a week? 1 1-2 2-3 3-4 4	Time 0.5 1.5 2.5 3.5 4.5
How many months a year? 1 1-3 4-6 7-9 9	Proportion 0.04 0.17 0.42 0.67 0.92

10. In comparison with others my own age, I think my physical activity during leisure time is

      much more more the same less much less   5 4 3 2 1

11. During leisure time I sweat

      very often often sometimes seldom never   5 4 3 2 1

12. During leisure time I play a sport

      never seldom sometimes often very often   1 2 3 4 5

13. During leisure time I watch television

      never seldom sometimes often very often   1 2 3 4 5

14. During leisure time I walk

      never seldom sometimes often very often   1 2 3 4 5

15. During leisure time I cycle

      never seldom sometimes often very often   1 2 3 4 5

16. How many minutes do you walk and/or cycle per day to and from work, school, and shopping?

      <5 5-15 15-30 30-45 >45   1 2 3 4 5

Calculation of the simple sport score (1_9):

(A score of zero is given to people who do not play a sport.)

$1_9 = \Sigma$ (intensity $\times$ time $\times$ proportion) = 0/0.01 <4.4 <8.8 <12 12   1 2 3 4 5

Calculation of scores of the indexes of physical activity:

Work index = $[1_1 + (6 - 1_2) + 1_3 + 1_4 + 1_5 + 1_6 + 1_7 + 1_8]/8$

Sport index = $[1_9 + 1_{10} + 1_{11} + 1_{12}]/4$

Leisure-time index = $[(6 - 1_{13}) + 1_{14} + 1_{15} + 1_{16}]/4$

Physical Activity Recall Items

Now we would like to know about your physical activity during the past 7 days. But first, let me ask you about your sleep habits.

1. On the average, how many hours did you sleep each night during the last 5 weekday nights (Sunday-Thursday)?_____hours

2. On the average, how many hours did you sleep each night last Friday and Saturday nights?____hours

Now I am going to ask you about your physical activity during the past 7 days—that is, the last 5 weekdays and last weekend, Saturday and Sunday. We are not going to talk about light activities, such as slow walking or light housework, or unstrenuous sports such as bowling, archery, or softball. Please look at this list, which shows some examples of what we consider moderate, hard, and very hard activities. [Interviewer: hand subject "Examples of Activities in Each Category" (page 334) and allow time for the subject to read it over.] People engage in many other types of activities, and if you are not sure where one of your activities fits, please ask me about it.

3. First, let's consider moderate activities. What activities did you do and how many total hours did you spend during the last 5 weekdays doing these moderate activities or others like them? Please tell me to the nearest half hour.____ hours

4. Last Saturday and Sunday, how many hours did you spend on moderate activities and what did you do? (Probe: Can you think of any other sports, job, or household activities that would fit into this category?)____hours

5. Now, let's look at hard activities. What activities did you do and how many total hours did you spend during the last 5 weekdays doing these hard activities or others like them? Please tell me to the nearest half hour.____ hours

6. Last Saturday and Sunday, how many hours did you spend on hard activities and what did you do? (Probe: Can you think of any other sports, job, or household activities that would fit into this category?)____hours

7. Now, let's look at very hard activities. What activities did you do and how many total hours did you spend during the last 5 weekdays doing these very hard activities or others like them? Please tell me to the nearest half hour.____hours

8. Last Saturday and Sunday, how many hours did you spend on very hard activities and what did you do? (Probe: Can you think of any other sports, job, or household activities that would fit into this category?)____hours

9. Compared with your physical activity over the past 3 months, was last week's physical activity more, less, or about the same?

   1. More

   2. Less

   3. About the same

Interviewer: Please list below any activities reported by the subject which you don't know how to classify. Flag this record for review and completion.

Activity (Brief description)	Hr: workday	Hr: weekend day
_____	_____	_____
_____	_____	_____
_____	_____	_____
_____	_____	_____
_____	_____	_____

Example of Activities in Each Categoy

Moderate Activity

- Occupational tasks: delivering mail or patrolling on foot; house painting; truck driving (making deliveries, lifting and carrying light objects)
- Household tasks: raking the lawn; sweeping and mopping; mowing the lawn with a power mower; cleaning windows
- Sport activities (actual playing time): volleyball; Ping-Pong; brisk walking (4.93 km/hr [3 miles/hr] or 12.5 min/km [20 min/mile]); golf, walking and pulling or carrying clubs; calisthenic exercises

Hard Activity

- Occupational tasks: heavy carpentry; construction work, physical labor
- Household tasks: scrubbing floors
- Sport activities (actual playing time): tennis doubles; disco, square, or folk dancing

Very Hard Activity

- Occupational tasks: very hard physical labor, digging or chopping with heavy tools; carrying heavy loads such as bricks or lumber
- Sport activities (actual playing time): jogging or swimming; singles tennis; racquetball; soccer

Note: Some examples of various levels of activity are given. However, the list in Appendix C is more comprehensive and can be used with moderate activities rated 3 to 5 METs; hard activities, 5.1 to 6.9 METs; and very hard activities, 7 METs and more.

Reprinted from Sallis et al. 1985.

Answers and Possible Solutions to Review Questions

Chapter 1 Bioenergetics

1. The calculations are as follows:

   a. To calculate the treadmill parameters, you need to first determine the $\dot{V}O_2$ on the cycle ergometer.

$$\dot{V}O_2 = 2 \text{ ml/kgm} \times 200 \text{ kgm} + (3.5 \text{ ml } O_2/\text{kg/min} \times 61 \text{ kg})$$

$$\dot{V}O_2 = 400 + 213.5 = 613.5 \text{ ml } O_2/\text{min}$$

$\dot{V}O_2$ is expressed in ml O_2/kg/min on the treadmill, so you need to convert 613.5 by dividing by his weight.

$$613.5/61 = 10.1 \text{ ml } O_2/\text{kg/min}$$

Now, to calculate the speed in m/min required to elicit a $\dot{V}O_2$ of 10.1 ml O_2/kg/min, you substitute in the treadmill equation, using speed as your unknown.

$$10.1 \text{ ml } O_2/\text{kg/min} = 0.1 \text{ ml } O_2/\text{m/min} \times \text{speed} + (\text{elevation factor}) + 3.5$$

Solving for speed, you get 66 m/min or 66/26.8 = 2.5 mph.

   b. Since 2.5 is too fast for this patient, you need to substitute 1.5 mph (40.2 m/min) in the equation and solve for elevation.

$$10.1 \text{ ml } O_2 / \text{kg} / \text{min} =$$
$$\frac{0.1 \text{ ml } O_2 / \text{kg} / \text{min}}{\text{m} / \text{min}} \times 40.2 \text{ m} / \text{min} +$$
$$\left(\text{elevation} \times 40.2 \text{ m} / \text{min} \times \frac{1.8 \text{ ml } O_2 / \text{kg} / \text{min}}{\text{m} / \text{min}} \right) +$$
$$3.5 \text{ ml } O2 / \text{kg} / \text{min}$$

Solving for elevation:

$$10.1 \text{ ml } O_2/\text{kg/min} =$$
$$4.02 \text{ ml } O_2/\text{kg/min} +$$
$$(\text{elevation} \times 72.36 \text{ ml } O_2/\text{kg/min}) +$$
$$3.5 \text{ ml } O_2/\text{kg/min}$$

$$6.6 \text{ ml } O_2/\text{kg/min} =$$
$$4.02 \text{ ml } O_2/\text{kg/min} +$$
$$(\text{elevation} \times 72.36 \text{ ml } O_2/\text{kg/min})$$

$$2.58 \text{ ml } O_2/\text{kg/min} =$$
$$\text{elevation} \times 72.36 \text{ ml } O_2/\text{kg/min}$$

elevation = 0.036 or 3.6% grade

   c. To calculate METs, simply divide the energy expenditure (10.1 ml O_2/kg/min) by 3.5 ml O_2/kg/min, which is approximately 3 METs.

   d. To convert Peter's O_2 consumption into energy expenditure, use the conversion factor 5 kcal/L O_2/min.

$$5 \text{ kcal/L } O_2 \times .6135 \text{ L } O_2/\text{min} =$$
$$3.07 \text{ kcal/min or } 61.35 \text{ kcal}$$
$$\text{for the 20 min session}$$

2. To compare the two athletes, you need to normalize the $\dot{V}O_2$ maximum by dividing it by each person's weight, in kg. Sue has a maximum exercise capacity to consume O_2 of 55.9 ml O_2/ kg/min, while Sara's is 58 ml O_2/kg/min; thus you will have to give Sara the edge.

3. 50% of the peak METs is 5 METs or 5×3.5 ml O_2/kg/min = 17.5 ml O_2/kg/min. 17.5 ml O_2/ kg/min 3 75 kg = 1312.5 ml O_2/min. Substituting this in the metabolic equation for the cycle ergometer, and assuming 6 m/revolution at 60 rpm:

$$1312.5 \text{ ml } O_2/\text{min} =$$
$$\text{kg} \times 2 \text{ ml } O_2/\text{kgm} \times 6 \text{ m/revolution} \times 60 \text{ rpm}$$

$$1312.5 \text{ ml } O_2/\text{min} = \text{kg} \times 720$$

$$\text{resistance} = 1.8 \text{ kg on the ergometer}$$

Chapter 2 Nutrition

1. Glucose is the primary form of carbohydrate used for energy production. Other forms of carbohydrate must be converted to glucose to be transformed to produce ATP.

2. Glycogen, a long-chained carbohydrate, is stored in the muscle and liver. In the muscle, glycogen is a form of glucose readily available to produce ATP during periods of high demand such as at the start of exercise or during high-intensity exercise. In the liver, this glycogen serves as a reservoir for glucose.

3. Lipids are transported as lipoproteins. A high concentration of lipid makes the lipoprotein low density; a low amount of lipid makes lipoprotein more dense.

4. Fatty acids are stored in the form of triglycerides. The fatty acids connected to these triglycerides can be mobilized from the stores and transported to the muscle where they are metabolized to produce ATP.

5. You should recommend that fat intake be reduced to below 30% of the total caloric intake. It is easy to see how to do this by looking at the food label and dividing the calories from fat by the total calories.

6. You should recommend that blood cholesterol levels be kept below 200 mg/dl.

7. A low-fat diet can reduce the risks by lowering cholesterol and the LDL levels. Exercise can reduce the risk by increasing the HDL levels.

8. A healthy, balanced diet that is meeting caloric needs generally supplies sufficient amino acids.

9. Meat is a good source of the amino acids we need in our diets. The ability to use ingested amino acids depends on having all of the essential amino acids available at the same time. If meat is eliminated from the diet, careful selection of foods must ensure that all of these essential amino acids are ingested through other sources.

Chapter 3 Energy Balance

1. In each case, the person's natural gait has been altered sufficiently to increase the energy expenditure. The walking economy has been decreased. You should tell all these people that the alteration in their gait has made them less economical and that the increased energy expenditure is leading to sensations of breathlessness and early fatigue.

a. In the case of the athlete, it is possible that further practice with the shorter stride length might shift her economy curve so that she begins to expend less energy. You should advise her coach to be aware that dramatic changes in gait to make someone run so they look "ideal" may make a person less efficient because each person's anatomy and muscle configuration is different.

b. A period of gait retraining for the stroke patient may make him a more economical mover. Measuring $\dot{V}O_2$ while the patient walks on a treadmill might allow you to identify the most economical gait.

c. The patient with the cast might find that using crutches or a cane makes her movement more economical.

2. Your mistake was simply telling him that he should lose weight. The critical issue is not his weight, but his overfatness. The more you can get him to concentrate on this, the greater his likelihood of success. You must now explain to him the difference between being overweight and overfat and the health implications of the latter.

3. You should consider using skinfold measures of body composition to estimate his % body fat.

4. Weight loss through severe dietary restriction is resisted by the body's set point. As energy intake decreases, the body responds by slowing metabolism to conserve energy. When the person goes off the diet, this slower metabolism leads to weight gain. Repeated cycles of dieting, slowing of metabolism, and weight gain can result in a gradual accumulation of fat. Exercise along with dietary restriction would be a better approach to weight loss.

5. The apparent ineffective (uneconomical) movement has likely increased the energy expenditure more than expected. You need to make an assessment of this increase and account for it in her diet.

6. Body mass index (BMI) is likely readily available from student health records. Simply access the weight (convert it to kg, if necessary) and divide it by the height, in meters, squared.

Chapter 4 Metabolism

1. $\dot{V}O_2$ and exercise

a. The energy deficit is being met through the ATP and CP stores and through glycolytic metabolism.

b. The primary fuel is carbohydrate in general and glucose in particular. Since this deficit phase also is a period when O_2 delivery is not meeting demand, glycogen is used particularly during the transition from rest to steady-state exercise.

c. At the onset of exercise, the stored ATP is used immediately but is quickly replenished from CP stores. Ca^{++}, which is released into the cell to initiate muscle contraction, also stimulates the conversion of phosphorylase b to phosphorylase a so that glycogen can be broken down for use in the glycolytic pathway. Its use in this pathway is enhanced because the ADP and Pi left over from ATP breakdown stimulate the key enzyme, PFK, to increase glycolysis. The end product of glycolysis, pyruvic acid, begins to enter the oxidative pathway. The pyruvic acid that can't enter this pathway is converted to lactic acid. At the same time, blood flow to the muscle increases, delivering increased lipids, which are then processed through β-oxidation and oxidation. The increased citric acid from the TCA cycle inhibits PFK, thus slowing down glycolysis. As a result of this process, eventually up to 80% of the energy is derived from oxidation.

d. No. Some of the oxidative energy comes from glucose that has passed through glycolysis first.

2. Respiratory quotient during exercise

a. Phase A is the onset phase, and the intensity does not appear to be too high because the RQ increases moderately. High-intensity exercise would be indicated by an RQ closer to, or over, 1.0. This assessment is confirmed in phase B in which RQ drops, indicating a shift toward fat as the primary fuel. Toward the end of phase B, the RQ increases, indicating an increase in intensity. Again, this reflects the greater reliance on carbohydrate as a fuel, always a glycolytic fuel and used during higher-intensity exercise.

b. Phase A: glucose/glycogen. Phase B: initially fat, then back toward glucose. Phase C: a return toward a resting RQ with fuel use of approximately 50% carbohydrate and 50% fat. It would appear that this might have been measured during a race in which the intensity was moderate, then picked up at the end. Phase C would represent the recovery phase.

c. The answer to question 1c describes the various basic metabolic control mechanisms.

Chapter 5 Metabolic Diseases

1. Exercise increases the sensitivity of the insulin receptors. In the case of type 1 diabetes, this means that the patient will make more effective use of the available insulin, whether that is produced by the body or injected. In type 2 diabetes, the exercise-related increase in receptor sensitivity will overcome the disease-related decrease in sensitivity.

2. A patient whose diabetes is not under control will likely start exercise with a higher-than-normal blood glucose level that will climb during exercise. The reason is a lack of insulin replacement. The normal response to exercise is an increased uptake of glucose during the early stages, resulting in a lower insulin level. The lowering of insulin enables the release of glucose from the liver via stimulation by glucagon. This response, of course, helps ensure maintenance of blood glucose levels. The lack of insulin in the person with poorly controlled diabetes means that there is less removal of glucose; at the same time, normal glucagon levels promote the release of liver glucose, thereby promoting an increase in

blood glucose levels. This is why exercise is contraindicated for the patient with poorly controlled diabetes.

3. Hypoglycemia in the patient with diabetes is a result of too much insulin relative to the exercise-induced increase in receptor sensitivity. The usual strategies to minimize the occurrence of hypoglycemia are to reduce insulin and to ensure adequate blood glucose availability, the latter through ingestion of a snack of fruit to provide fructose that can be converted to glucose in the body.

4. McArdle's disease is manifested in a reduced ability to process glycogen through glycolysis. Processing of less substrate through the glycolytic pathway results in less pyruvic acid production.

Chapter 6 Neural Control of Movement

1. The resting membrane potential is determined largely by the separation of sodium (higher concentration on the outside) and potassium (higher on the inside) across the cell membrane. The cell membrane is more permeable to K^+ than to Na^+, so more K^+ diffuses out than sodium diffuses in, and the inside of the cell becomes more and more negative until an equilibrium potential is reached, around −70 mV. The separation of charges is maintained by a Na^+-K^+ pump that pumps NA^+ out of the cell and K^+ back in.

2. The cell is depolarized when the membrane permeability, first to Na^+ and then to K^+, increases and the membrane potential is driven in the positive direction. Basically this process is a membrane event, involving the movement of relatively few ions, which enables the resting membrane potential to be restored very quickly.

3. When the cell membrane is hyperpolarized, the cell cannot be easily depolarized. If the period of hyperpolarization that typically follows an action potential is short, the nerve can have a high discharge rate; if the afterhyperpolarization is relatively long, it is possible to generate action potentials at only a slow rate. This fact helps distinguish between slow and fast motor units in that the discharge rate of fast units is high and that of slow units is low.

4. In unmyelinated nerve axons the current is spread passively down the axon, so transmission speed is slow. Myelinated fibers have gaps of exposed membrane; this enables the current to jump from gap to gap and speed up transmission.

5. Receptor potentials are generated in sensory nerve endings by such stimuli as mechanical deformation, stretch, and chemical substances that are then passively transmitted to a segment of the axon called the trigger zone. Stimuli to motor nerves are received either by dendrites or directly by the soma, and can be inhibitory or excitatory. These inputs are summed at the axon hillock, and if they are of sufficient magnitude to reach threshold, an action potential is generated. In both cases, the number of action potentials developed (discharge rate) is proportional to the magnitude of the input stimulus.

6. The knee-jerk reflex is initiated by tapping (or pulling) on the patellar tendon. This action stretches the rectus femoris muscle and activates the muscle spindles, which lie in parallel with the muscle fibers. The stretch of the spindle is the input signal that is picked up by the sensory nerve endings, which generate a graded receptor potential. The receptor potential is passively transmitted to the trigger zone of the axon; and a series of action potentials is generated and transmitted to the axon terminal, which synapses with a motor neuron in the spinal cord. The motor neuron develops a graded synaptic potential, which is transmitted to the trigger zone and results in the generation of a number of action potentials; these are delivered back to the muscle where the input signal originated. The muscle fibers contract with an intensity that is exactly proportional to the magnitude of the original input. If the input (stretch) increases in magnitude, the receptor potential is bigger, more action potentials are generated, and the resulting muscle contraction is greater.

7. Sensory nerves tend to distribute their information by divergence, whereby a single sensory nerve has an impact on many different target cells. This enables a single piece of information to be delivered to many areas and increases the likelihood that it will be interpreted correctly. It also enables the delivery of signals by different pathways in the event that one or more pathways are destroyed. Motor nerves, on the other hand, tend to receive convergent information from many different sources. This allows information to be processed or modified at many different levels before the motor neuron is impacted, and helps ensure the proper selection of muscles or motor units and the correct timing sequence.

8. Parallel processing is a means by which similar information is delivered to or from a brain center via parallel pathways. If one pathway is eradicated, this process provides an alternate pathway through which the movement behavior *may* be relearned.

9. Structures of the CNS are arranged in a structural and functional hierarchy whereby the lower levels (e.g., spinal cord) regulate simple movements with little or no input from the higher levels. As the movement becomes more complex, the higher levels become more and more involved.

10. The quality of any movement is affected by motivation or emotion, which are regulated by the limbic system. The limbic system has direct connections to the motor cortex and can influence motor behavior directly. It also connects to the hypothalamus, which modulates the body's state of readiness. The hypothalamus, through the sympathetic nervous system, regulates heart rate, blood flow, and hormone release, among other things. A person who is highly motivated will perform a movement behavior, especially gross movements, more effectively than someone who is bored or uninterested.

11. Basal ganglia, through connections with association areas of the cerebral cortex, help determine which patterns of movement will be used together and in what sequence. The cerebellum compares outgoing motor information with incoming sensory input to determine whether or not the movement is proceeding as planned. It also helps the motor cortex plan the next movement sequence based on the backdrop of the incoming information.

12. Muscle spindles lie in parallel with muscle fibers and respond to stretch, while GTOs reside in the tendon and respond to changes in muscle force. Muscle spindles help regulate muscle length by responding to both the rate and magnitude of length change, causing the muscle to contract and antagonistic muscles to relax. The GTO is much more responsive to active force changes than to passive stretch, and acts as a "governor" by inhibiting the contracting muscle and exciting the antagonist. This prevents high forces from damaging muscle, tendon, and ligaments.

13. The motor, sensory, and motivational systems must work together for precise movement control. The motor system relies on a constant supply of sensory information to select the appropriate muscles and to time the sequences of muscle activation and inhibition. The motivational system acts directly on the motor cortex to influence muscle recruitment, but also acts indirectly through the hypothalamus to control the body's state of readiness.

14. The motor neuron cell body of slow motor units is relatively small, has a low resistance, and therefore can be easily activated. In addition, the slow motor neuron has a long afterhyperpolarization, which limits the discharge rate. The somas of fast motor units have a high resistance and a short afterhyperpolarization period. Thus, slow motor neurons are easily activated and have slow discharge rates, while fast neurons need a greater stimulus to depolarize and are capable of generating high frequencies of action potentials. These two characteristics more or less outline the differences between fast and slow motor units—slow units are used most frequently and are stimulated at low frequencies, while fast units are used infrequently and have high discharge rates. The axons of these two motor units are also different, again reflecting their basic characteristics. Fast units have axons with large diameters (high conduction velocity), and slow units have small-diameter axons (slow conduction velocities). Regardless of these differences, both fast and slow units deliver action potentials to the muscle fiber via the NMJ. Here, ACh is released from the axon terminal and diffuses to the muscle cell membrane, where an equivalent number of action potentials is generated. In this manner the original signal is delivered, in its entirety and without fail, from the motor neuron to the muscle fiber.

15. When a muscle contracts, various metabolites form and then diffuse out of the muscle cell. Adenosine is formed when phosphates are cleaved from ATP to release energy. As the concentration builds up over time, adenosine diffuses out of the cell and interferes with the release of ACh from the presynaptic terminal. Hydrogen ion, from the breakdown of lactic acid, is associated with short-term high-intensity exercise and will interfere with the binding site for ACh on the muscle fiber. Thus, adenosine is associated with fatigue over longer-duration exercise, and H^+ over shorter-term exercise.

16. Fiber type is largely determined by the pattern of activity of the motor nerve. This pattern of activity is related to factors covered in question 14. That is, slow motor nerves are capable of delivering action potentials at a low frequency (low discharge rate), whereas fast motor nerves have a high discharge rate. These rates somehow determine which types of proteins will be synthesized within the muscle fiber. A particular type of training can change the characteristics of the fiber to some extent (e.g., more oxidative, or more glycolytic), but it is unlikely the muscle fiber type can be altered unless the motor nerve undergoes changes as well.

17. The key principle here is that most muscles in humans have a spectrum of fiber types from slow to fast and from oxidative to glycolytic, rather than just three or four. This enables us to perform a wide variety of movements ranging from fast to slow and from short duration, high intensity to long duration, low intensity.

18. The vomiting and diarrhea caused a considerable loss of extracellular potassium, which made the inside of nerve (and muscle) cells more negative. This hyperpolarization made it more difficult to generate an action potential and resulted in Karen's muscle weakness.

19. Anatomically, a dorsal rhizotomy involves the transection of the spinal nerve roots to relieve pain and spasticity. It entails identifying and cutting sensory nerve fibers just dorsal to the spinal cord. A selective dorsal rhizotomy is a more specific procedure whereby the nerve fibers that are cut are ones that have been identified as generating high activity when stimulated.

Physiologically, the procedure is usually performed to reduce tone in the spastic muscles of the lower extremities. With part of the sensory nerve transected, muscle tone is decreased. The hope is to induce a reduction in the amount of random muscle recruitment to improve ambulation and increase range of motion. After surgery, physical therapy is needed to optimize results.

20. The higher ventilation with use of the arms is explained by the greater tension developed by the smaller muscle mass of the arms, relative to the lower tension developed by the greater mass of the legs doing the same work (i.e., the arms are working harder than the legs). Because of the greater tension, joint receptors in the arms send more signals to the respiratory center and stimulate greater ventilation. Since the work is the same and the ventilation is greater with the arm work, the person breathes more than necessary, causing a decrease in blood CO_2 and dizziness.

Chapter 7 Neuropathies of the Nervous System

1. The rugby player

a. Passive and active ROM exercises should be prescribed to maintain joint and muscle flexibility, and bed mobility activities should be prescribed to minimize the loss of function.

b. Demyelination of the phrenic nerve may result in respiratory paralysis.

c. Demyelination of the peripheral nerve produces loss of alpha motor neuron transmission to the muscle.

d. Schwann cells.

e. Prognosis is very good, with potential to return to professional rugby.

2. Tom

a. The region of the spinal cord that causes weakness is the anterior part of the cord, specifically the corticospinal tract in the spinal cord.

b. Tom experiences sensory loss because the inflammatory process involved the entire spinal cord, the anterior (motor) and posterior (sensory) regions.

c. Deep tendon reflexes are intact. All muscles are actively contracting; therefore alpha motor neurons are intact. Numbness exists but does not have an impact on the Ia afferent fiber from the muscle spindle, which composes the other half of the monosynaptic reflex arc.

d. Prognosis, similar to that for encephalitis, is good as long as the condition is medically managed and resolved.

3. Emma

a. A possible diagnosis is CVA, specifically lack of blood supply to the brainstem region limiting function of cranial nerves.

b. The ischemic condition from the stroke did not impair the descending or ascending brainstem/spinal tracts for motor and sensory function.

c. Emma will be deconditioned as a result of her hospitalization and should begin a graded exercise program under close supervision, including monitoring of vital signs.

Chapter 8 Skeletal Muscle Structure and Function

1. In adult muscle, different fiber types are interspersed, forming a mosaic pattern of distribution. This pattern is established during development as primary myotubes separate after being innervated, form a motor unit, and then serve as a scaffolding for the formation of secondary myotubes. After these secondary cells are innervated by their own nerves they separate, and other cells form on them. This process continues until all motor units within the muscle are established.

2. During the process described in the answer to question 1, some myoblasts become trapped between the basal lamina and the cell membrane of the newly formed cells. These myoblasts remain trapped and in their primitive state until a signal is received (through damage or mechanical stress) and they begin to divide, fuse, and either repair damaged fibers or add to existing protein already present (hypertrophy).

3. Some muscles have fibers arranged in parallel to the long axis, which means that all force is transmitted directly to the tendon. Others have fibers arranged at an angle to the long axis, which reduces the force transmitted to the tendon by each fiber but enables more fibers to be packed into the muscle. This results in more total force.

4. Slow postural muscles benefit from the increased tensile strength and stiffness provided by the greater amount of connective tissue surrounding them. This helps them provide continuous support of the body. Fast muscles have less connective tissue, and this facilitates efficiency and speed of contraction.

5. Sarcomeres are held in place by an infrastructure of support proteins that connect myosin molecules together, anchor actin, and provide a link to the cell membrane. Even when forceful contractions distort the sarcomeric organization, these support structures help return sarcomeres to their original positions.

6. When muscle fibers are activated, calcium release from the SR triggers myosin cross-bridge cycling (attachment and detachment of the myosin heads). Continual cycling of the myosin heads moves the actin relatively long distances and enables the muscle to move the bony lever through the entire range of motion. Because the cross-bridges cycle independently of one another, a smooth (rather than ratchet-like) movement is produced.

7. When a phosphate is removed from ATP to produce ADP, the energy released is used to power the swiveling of the myosin head (power stroke). The ADP and Pi are released, and another ATP molecule attaches and dissociates actin from myosin. If ATP were depleted, or even significantly reduced, myosin would remain attached to actin and the muscle would be in a state of rigor (very stiff).

8. The rate of force development is directly related to the speed of cross-bridge cycling, which is fast in type II fibers and slow in type I and is controlled largely by the activity of myosin ATPase (the enzyme that cleaves the phosphate bond). The rate of relaxation is related to the speed with which calcium is taken back into the SR, which is regulated by the calcium-activated SR ATPase. Thus, fast fibers contract and relax quickly, and slow fibers do so slowly, largely on the basis of these two processes.

9. When a fiber is activated by a single action potential, enough calcium is released to saturate all or most of the troponin. Since calcium reuptake begins immediately, calcium concentration falls, and the amount of cross-bridge cycling decreases accordingly. The mechanical response of the muscle is called a twitch. When several action potentials arrive at the muscle fiber, it doesn't have time to completely relax between stimuli, so forces of the individual twitches summate. If the action potentials are close enough together, maximum tetanic force is achieved and is about three to four times higher than the peak twitch force. During a twitch there is not enough time to stretch the elastic components of the muscle, so the external force (that transmitted to the tendon) is less than the internal force (that developed by the sarcomeres). During a tetanic contraction, the elastic components are stretched to their maximum, so external force equals that generated internally.

10. Endurance performance is linked directly to stores of muscle glycogen. If stores have been reduced from prior activity, endurance time will decrease. Conversely, if a person overloads glycogen stores by manipulating exhaustive exercise and diet, endurance time can be greatly increased. Dehydration is often accompanied by

electrolyte imbalance. Low levels of body water affect both cardiovascular and muscle function. An electrolyte imbalance affects membrane excitability. Cold muscles also have impaired excitation, whereas muscles at a high temperature have increased ATP utilization and decreased ATP production.

11. Isometric contractions are the most fatiguing because blood vessels are compressed by the contracting muscle fibers and flow is cut off. This makes the muscle hypoxic and results in a buildup of metabolites, causing rapid fatigue. Dynamic contractions are less fatiguing because blood flow is interrupted only intermittently. Eccentric contractions are less fatiguing than concentric contractions because less energy is used at any submaximal force level.

12. Your figure could look something like the following.

The first panel (*a*) represents the mechanical response to a single stimulus—a twitch. Notice that

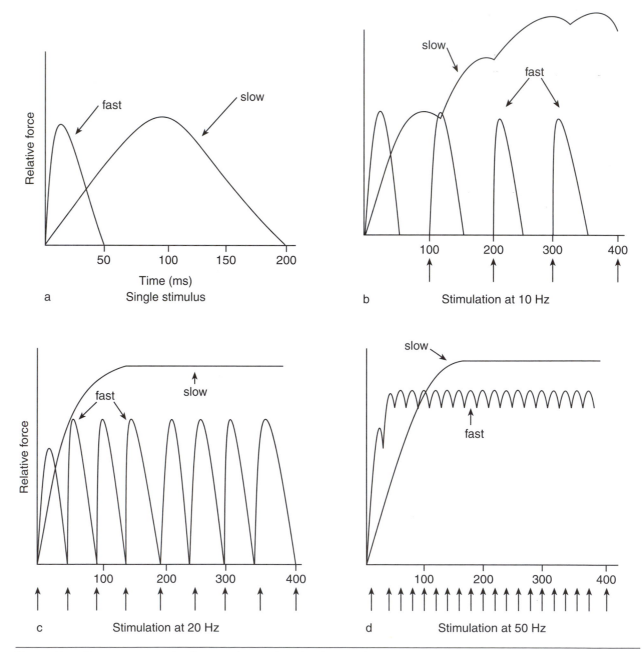

Figure F.1 Responses to a single stimulus and multiple stimuli

the fast fiber has faster rates of both force development and relaxation (see question). At 10 Hz (*b*) the action potentials are 100 ms apart, so the fast fiber has time to contract and relax between stimuli; but because sarcoplasmic calcium concentration is still high in the slow fiber when the next stimulus arrives, the forces summate. At 20 Hz (*c*) the force of the fast fiber is summated slightly, and the slow fiber has reached maximum force—a tetanic contraction. At 50 Hz (*d*) the stimuli are arriving only 20 ms apart, so the forces of the fast fiber are now summating and the slow fiber is still producing maximum force. No further increases in force are observed once the maximum force potential has been reached.

13. In terms of muscle mechanics, the worker has a large angle at the elbow, which affects the angle of force application of the muscle on the bone (moment arm). An angle greater or less than optimum decreases the effectiveness of the force transfer to the bone. Also, at this angle the muscle is stretched; this pulls the actin away from the myosin and reduces the number of cross-bridges that are able to attach. This principle is the length-tension relationship. The cause of fatigue relates to the type of contraction. The work involves an isometric contraction, meaning that blood flow to the muscle is reduced or even shut off completely. The muscle fatigues very quickly because of a rapid buildup of waste products such as lactic acid. As a remedy, you can have the industrial engineer redesign the work space so that the worker does not have to reach across the table, and you can instruct the worker in proper lifting and carrying techniques (holding the object closer to the body.)

14. Sedentary people most frequently recruit slow motor units, as these are used mostly for activities of daily living, and rarely recruit fast motor units (high force or power). When a limb is casted, the fiber types used the most will be affected the most—in this case the slow fibers. In the sprinter, the fast fiber types would be more developed, and these would be most affected relative to the decline in mass with the sedentary person.

15. The worker becomes exhausted because of a large production and accumulation of lactic acid, which causes fatigue in many ways (see page 143). Increasing the rest time not only may not eliminate the fatigue problem; it will reduce productivity because the worker will be taking more time to finish the job. A better solution is to keep the work:rest ratio the same but reduce the work time, say to 2 min. Lactic acid production will be much less, and the worker will be able to finish the job without becoming fatigued.

16. The workers are experiencing low-frequency fatigue, which is selective for slow motor units and has a long recovery, possibly lasting into the next day. You may suggest that the workers be given different tasks on alternate days. Even though fatigued from low-intensity, long-lasting work tasks, they should be able to effectively perform high-intensity, short-duration activities, which use different motor units.

Chapter 9 Muscle Plasticity

1. Large gains in strength can be achieved with isometric resistance training, but the application to real-life movements is limited because strength gains are specific to the joint angle at which the training has occurred. Some weight lifters have used isometrics at a specific joint angle to overcome limitations at that angle (a "sticking point") during a dynamic lift. Dynamic resistance training is more applicable to real-life movements.

2. An isokinetic exercise machine allows the exerciser to move a limb through an entire range of motion at a constant angular velocity. During concentric contractions, you can control the force production by changing the velocity. At high velocities, force production is low; at low velocities, it is possible to generate more force. This is appealing for rehabilitation purposes. The same does not hold true for eccentric contractions, however. Large forces are produced at almost all velocities, and this may not be desirable when the patient has muscle weakness or is recovering from an injury. In addition, training at the relatively slow speeds of the isokinetic machine does not translate into performance at the high velocities needed in many powerful movements.

3. Plyometric training takes advantage of elastic recoil and the stretch reflex to improve jumping ability. This occurs when an eccentric contraction is immediately followed by a concentric contraction—the latter is much more powerful than a concentric contraction performed from a resting state. The increase in jumping power is attributable largely to neural adaptations, which result in greater activation of the involved muscles.

4. Specificity of training means that muscles should be conditioned in a manner similar to that in which they are to perform. With repetition of a specific movement pattern, adaptations of the entire musculoskeletal system (neural, muscular, connective tissue) occur, and the best improvement is seen in that specific movement. Muscles should be overloaded gradually, with plenty of rest between workouts. This allows the muscle to recover from each workout and to synthesize proteins necessary for growth. If the person overtrains, he or she becomes much more susceptible to injury.

5. The statement should be refuted. Fast oxidative motor units are used primarily for activities that require sustained power for relatively long periods of time. The fast contraction speed and large cross-sectional area provide high levels of power, and the high oxidative potential provides the ability to supply energy through oxidative metabolism.

6. Eccentric contractions produce higher forces per cross-sectional area than either isometric or concentric contractions and therefore a greater stimulus for protein synthesis. In addition, lengthening the fibers while they are activated may cause more sarcolemmal wounding than either concentric or isometric contractions. These microtears in the cell membrane are linked to the release of growth factors that regulate the hypertrophic response.

7. As mentioned in the answer to question 6, the stimulus is sarcolemmal wounding, which first stimulates the release of FGF from the cell membrane. Fibroblast growth factor activates satellite cells to divide and fuse, adding protein to that already existing in the fiber. Insulin-like growth factor I is synthesized within the fibers, and it increases amino acid transport into the fiber.

8. The fast glycolytic fiber type is best suited for producing explosive power because of the high velocity of force generation combined with the large cross-sectional area of the fiber, which provides high levels of force.

9. The first stage of muscle damage is related to the high stress and strain on the fiber, usually associated with eccentric contractions. In combination, the stress and strain produce a mechanical damage that occurs initially with disruption of the Z line. After two or three days, further cell damage, associated with the inflammatory process, occurs. This latter phase of injury is accompanied by muscle soreness (delayed-onset muscle soreness).

10. More injury occurs during eccentric contractions than during concentric ones, as well as when a movement begins with the muscle in a lengthened position versus a neutral or shortened position. Thus, lowering the weight (eccentric contraction) from 90° to full extension would result in the most muscle damage.

11. Eccentric muscle contractions produce more muscle damage than other types of contractions because of the associated high stress and strain. In particular, this type of training causes widespread damage in people who are unaccustomed to these movements. The extensive muscle damage causes cellular debris (mostly protein) to be released into the blood. In its attempt to filter the debris, the kidney may fail. Your advice should be to begin a resistance-training program at low resistance (50-60% of 1 RM), allow plenty of rest between lifting sessions (lifting no more than three times per week for one muscle group), and increase resistance gradually. A combination of concentric and eccentric movements should be used initially.

Chapter 10 Muscular Diseases

1. The selection of exercises depends on the muscles and specific fibers affected. For instance, exercise for a person whose disease progresses from proximal to distal muscle groups should emphasize proximal muscle groups early in the rehabilitation process. In terms of the specific muscle fibers, see the answers to questions 5 and 6.

2. Many patients with muscular disease feel socially isolated. Activity can be an important means by which children can gain a positive social self-image. Those planning activity programs should consider incorporating unusual activities, for example, karate, boxing, and tai chi, that will make participants feel "special."

3. Regular exercise is not harmful to these patients and may provide benefits in terms of slowing disease progression and maintaining function for a longer time.

4. Stretching helps prevent the muscle contractures associated with these diseases.

5. Lower-resistance exercises done at somewhat higher velocities will stimulate most fibers in the

muscle and may promote the development of type II fibers.

6. Since corticosteroids affect primarily type II fibers, you might consider exercise programs that will ensure recruitment of these fibers. Again, such programs might include the low-resistance, high-velocity (power) type of resistance exercises.

Chapter 11 Pulmonary System

1. The restriction of the chest wall movement would make the elastic forces of breathing so high that the worker would have to breathe at high frequencies to accommodate the increased exercise ventilation. With increasing frequency, however, the Vd/Vt rises. The greater wasted ventilation means that less air is going into the alveoli so that actual hypoventilation might occur.

2. A low PAO_2 means that there is less driving pressure to assist the movement of O_2 across the alveolar membrane into the capillary.

3. $\dot{V}_A/\dot{Q}c$ is better in the supine position because the gravity-dependent unevenness in the distribution of both $\dot{V}_A$ and $\dot{Q}c$ is smaller. Instead of an unevenness going from head to toe, the unevenness is from anterior to posterior only.

4. The high volume of fat on the chest wall restricts the movement of the chest wall and dramatically increases the elastic work of breathing. To accommodate this, the person increases breathing frequency. This results in the same hypoventilation as described in the answer to question 1.

5. When the small muscle mass of the arms performs the same amount of work as the larger muscle mass of the legs, the result is greater stimulation of the neural receptors that in turn stimulate breathing. In addition, more lactic acid may form, further increasing ventilation. This heightened ventilatory response results in hyperventilation and the sensation of dizziness.

6. The pulmonary patient might have a low PaO_2 that will result in a lower SO_2. As the patient breathes higher inspired O_2, the PaO_2 increases, thereby dramatically increasing the levels of SO_2 and O_2 being delivered to the body. The healthy person is already carrying about 96% of the potential O_2 on the Hb, so breathing high O_2 will provide little benefit.

7. The greater mitochondrial capacity to accept pyruvic acid into oxidation means that less of the pyruvic acid being produced will have to be transformed into lactic acid.

Chapter 12 Pulmonary Diseases

1. The low FVC and flow rates are indicative of obstructive disease, but obstructive diseases make it difficult for people to get much of their air out very quickly, resulting in a low FEV_1/FVC ratio. The fact that this patient was able to exhale almost all of her vital capacity in 1 s indicates that she was not obstructed but has a restrictive disease. Imagine inflating a very stiff balloon. The volume that you could get in and out would be small (equivalent to the low FVC). The volume let out in 1 s also would be low (equivalent to the low FEV_1), but the airflow would be extremely fast because the stiffer balloon "wants" to collapse to a lower volume so that almost all of the air would come out quickly. This is the same thing that happens in restrictive diseases. The stiffer system has a low FVC and consequently a low FEV_1, but returns to a lower volume very quickly as reflected in the high FEV_1/FVC ratio.

2. The problem is that EIB is caused by the loss of heat and/or water from the airways. Exercising in the cold, dry air of winter would exacerbate EIB. You might first ask him to consider a different sport. On the other hand, EIB is different from asthma triggered by other factors in that it results from a bronchospasm that is preventable or very treatable with medication. He should consult with his physician to optimize medical management of the asthma and determine the best way to manage EIB.

3. You should tell her that smoking is slowly eroding her lung capacity but that because of the presence of an "excess" or reserve she does not notice the loss of this capacity, especially if she does little exertion. As the capacity becomes more eroded, however, she will begin to notice shortness of breath when she does exert herself. She should know that when enough capacity is gone that she feels the effect doing routine activities, she will have lost a considerable part of her lung capacity. Most, if not all, of this capacity is unrecoverable even if she stops smoking at that time. You might emphasize that very little rehabilitation is possible once a large part of the lung function is lost.

4. The hyperinflation associated with COPD makes it very difficult—and would cost a lot of

energy—to increase volume any further. Thus, even though breathing faster through obstructed airways costs more energy, it is still the best strategy for breathing.

5. You shouldn't be surprised, because the system has found the most energy-efficient combination of frequency and tidal volume. Your alteration of this combination requires more energy—something the system will resist despite the fact that arterial blood gases (PaO_2 and $PaCO_2$) might deteriorate with the "chosen" combination of frequency and tidal volume.

Chapter 13 Cardiovascular System

1. Skeletal muscle at rest has very low energy needs, so blood flow is very low. Even though muscle makes up approximately 40% of total body mass, only about 20% of available blood is flowing through muscle at rest. During maximum exercise, the metabolic needs in the working muscle of a trained athlete increase tremendously, so cardiac output increases dramatically. The sympathetic nervous system increases heart rate and force of contraction to increase stroke volume. An increase in venous return fills the heart more and helps increase stroke volume. In order for about 90% of cardiac output to be diverted to the working muscle, other areas of the body, even nonworking muscle, must have blood flow reduced. This is accomplished by sympathetic vasoconstriction in these areas while arterioles in contracting muscle are dilating.

2. Large arteries, which are basically conduits, propel the blood through the vascular tree by elastic recoil. Therefore, they have a large elastic modulus and not much sympathetic innervation of their smooth muscle layer. The smaller feed arteries and arterioles that lie within skeletal muscle are responsible for increasing or reducing blood flow on the basis of metabolic demand. As such, these smaller vessels are richly innervated with sympathetic vasoconstrictor nerve fibers and can change diameter over a wide range. Also, they have a small elastic component.

3. The left ventricle has a much larger mass because the resistance (systemic circulation) that it must pump against is much larger than that on the right (pulmonary circulation). Muscle mass of the heart chambers increases or decreases in much the same manner skeletal muscle does; such changes mainly depend on resistance or load.

4. Vasodilator substances are released from contracting muscle and cause arterioles in that area to dilate. This dilator signal is propagated back up the vascular tree to the feed arteries, which regulate blood flow into the muscle in the exact amount dictated by the degree of vasodilation within the muscle.

5. During rhythmic/dynamic exercise, stroke volume is increased largely through an increase in venous return. This is assisted through venoconstriction, which reduces the blood volume in the venous system; the muscle pump, which helps move the blood out of the muscle and back to the heart; and increased breathing, which decreases intrathoracic pressure and facilitates filling of the heart.

6. Training for the three women

a. Even though their maximum capacities vary widely, all three have approximately equal resting cardiac outputs because they are the same age and sex and have equivalent body mass and composition. Their maximum cardiac outputs, though, are radically different. The ACT has the highest because she is moderately trained, followed by the SED and then the MS. How they achieve resting and maximum cardiac outputs relates to differences in stroke volume. The MS is limited by a mechanical filling deficiency, so her stroke volume is low at rest and cannot increase much during exercise. She has the highest resting heart rate (probably around 100). The SED is unconditioned, so her resting stroke volume is moderately low, as is her resting heart rate (around 75). The ACT has the highest resting stroke volume and therefore the lowest resting heart rate—let's say 65. All three have approximately the same maximum heart rate (based on sex and age), so it is evident their maximum cardiac outputs are limited by stroke volume. If the SED adopted the training program of the ACT, she would see an increase in her stroke volume and a subsequent decline in resting heart rate. The decline in heart rate is mediated by an increased input from the parasympathetic nervous system. The MS would have a very limited conditioning response because of the limitations in ventricular filling. Most of her adaptations would be peripheral (in the muscle).

b. The SED has the greatest potential to increase $\dot{V}O_2$max. ACT is already trained, so the percent increase will be low, and MS is limited by the ventricular filling deficiency, so won't get much of an adaptation of the heart.

7. Endurance training produces a volume overload—small increase in LV mass and large increase in size of chamber, and an increase in plasma volume. This produces large increases in cardiac output and an increase in CV fitness. Resistance training produces a pressure overload with moderate increases in LV mass and little or no change in chamber size. This brings about a small increase in stroke volume and not much increase in cardiac output, which limits CV function. You would recommend endurance training for CV benefits.

Chapter 14 Cardiovascular Diseases

1. You can make the relationships clearer if you present information about smoking and information about cardiovascular disease separately. In addition, your explanation should always give the patient some indications of how to implement recommendations to reduce risks. Let's start with smoking. You need to stress that smoking has the short-term effect of reducing the amount of O_2 carried in the blood and the long-term effect of destroying the lungs' ability to exchange gases. Stopping smoking will immediately reverse these effects to varying degrees. Now you can move to the cardiovascular system. Since less O_2 is being carried in the blood, the heart must work harder to deliver the necessary O_2. In addition, the chemicals in cigarette smoke cause vasoconstriction, further increasing the work of the heart. Finally, these chemicals also injure the vessels, leading to potential plaque formation.

2. The answer to this question is really an extension of the answer to question 2. The increased work of the heart, under relatively low O_2 conditions, can lead to an inability of the heart to meet demand, and thus heart failure.

3. The stepping device, because of the increased stress on small muscles of the leg, would most likely lead to the conditions that promote intermittent claudication, that is, low blood flow and more non-oxidative metabolism.

4. Exercise program patient

a. Her peak exercise capacity for the purposes of establishing her exercise program is >3 and <5 METs.

b. This patient would be classified as cardiovascular functional status III (see table 14.1).

c. This information would characterize the patient as starting in cardiac rehabilitation phase I or at the very beginning of phase II (see figure 14.4).

d. She should begin at a rate of <125 and progress, with ECG monitoring, to a heart rate approaching 130. Of course, the exercise intensity at this heart rate should increase as she becomes conditioned. She should start with active range of motion and light walking and progress to walking at about 5 to 6.5 km/hr (3 to 4 mph) and slow cycling. Activities of daily living may be limited at first, but the patient should be able to increase the number and intensity of these activities over the rehabilitation period. You should be able to find activities from the compendium (appendix C) that are at an appropriate intensity for her condition.

e. The previous answer encompasses the answer to this question.

f. Diet and exercise are the most obvious educational choices.

Chapter 15 Pediatrics

1. Age-related factors that can explain differences in exercise capacities include muscle mass, size of the lungs and heart, and the metabolic properties of the muscles.

2. Sex-related factors that can explain differences in exercise capacities between growing boys and girls include the loss of red blood cells and therefore Hb, and the increased % body fat, in females after adolescence. The % lean body mass increases to a greater degree in growing boys than in girls. Although this cannot account for all of the differences in aerobic capacity, it can explain most of the sex differences in strength. The effects of culture, that is, fewer opportunities for females, cannot be discounted.

3. Differences in exercise capacities between young people and adults, both male and female, can be accounted for by differences in volumes, that is, lung and heart volume, Hb-carrying capacity, and muscle mass.

4. During exercise under hot, humid conditions, blood flow is shunted from the working muscles to the skin to promote cooling. Adults compensate partially for this loss of blood flow by increasing O_2 extraction. Children, because they are already extracting more, have less capacity to draw on this mechanism.

Chapter 16 Aging

1. A cross-sectional study compares measured variables between a group of people of approximately the same age and another group of people of a different age. This means that the two groups have members who are genetically biased, which may confound the results of the study. A longitudinal study, in which people are studied over long periods, provides more accurate data regarding the true effect of physical activity on aging. The drawbacks of longitudinal studies are the presence of confounding variables, such as lifestyle and environmental factors, and the difficulty of keeping subjects in a study that lasts so long.

2. Any given person does not have a linear decline in functional capacity over his or her life span. For example, a highly trained athlete will have a sharper decline in $\dot{V}O_2$max over a period of years if training is reduced or stopped than one who maintains a high level of training. A person who develops HTN or CAD is another example. Both these conditions reduce the functional capacity of the cardiovascular system and would result in a greater decline in $\dot{V}O_2$max than normal. As yet another example, a sedentary older adult who begins a training program will have an upswing in functional capacity.

3. Older adults typically have a decline in strength of respiratory muscle and decreased elasticity of the lungs and chest wall. These changes decrease total lung capacity and increase the work of breathing, which limits work capacity and leads to early fatigue. The decreased surface area for gas exchange and reduced capillarity produce a mismatch in ventilation to perfusion, causing a reduced arterial partial pressure of O_2.

4. Aging women are more susceptible to osteoporosis than men because they tend to have a lower calcium intake as adults and tend to ingest less calcium with aging.

5. Mechanical stress produced by weighted exercise induces increased bone density, just as bed rest or unweighting will result in the loss of bone.

6. After age 30, people tend to increase body fat and reduce lean body mass. This change in body composition reduces resting metabolic rate because fat is less metabolically active than muscle. The reduced muscle mass also reduces O_2 extrac-tion capabilities at maximum exercise, thereby reducing $\dot{V}O_2$max.

7. There are two explanations here. There is an increased stiffness of peripheral blood vessels with aging per se that results in a slight LV hypertrophy. Then, with the onset of borderline HTN, the LV hypertrophies more. The R wave is a measure of ventricular depolarization, and the current flow (electric axis) is normally biased toward the LV (points directly at the V5 lead) because of its greater mass compared to the right ventricle. Normally, the highest R wave is seen in the V5 lead because of the direction of the electric axis. With selective LV hypertrophy, we see a left axis deviation, so the electric axis is now pointing more toward lead V6—hence the increased amplitude in this lead.

Chapter 17 Exercise in Various Environments

1. Central and peripheral temperature-sensitive receptors provide feedback about core and skin temperature, respectively. If body temperature increases, various heat-dissipating mechanisms (e.g., increased skin blood flow, sweating) are enacted to dissipate heat. Conversely, when body temperature drops, heat is conserved through decreases in skin blood flow. In this manner, body temperature varies around a set point that is determined by the temperature-regulating center in the hypothalamus.

2. Under thermoneutral conditions and during light activity, the body can effectively thermoregulate through conduction, convection, and radiation. Altering skin blood flow is the major heat-regulating mechanism.

3. Skin blood flow is regulated by the sympathetic nervous system. Vasoconstrictor nerve fibers reduce skin blood flow when heat needs to be conserved, and vasodilator nerves increase skin blood flow when heat needs to be lost.

4. When core temperature increases under resting conditions compared to exercise conditions, skin blood flow rises at a faster rate. When a person begins to exercise under thermoneutral conditions, skin blood flow decreases at first and then increases slowly as core temperature rises.

5. Exercise in a hot environment results in a very high skin blood flow, creating a competition

for blood between the working muscle and the skin. Consequently venous return is decreased, central blood volume is low, and stroke volume decreases. Heart rate must increase in an effort to maintain cardiac output. During exercise in cold weather, this situation is reversed. Skin blood flow is low, so central blood volume is high, filling of the heart is optimal, and stroke volume is high. As a result, heart rate remains low.

6. Prolonged exercise in a thermoneutral environment is typically accompanied by a slow decrease in stroke volume—a result of an increase in skin blood flow. Core temperature rises slowly because of the internal production of heat, causing the rise in skin blood flow and a concomitant decrease in venous return. To maintain cardiac output, heart rate must increase by the same percentage that the stroke volume decreases. Because cardiac output remains the same but skin blood flow is increasing, skeletal muscle blood flow is likely maintained by increased vasoconstriction in nonessential areas of the body. The increased O_2 consumption that occurs can be explained by a temperature-related increase in metabolism and ventilation, uncoupling of oxidative phosphorylation, and an increased heart rate.

7. An acute exposure to altitude results in an increased ventilation, which helps maintain PaO_2, and an increased cardiac output, which enhances delivery. Long-term exposure increases red cell production and reduces plasma volume, which increases the O_2-carrying capacity of the blood. Also, bicarbonate is excreted, which reduces blood pH toward normal.

8. Chronic exposure to heat produces an increase in plasma volume, whereas long stays at altitude decrease plasma volume. These differing adjustments benefit the body in different ways. An increased plasma volume helps maintain venous return during exercise, keeping stroke volume high. The increased plasma volume also enables perfusion of both the skin and muscle while maintaining blood pressure. The reduced plasma volume at altitude increases O_2-carrying capacity of the blood and helps deliver adequate O_2 to the muscles during exercise.

9. Since you are unacclimatized to the heat, when exercising you will have problems losing enough heat. Skin blood flow will be higher un-

der these conditions than when you run in Flint, so central blood volume will be lower, ventricular filling less, and stroke volume lower, requiring an increase in heart rate. Increased body temperature is a stimulus for ventilation, which is why the breathing will be increased. Sweating is the major heat-dissipating mechanism here—the other three mechanisms are inadequate under these conditions to dissipate enough heat.

To avoid becoming dehydrated, you need to drink fluids before your run and then take in small amounts at frequent regular intervals during the run. After running, you should drink more volume than estimated to be lost through sweating. You can make a rough estimate by weighing yourself before and after the run. For every kilogram (2.2 lb) lost through sweating, you should take in approximately 1.5 L of fluid (1 L = 2 pints). Because of the heat load built up during the run, it is advisable to do the first half of the run with the wind, then for the second half turn and run into the wind so that convective wind currents help evaporate the sweat. If you did it the other way around, your speed on the return would be the same as the wind speed, so convection would be negligible. Because of the internal heat load acquired during the first part of the run, the added environmental heat stress would create heat stress–related problems.

If you ran in the heat for six days, you should benefit from acclimatization. You would sweat more, have an increased plasma volume, and have a lower threshold for sweating. These changes combine to decrease skin blood flow, which increases central blood volume and helps maintain stroke volume. Your performance should be close to what was normal for you before you left Flint to travel to Miami.

Chapter 18 Exercise Testing and Prescription

1. This target heart rate range was calculated on the basis of her resting and peak exercise heart rates. The difference between these two rates, 115 minus 82, was 33 beats. To exercise at 60% of her peak capacity, the target heart rate was calculated as 0.6 multiplied times 33; and the resting heart rate, 82, was then added to that. This set the lower heart rate limit at 101 beats/min. Based on the results of the exercise test, her exercise prescription

was designed to keep her heart rate below 115, since this was the level at which she first noticed angina. She exercised on a cycle ergometer with the load adjusted to keep her heart rate between 101 and 115 beats/min.

2. You should include information about heart disease, diet, stress, medications, lifestyle, sex, exercise, and diabetes.

3. She will have to decrease her insulin injections.

4. You will need to know something about her past activity patterns, her likes and dislikes, and activities that she will most likely pursue on a regular basis. You might find out whether her grandchildren can be part of her activity program. Can she walk with the children or do other activities with them on a regular basis? You can use her answers to these and other similar questions to optimize the possibility that your client will remain compliant to her program.

Credits

Figure 2.4—Reprinted, by permission, from M. Houston, 1995, *Biochemistry primer for exercise science* (Champaign, IL: Human Kinetics).

Figure 3.1—Reprinted, by permission, from H.B. Falls and L.D. Humphrey, 1976, "Energy cost of running and walking in young women," *Medicine and Science in Sports* 8:9.

Figure 3.4—Adapted, by permission, from W. Westcott, 1996, *Building strength and stamina* (Champaign, IL: Human Kinetics), 15.

Figure 4.2—Reprinted, by permission, from Stegeman, 1977, *Exercise Physiology* (New York: Thieme Publishing).

Figure 4.15—Adapted, by permission, from J. Keul, E. Doll, and D. Keppler, 1972, Energy metabolism of human muscle. In *Medicine and Sport Series* (Farmington, CT: S. Karger Publishers).

Figure 5.1—Reprinted, by permission, from J.S. Skyler, 1998, Insulin treatment. In *Therapy for diabetes mellitus and related disorder*, 3rd ed. (Alexandria, VA: American Diabetes Association), 190.

Figure 5.2, 5.3—Adapted, by permission, from M. Berger, P. Berchtold, H.J. Cuppers, H. Drost, H.K. Kley, W.A. Muller, W. Wiegelmann, H. Zimmermann-Telschow, R.A. Gries, and H.L. Kruskemper, 1977. "Metabolic and hormonal effects of muscular exercise in juvenile type diabetes," *Diabetologia* 13:355.

Figure 5.4a, 5.4b—Reprinted, by permission, S.S. Sternberg, 1992, *Histology for pathologists* (New York: Raven Press).

Figures 6.1, 6.2, 6.9, 6.10, 6.11, 6.14, 8.11, 8.17, 8.18—Adapted, by permission, from A. Vander, J. Sherman and D. Luciano, 1994, *Human physiology*, 6th ed. (New York: McGraw-Hill Companies),191, 194, 356, 356, 357, 318, 323, 333, 334.

Figures 6.3, 6.5, 6.6, 6.7, 6.8, 6.12, 8.13—Adapted, by permission, from E. Kandel, J. Schwartz, and T. Jessell, 1991, *Principles of neural science*, 3rd ed. (New York: McGraw-Hill Companies), 23, 28, 25, 280, 583, 281, 553.

Figure 6.16, 8.21, 13.13—Adapted, by permission, P. Astrand and K. Rodahl, 1986, *Work Physiology*, 3rd ed. (New York: McGraw-Hill), 105, 306, 193.

Figure 7.2—Reprinted, by permission, from J.H. Martin and T.M. Jessell, 1992, Anatomy of the somatic sensory system. In *Principles of neural science*, 3rd edition, edited by E.R. Kandel, J.H. Schwartz, and T. M. Jessell (New York: McGraw-Hill Companies), 59.

Figure 7.6—Reprinted, by permission, from J. Wilmore and D. Costill, 1994, *Physiology of sport and Exercise* (Champaign, IL: Human Kinetics).

Figure 8.1—Adapted, by permission, from A. Kelly and N. Rubinstein, 1986, "Development of neuromuscular specialization," *Medicine and Science in S ports Exercise* 18:292-298.

Figure 8.3, 8.14, 8.16—Adapted, by permission, from R. Leiber, 1992, *Skeletal muscle structure and function* (Baltimore, MD: Lippincott, Williams, and Wilkens), 41, 24, 118.

Figure 8.4, 8.10—Adapted, by permission, from A. Guyton, 1991, Contraction of skeletal muscle. In *Medical physiology*, 8th ed. (Philadelphia, PA: W.B. Sanders), 68, 75.

Figure 8.5—Adapted, by permission, from C.M. Waterman-Storer, 1991, "The cytoskeleton of skeletal muscle: is it affected by exercise?" *Medicine and Science in Sports and Exercise* 23:1249.

Figure 8.6, 8.7—J. Squire, *Molecular mechanisms in muscular contraction*, 1990, MacMillan Press, adpated with permission of Palgrave.

Figure 8.8—Adapted, by permission, W. McCardle, F. Katch, and V. Katch, 1996, "Skeletal muscle: structure and function," In *Exercise physiology: energy, nutrition, and human performance*, 4th ed. (Baltimore, MD: Lippincott, Williams, and Wilkens) 323.

Figure 8.15—Adapted, by permission, from R. Enoka, 1992, *Neuromechanical basis of kinesiology*, 2nd ed. (Champaign, IL: Human Kinetics) 134.

Figure 8.20—Adapted from J. Faulkner, 1983, "Fatigue of skeletal muscle fiber," *3rd Banff International Hypoxia Symposium*.

Figure 9.3—Reprinted, by permission, A. Thorstensson et al. 1976, "Effect of strength training on enzyme activities and fibre characteristics in human skeletal muscle," *Acta Physiologica Scandinavica* 96:392-398.

Figure 9.4—Adapted, by permission, from D. Sale, 1988, "Neural adaptation to resistance training." *Medicine and Science in Sports Exercise* 20:S135-S145.

Figure 9.6, 9.7—Reprinted, by permission, from J. Faulkner et al.,1984, Human muscle power. In *Proceedings* (Champaign, IL: Human Kinetics), 64, 86.

Figure 9.11—Adapted from J. Faulkner, 1989, "Injury to skeletal muscles of mice by forced lengthening during contractions," *Quarterly Journal of Exercise Physiology,* 74:661-670.

Figure 11.20—Reprinted, by permission, from J. Wilmore and D. Costill, 1999, *Physiology of sport and Exercise,* 2nd ed. (Champaign, IL: Human Kinetics), 254.

Figure 12.4—Adapted, by permission, from F. Haas and K. Axen, 1991, *Therapy and rehabilitation,* 2nd ed. (Baltimore, MD: Lippincott, Williams, and Wilkens).

Figure 13.1—Adapted, by permission, from L. Rowell, 1986, Circulatory adjustments to dynamic exercise. In *Human circulation–regulation during physical stress* (New York: Oxford University Press, Inc.), 240.

Figure 13.2—Reprinted, by permission, from L. Rowell, 1986, Circulatory Adjustments to dynamic exercise. In *Human circulation–regulation during physical stress* (New York: Oxford University Press, Inc.), 235.

Figure 13.3—Adapted, by permission, from A. Guyton, 1991, Heart muscle: the heart as a pump. In *Medical physiology,* 8th ed. (Philadelphia, PA: W.B. Saunders), 99.

Figure 13.4—Reprinted, by permission, from R.C. Little, 1977, Physical characteristics and functional significance of cardiac structure. In *Physiology of the heart & circulation* (Chicago: Yearbook Medical Publishers), 38.

Figure 13.6—Adapted, by permission, from S. Segal, 1992, "Communication among endothelial and smooth muscle cells coordinates blood flow control during exercise," *News in Physiological Sciences* 7:152-156.

Figure 13.9, 13.10, 13.11—Reprinted, by permission, from M. Thaler, 1999, *The only EKG book you'll ever need,* 3rd ed. (Balitmore, MD: Lippincott, Williams, and Wilkens), 35, 41, 53.

Figure 13.12—Adapted, by permission, from J. McDougall et al., 1985, "Arterial blood pressure response to heavy resistance exercise," *Journal of Applied Physiology* 25:785-790.

Figure 13.15—Reprinted, by permission, from N. Secher et al., 1977, "Central and regional circulatory effects of adding arm to leg exercise," *Acta Physiologica Scandinavica* 100:288-297.

Figure 15.1, 15.2—Reprinted from P.V.V. Hamill, T.A. Drizd, C.L. Johnson, R.B. Reed, A.F. Roche, and W.M. Moore: Physical growth: National Center for Health Statistics percentiles. *American Journal of Clinical Nutrition* 32:607-629, 1979. Data from the National Center for Health Statistics (NCHS), Hyattsville, Maryland.

Figures 15.3, 15.4, 15.5, 15.6, 15.7—Adapted, by permission, from O. Bar-Or, 1983, *Pediatric sports medicine for practitioner* (New York: Springer-Verlag).

Figure 16.4—Adapted, by permission, from G. Gersenblith et al., 1987, "Cardiovascular response to exercise in younger and older men," *Journal of Federation Proceedings* 46:1835.

Figure 17.1—Adapted, by permission, from A. Guyton, 1991, Body temperature, temperature regulation, and fever. In *Medical physiology,* 8th ed. (Philadelphia, PA: W.B. Saunders), 801.

Figure 17.2—Adapted, by permission, from L. Rowell, 1986, Thermal stress. In *Human circulation–regulation during physical stress* (New York: Oxford University Press, Inc.), 196.

Figure 17.3—Adapted, by permission, from L. Rowell, 1986, Dynamic exercise and heat stress. In *Human circulation–regulation during physical stress* (New York: Oxford University Press, Inc.), 368.

Figure 17.4—Adapted, by permission, from L. Rowell, 1986, Dynamic exercise and heat stress. In *Human circulation–regulation during physical stress* (New York: Oxford University Press, Inc.), 366.

Figure 17.5—Reprinted, by permission, from J. Wilmore and D. Costill, 1999, *Physiology of Sport and Exercise,* 2nd ed. (Champaign, IL: Human Kinetics), 323.

Figure 17.7—Adapted, by permission, from A. Lind and D. Bass, 1963, "Optimal exposure time for development of acclimatization to heat," *Journal of Federation Proceedings* 22:704.

Figure 17.9—Adapted, by permission, from J.B. West, S.J. Boyer, D.J. Graber, P.H. Hackett, K.H. Maret, J.S. Milledge, R.M. Petters, C.J. Pizzo, M. Samja, R.H. Sarnquist, R.B. Schoene, and R.M. Winslow, 1983, "Maximal exercise at extreme altitudes on Mount Everst," *The Journal of Applied Physiology* 55:688-698.

Figure 18.5—Adapted, by permission, from W.H. Martin and E.F. Coyle, 1986, "Effects of physical deconditioning after intense training on left ventricular dimensions and stroke volume," *Journal of American College Cardiology* 7:982-989.

Figure 18.8a—G. Borg, 1998, *Borg's Perceived Exertion and Pain Scales* (Champaign, IL: Human Kinetics), 47.

Figure 18.8b—Modified G. Borg, 1998, *Borg's Perceived Exertion and Pain Scale* (Champaign, IL: Human Kinetics), 50.

Table 2.4—Adapted from E.N. Whitney and R.S. Sizer, 1994, *Hamilton and Whitney's nutrition concepts and controversies,* 6th ed. (Pacific Grove, CA: Thomson Learning).

Table 3.1—From *Lifetime Physical Fitness and Wellness,* 5th edition, by W.W.K. Hoeger ©1998. Reprinted with permission of Wadsworth, an imprint of the Wadsworth Group, a division of Thomson Learning, Fax 800-730-2215.

Table 14.1—Adapted from American Heart Association.

Table 15.1—Adapted, by permission, from O. Bar-Or, 1983, *Pediatric sports medicine for practitioner* (New York: Springer-Verlag).

Photo 10.3c—Adapted, by permission, from I. Damjanov and J. Linder, 1996, *Anderson's pathology,* 10th ed. (St. Louis, MO: Mosby-Yearbook Publishing).

Appendix C—Reprinted, by permission, from B.E. Ainsworth, W.L. Haskell, A.S. Leon, D.S. Jacobs Jr., H.J. Montoye, J.R. Sallis, and R.S. Paffenbarger, 1993, "Compendium of physical activities," *Medicine and Science in Sports and Exercise* 25(1):71-80.

Appendix E Baecke Questionnaire—Reprinted, by permission, from J.A.H. Baecke, J. Nurema, and E.R. Fritters, 1982, "Baecke Questionnaire," *American Journal of Clinical Nutrition* 36:936-942.

Appendix E Physical Activity Recall Items—Reprinted, by permission, from J.R. Sallis, W.L. Haskell, P.D. Wood, S.P. Fortman, T. Rogers, S.N. Blair, and R.S. Raffenbarger, Jr., 1985, "Physical activity assessment methodology in the five city project," *American Journal of Epidemiology,* 121:91-106.

Glossary

absolute and normalized energy expenditure—Absolute energy expenditure is expressed in terms of oxygen consumption expressed in L O_2/min while normalized energy expenditure is expressed in terms of the oxygen consumption for each kg of body mass or ml O_2/kg/min.

acclimatization—Physiological adaptations that occur after prolonged exposure to different environmental conditions, such as heat and altitude.

acetazolamide—A drug that inhibits the action of carbonic anhydrase in the kidney causing an increase in urinary excretion of sodium, potassium, and bicarbonate and a decreased excretion of ammonium. This leads to greater production of urine (loss of body water) and lowered blood pH.

acetylcholine—The neurotransmitter between the motor nerve and muscle fiber.

acetylcholinesterase—The enzyme that breaks down acetylcholine.

actin—One of the contractile proteins in skeletal muscle, which contains attachment sites for the globular head of myosin.

action potential—An electrical signal generated in excitable cells by depolarization of the cell membrane. Transmits information between excitable cells.

active transport—The use of energy (ATP) to transport an ion against its concentration gradient.

acute mountain sickness—Hypoxia (common at high altitude) associated illness characterized by nausea, headache, breathlessness, lethargy, and inability to sleep. May be associated with subclinical or overt pulmonary edema and severe respiratory distress. Treatment is elimination of hypoxia by administering oxygen or moving to a lower altitude.

adenosine—A nucleoside consisting of adenine and ribose that forms the basic structure of the primary energy transferring molecule in cells—adenosine triphosphate.

adenosine diphosphate (ADP)—An adenosine with two phosphoric acids. A precursor to the formation of adenosine triphosphate, the primary source of energy in the body.

adenosine triphosphate (ATP)—An adenosine molecule with three phosphates attached. The energy from these three phosphates is released to support energy-requiring processes in the body, including muscle contraction.

adipocytes—Fat-containing cell.

aerobic capacity—The peak capacity to consume oxygen during exercise. Also called the $\dot{V}O_2$ maximum if certain criteria are met during its measurement.

aerobic metabolism—Metabolic process of forming ATP through the process of oxidation—the transfer of an electron or a hydrogen. The final hydrogen acceptor is oxygen to form H_2O. Both carbohydrates and fats are used as substrate for aerobic metabolism.

afterhyperpolarization—The period of time after an action potential when the cell membrane is hyperpolarized and is unable to generate an action potential.

afterload—Resistance to flow from the heart.

airway resistance (Raw)—A measure of the resistance in the airways.

akinesia—Loss of voluntary movement.

aldosterone—A hormone that increases retention of sodium by the kidneys.

alpha motor neuron—Innervates skeletal muscle fibers.

alveolar ventilation—The air breathed in and out that reaches the gas exchange units, the alveoli. Air that does not reach the alveoli is part of the dead space ventilation.

alveolus—The terminal sac-like portion of the branching structure at the end of the branchings of the pulmonary airways. This is the gas-exchange unit of the lung where the alveolar air comes into contact with the pulmonary capillary blood.

Alzheimer's disease—A dementia characterized by a loss of memory and, eventually, body functions. Caused by the progressive accumulation of insoluble fibrous material (amyloid) in the brain.

amino acid—An organic acid with an NH_3 on one carbon. Proteins are made up of amino acids.

amyotrophic lateral sclerosis (ALS)—Progressive motor neuron disease of unknown origin. Degeneration of motor neurons is found in the cerebral cortex, brainstem and spinal cord.

anabolism—Rejuvenation of the cell.

anaerobic—The metabolic process of forming energy (ATP) without oxygen. This process is also called glycolysis since the substrate is carbohydrate in the form of glucose.

anaerobic threshold—Originally defined as the exercise intensity at which anaerobic metabolism begins to result in the formation of lactic acid. The threshold is identified by the point where lactic acid accumulation begins to stimulate an increase in ventilation beyond that expected for aerobic metabolism. Since anaerobic metabolism takes place all the time, this point of inflection is better described as the ventilatory or lactate threshold.

aneurysm—A weakness in a vascular wall evidenced by the wall ballooning into the surrounding interstitial space.

angina—Chest pain caused by inadequate blood flow in the heart muscle.

angina pectoris—Pain associated with complete or partial blockage of a vessel in the myocardium.

antidiuretic hormone—Also known as vasopressin; reduces urine production.

aortic valve—Between the left ventricle and aorta.

aphasia—Inability to communicate by speaking, writing, or signs.

apoptosis—Programmed cell death.

arterioles—Small, very reactive arteries that distribute blood within organs.

arteriosclerosis—Hardening of the arterial wall.

arteriosclerosis obliterans (ASO)—A vascular disease associated with intermittent spasm of the vessels that blocks blood flow to an area.

asthma—An intermittent narrowing of the airways caused by a variety of stimuli or triggers.

asynchronously—Occurring at different times.

ataxia—Inability to control muscular activity.

atherosclerosis—A vascular disease resulting in a progressive narrowing, and eventually blockage, of arterial vessels.

atrophy—Decrease in size.

augmented—The signals from these ECG leads are amplified to get an adequate recording.

auscultation—Listening to sounds of internal organs.

autonomic nervous system (ANS)—Modulates the body's state of readiness.

autosomal—Related to a chromosome other than a sex chromosome.

axial growth—An increase in size related to the longitudinal axis (i.e., increase in length).

axon hillock—Trigger zone for motor nerves.

β-oxidation—The process by which long chain carbon fats are broken down into two carbon acetyl groups. The process results in the formation of electrons that can be processed to produce ATP in the electron transport chain. The acetyl groups enter the tricarboxylic acid cycle to produce energy. All of these processes take place in the mitochondria.

Balke protocol—One of the standard graded, incremental tests for measuring exercise or aerobic capacity. The workload or power output is increased in increments, usually every two minutes, until the subject/patient can no longer continue.

basal ganglia—Groups of neurons lying deep within the cerebrum, which project on to motor and premotor cortices to help plan and update complex movements.

basal lamina—A thin layer of tissue on the outer surface of the sarcolemma.

bicarbonate—A major buffer agent in blood.

bifurcate—To divide into two branches.

blastula stage—Early stage of an embryo where cells form a hollow sphere.

body mass index—Calculated as the weight in kilograms divided by the square of the height in meters. Useful in determining nutritional status and as an indication of overweight.

bradykinin—Present in blood, a potent vasodilator associated with the inflammatory response as a result of skeletal muscle injury.

breathing frequency—The number of breaths taken per minute.

bronchitis—An inflammation of the airways that many times will become chronic.

Bruce protocol—A standard graded, incremental treadmill test for measuring exercise of aerobic capacity in patients with cardiac disease.

Brudzinski's sign—A test for meningitis. An involuntary flexion of the lower limbs when a supine patient attempts to raise the head and legs with the hands cupped behind the head. This will reduce pain from inflamed meninges and can indicate meningitis.

cable properties—Diameter of the axon affects speed of signal transmission.

calorie (cal), kcal, and Cal—The heat required to raise 1 g of water 1° C. An expression of energy. 1 kcal and 1 Cal = 1000 cal.

cardiac output—Blood flow from the heart in liters per minute.

cardiovascular drift—The slow steady increase in heart rate observed during prolonged steady state exercise. The increased heart rate is related to a decline in stroke volume resulting from increased skin blood flow in response to a slow, steady increase in core temperature.

carnitine—A carrier of lipids in the muscle cell and into the mitochondria.

catabolism—Cellular breakdown.

central fatigue—Decrement in performance related to decreased excitation of motor neurons.

central nervous system (CNS)—All nerves and pathways lying within the brain and spinal cord.

cerebellum—Lies behind the pons and helps determine timing sequences and pattern of muscle recruitment during movement.

cerebral cortex—Has four lobes: frontal—planning movements; parietal—somatic sensation; occipital—vision; temporal—audition, learning, and memory.

cerebrovascular accident (CVA)—A stroke.

cerebrum—Contains the cerebral cortex and basal ganglia. Controls more complex functions like perception, cognition, and higher motor skills.

chemoreceptors—Receptors that sense chemical changes in the body. Breathing is controlled, in part, by chemoreceptors that sense the level of O_2 (carotid body receptors) and of CO_2 (central or medullary receptors) in the blood and cerebral spinal fluid respectively.

cholesterol—A sterol found in animal cells implicated in the development of atherosclerosis.

chorea—Involuntary movements of the limbs or facial muscles.

chronic obstructive pulmonary diseases—Diseases of the lung that are associated with airway blockage, or obstruction, including chronic bronchitis, emphysema, and cystic fibrosis.

chylomicron—A complex composed of lipid and protein that is responsible for transporting the water-insoluble lipids in the blood from the intestines to storage sites. Other lipoproteins include the high-density and low-density lipoproteins.

citric acid—The first product in the mitochondrial tricarboxylic acid cycle.

citric acid cycle—A mitochondrial metabolic cycle that produces energy (ATP) and high energy compounds (NADH and FADH$_2$) that can be processed through the electron transport chain to produce ATP. (Also tricarboxylic acid cycle or Kreb's cycle.)

coactivated—When two different muscles are activated simultaneously.

collagen—The major protein found in connective tissue.

collateral circulation—Development of new blood vessels that results in increased circulation in tissue.

colloid osmotic pressure—The pressure exerted on a semipermeable membrane by an unequal concentration of particles on either side. When the concentration of particles is greater on one side, fluid is drawn from the other side to equalize the concentration.

complements—Nutrients that enhance the absorption of other nutrients when ingested together.

compliance (as related to attendance at treatment sessions)—The level at which a target number of sessions or treatments is met. It is usually expressed as the total number of sessions/treatments completed relative to the total possible treatments.

compliance (as related to tissue elasticity)—A measure of the ease with which a structure can be deformed.

concentric contraction—Skeletal muscle force exceeds the external load and muscle shortens.

conduction—A method of heat exchange whereby heat energy moves between adjacent molecules.

confounding factor—A factor that, if not separated from the measured variable, influences the outcome of the study.

congenital myopathies (central core disease [CCD], minicore/multicore disease, the nemaline [rod] myopathies, and myotubular myopathy)—A class of muscular diseases characterized by progressive weakness and alterations in and loss of muscle fibers.

contractile elements—The major proteins in skeletal muscle, actin and myosin, which interact to produce force.

convection—A method of heat dissipation whereby air or water next to the body is heated, moves away and is replaced by a cooler medium. Effectiveness is related to the velocity of the medium.

convergence—Information from many sources is delivered to a single target cell.

Cooper protocol—One of the standard field tests for estimating aerobic capacity. The test is usually performed for 12 minutes and the distance covered compared to norms.

core temperature—Refers to the deep tissues of the body.

coronary artery bypass graft surgery—Surgery that grafts a healthy vessel onto a coronary vessel that is blocked so that the blocked area is bypassed.

coronary artery disease (CAD)—Buildup of atherosclerotic plaque in arteries of the heart, which leads to decreased blood flow.

coronary heart disease (CHD)—Diseases of the coronary vessels. *See* atherosclerosis.

corticosteroids—A steroid produced in the adrenal gland.

costamere—A complex of proteins embedded in the cell membrane of skeletal muscle fibers that connects intracellular and extracellular structural components.

creatine phosphate (CP or PC)—Creatine combined with phosphoric acid. A source of cellular energy that supplies a phosphate for the formation of ATP from ADP.

cross bridges—In skeletal muscle, the connection between the myosin head and actin.

cross-sectional study—An examination of a population at one particular time.

cystic fibrosis—A congenital disease associated with a defect in the fluid transfer in exocrine glands, including the sweat, pancreatic, and lung mucous glands.

cytoplasm—The interior protoplasm of a cell containing the nucleus and other organelles, including the mitochondria.

cytoskeleton—Complex array of intracellular proteins that provide an infrastructure of support within a skeletal muscle fiber.

cytosolic—Pertaining to the cytosol, the fluid portion of the cell.

dead space—That part of the ventilatory system where air is moved but no gas exchange takes place. The conducting airways are dead space, but any alveolus that receives ventilation and does not have sufficient blood flow for gas exchange also is dead space.

dehydration—Reduction of body water.

delayed-onset muscle soreness (DOMS)—Related to an inflammatory process that accompanies skeletal muscle damage. Soreness peaks 2-3 days after injury occurs.

dementia—Loss of cognitive and intellectual capacity.

demyelination—Loss of the myelin sheath surrounding nerve fibers.

dendrites—Branches from neuron cell body that serve as receptive surfaces for the nerve.

depolarize—To alter the polarization of electrical charges across a cell membrane by movement of sodium into the cell and potassium out. This occurs rapidly.

desaturation—A decrease in the saturation of the Hb with oxygen. The level of saturation or desaturation is dependent on the PO_2 of the blood.

desmin—A protein in skeletal muscle that joins Z lines between adjacent myofibrils and links myofibrils to the costamere, a large structure in the cell membrane.

detrained—A physiological process where the changes induced by regular exercise training were lost during a subsequent period of rest.

diabetes—A disease associated with an inability to regulate blood glucose. Type 1 is associated with a lack of insulin production while type 2 is associated with a decreased sensitivity of the insulin receptors.

diastole—Heart is relaxed, period of filling.

diastolic blood pressure—Heart is relaxing, pressure is at its lowest point.

diencephalon—A major region of the brain containing the thalamus, which processes and relays most sensory input from the lower regions on the way to the cerebral cortex, and the hypothalamus, which regulates the autonomic nervous system and hormonal secretions of the pituitary gland.

diplopia—Double vision. Two objects are seen instead of one.

discharge rate—Rate at which action potentials are generated.

disinhibition—The removal of an inhibitory influence.

diuretic—Any agent that increases the amount of urine production.

divergence—A single piece of information is delivered to many target cells through parallel pathways.

dorsal horn—Gray matter of the spinal cord lying toward the back.

Duchenne's, Becker's—Muscular dystrophies (MD) associated with muscle loss and weakness due to a lack of production of the dystrophin protein.

dynamic contraction—A muscle contraction that involves movement.

dysarthria—Speech and language dysfunction due to brain injury or paralysis.

dysphasia—Swallowing difficulty.

dyspnea—Shortness of breath.

dyspneic—Having shortness of breath.

dystrophin—A membrane-stabilizing protein found in the costamere of skeletal muscle fibers.

eccentric contraction—The external load exceeds the force generated by the muscle, so the muscle lengthens.

effector—Target cell or organ that "effects" the desired response.

ejection fraction—The difference between the end systolic volume and the end diastolic volume expressed as a fraction of the systolic volume.

elastic forces—The forces required to overcome the elastic (e.g., lung and chest wall) tissues to effect ventilation.

elastic "work"—The energy cost, or work, associated with moving the elastic structures of the lung and chest wall.

electric axis—A vector representing the magnitude and direction of the average current flow though the ventricles of the heart.

electrical potential—Separation of electrically charged ions (potassium and sodium) across a cell membrane.

electron transport chain—A series of electron transporters in the mitochondria from which high energy ATP is formed with the final electron acceptor being oxygen.

embolism—A blockage of a vessel by a detached thrombus or foreign body.

emphysema—A chronic obstructive pulmonary disease characterized by alveolar destruction.

encephalitis—Inflammation of the brain.

end plate potential—Depolarization of the motor end plate region.

endomysium—Layer of connective tissue surrounding each skeletal muscle fiber.

end-diastolic volume—Amount of blood remaining in the ventricles at the end of diastole.

endocardium—The innermost layer of the heart.

endogenous—Within the body or tissue (i.e., glycogen is an endogenous glucose; insulin secreted by the pancreas is endogenous).

end-systolic volume—Amount blood remaining in the ventricles at the end of systole.

endurance—The duration a process can be continued. Usually refers to the time a particular exercise can be maintained.

epimysium—Layer of connective tissue surrounding skeletal muscle.

epinephrine—A catecholamine that is a major neurotransmitter in the adrenergic system. Epinephrine also stimulates several enzyme systems including lipase and phosphorylase.

epineurial sheath—Connective tissue layer that surrounds one or more nerve trunks.

equilibrium potential—The point at which the concentration gradient for an ion (e.g., K^+) and the electrical potential difference are at equilibrium and ions do not move in or out of the cell.

ergogenic aids—Any substance or treatment that improves exercise performance.

ergometer—A device of known calibration that is used to test exercise capacity, muscle strength, and other measures of exercise capacity.

essential amino acids—Amino acids that can not be produced by the body and therefore must be part of the diet.

essential fat—Fats that must be included in the regular diet as they can't be stored.

essential fatty acids—Fatty acids that must be ingested to be available.

evaporation—A method of heat dissipation whereby heat is required to transform water on the surface of the body from water to a gas. For each gram of water that evaporates, 0.58 Cal of heat is lost.

excitatory postsynaptic potential—Increase in membrane potential in a postsynaptic cell as a result of an increase in permeability to sodium, induced by a neurotransmitter released from a presynaptic cell.

exercise-induced asthma—Airway narrowing triggered by exercise or high levels of ventilation for 5-7 minutes in individuals who have asthma.

exercise-induced bronchospasm—Another term for exercise-induced asthma.

exogenous—From outside the body or tissue (i.e., ingested glucose is an exogenous glucose as opposed to glycogen) but stored endogenously within the tissue (injected insulin is exogenous insulin).

extrafusal—Outside the muscle spindle.

fasciculations—Involuntary twitches of groups of muscle fibers.

fatty acids (saturated and unsaturated)—Hydrolyzed fats with carbon chains of varying lengths. A fatty acid that has no unsaturated carbon-to-carbon bonds is saturated. Unsaturated fatty acids with a single unsaturated bond are called mono-unsaturated, those with multiple unsaturated bonds are called poly-unsaturated fatty acids.

fiber—Plant polysaccharides resistant to digestion.

fiber type—Basic characteristics of a skeletal muscle fiber—slow or fast, oxidative or glycolytic.

fibroblast growth factor (FGF)—Released from skeletal muscle cell membranes in response to sarcolemmal wounding. Activates satellite cells to divide and proliferate as part of muscle hypertrophy.

field tests—Exercise performance tests that can be performed outside of a sophisticated laboratory setting.

flexibility—The ability to stretch a muscle or muscle groups.

flow-resistive forces—The forces required to overcome the resistance to airflow in the airways.

flow-resistive "work"—The energy cost, or work, associated with moving air through the airways.

forced expiratory flow in mid-vital capacity ($FEF_{25-75\%}$)—Air flow measured during the middle 50% of a forced expired volume.

forced expired volume in one second (FEV$_1$)—Airflow measured in the first second of a forced vital capacity maneuver.

forced vital capacity (FVC)—The lung capacity measured from the highest lung volume, total lung capacity, to the lowest, residual volume.

functional capacity—Another term for exercise capacity that refers specifically to the peak capacity of the individual to function at a particular task.

functional reserve capacity—Maximum minus basal function of any physiologic system.

functional residual capacity—The lung capacity that represents the air left in the lungs at the end of a normal exhalation.

fused tetanus—Action potentials are delivered to muscle at a high enough rate that individual muscle twitches cannot be distinguished.

gamma motor neuron—Excites intrafusal muscle fibers in muscle spindle.

gating—An inhibitory command from higher brain centers prevents a peripheral sensory signal from discharging on a motor neuron.

glucagon and growth hormone—Metabolic regulatory hormones. They are called counter regulatory hormones because they oppose the glucose-uptake action of insulin. Glucagon stimulates the release of glucose from the liver while growth hormone stimulates fatty acid mobilization and utilization, which inhibits glucose utilization.

GLUT-1 and GLUT-4—Transporters involved in moving glucose from the cell exterior into the interior. Stimulated by insulin and exercise.

glycogen—Long, straight or branched chain glucose molecule that serves as a storage form of glucose in muscle and liver cells.

glycolipid—A lipid attached to one or more sugar.

glycolysis—The metabolic pathway where glucose is manipulated to produce energy (ATP) and hydrogen that is further processed in the electron transport chain to produce ATP.

golgi tendon organ—Encapsulated structure located in the myotendinous junction and sensitive to changes in muscle force. Through connections with motor neurons in the spinal cord it inhibits the contracting muscle and excites the antagonist.

graded exercise test (GXT)—The generic term for exercise tests designed to measure exercise capacity.

growth hormone—A hormone, released from the pituitary, that stimulates protein synthesis. Growth hormone also is involved in stimulating fat metabolism by promoting the release of lipid from adipocyte, or fat storage, cells.

Guillian-Barré disease (acute inflammatory demyelinating polyradiculoneuropathy)—An autoimmune-mediated disease resulting in demyelination of the peripheral nerves.

HbA$_1$c—A glycosylated hemoglobin that reflects the patient's serum glucose levels over the past several months. Used to monitor effectiveness of diabetes treatment.

heart failure—A condition where a diseased heart cannot generate sufficient pressure to maintain blood pressure and blood flow.

heat cramps—Involuntary cramping of exercising skeletal muscle related to fluid and electrolyte loss.

heat exhaustion—Inability of the circulation to meet the demands of exercising muscle and need to increase skin blood flow to dissipate heat. Related to dehydration that accompanies exercise in the heat. Characterized by profuse sweating and hypotension.

heat stress—Overload of the heat dissipating mechanisms produced by a combination of internal heat production by contracting muscles and hot/humid environments.

heat stroke—Complete breakdown of the body's heat regulating mechanisms. A person stops sweating and body temperature shoots up.

heme—The iron-containing part of hemoglobin to which O$_2$ is attached for transport in the blood.

hemianesthesia—One-sided anesthesia or loss of sensation.

hemoglobin—The O$_2$ carrying molecule in the red blood cells. Hemoglobin (Hb) allows the blood to carry significantly more O$_2$ than if the O$_2$ were just dissolved in the plasma.

Henneman principle—The size principle of motor unit recruitment. In general, motor units are recruited in order, from slow to fast.

hexose kinase (HK)—An enzyme of the glycolytic pathway of metabolism. HK catalyzes the transfer of a phosphate.

high-frequency fatigue—Decrement in muscle force related to high-frequency stimulation. Fatigue is specific to fast fibers and recovery of force generating capacity is relatively quick.

histamine (in lungs)—A powerful stimulator of the smooth muscle in the airways, causing airway constriction.

histamine (in skeletal muscle)—An inflammatory mediator that dilates arterioles and increases leakiness of capillaries. Associated with skeletal muscle injury.

homonymous hemianopsia—Loss of one half of the vision in one or both eyes.

humoral control of ventilation—Ventilatory control exerted by receptors that sense the levels of O_2 and CO_2 in the blood. *See* chemoreceptors.

Huntington's disease—Progressive hereditary disease characterized by abnormal movement, changes in personality, and dementia.

hyperglycemia—Blood glucose levels above normal.

hyperinflation—An overinflation of the lungs seen with obstructive lung diseases.

hyperpnea—An increase in ventilation that is appropriate to the amount of CO_2 being produced, such as during exercise. This is not to be confused with hyperventilation.

hyperpolarized—Inside of the cell becomes more negative, making it harder to generate an action potential.

hypertension (HTN)—An increase in blood pressure above that considered normal (i.e., 120 systolic and 80 diastolic).

hypertrophy—Increase in size.

hyperventilation—An increase in ventilation that is inappropriately high for the level of CO_2 being produced such that arterial blood CO_2 decreases. Hyperventilation is defined by a decrease in arterial blood CO_2. Since the increase in ventilation at mild to moderate levels of exercise intensity is appropriate to the amount of CO_2 produced, this increased ventilation is called hyperpnea.

hypoglycemia—Blood glucose levels below normal.

hypohydration—Insufficient fluid intake.

hypokinesis—Diminished or slow movement.

hypo-osmolarity—Low concentration of solutes.

hypothalamus—Regulates the autonomic nervous system and hormonal secretions of the pituitary gland.

hypoventilation—An increase in ventilation that is inappropriately low for the level of CO_2 being produced such that arterial blood CO_2 levels increase.

impedance—The resistance to electrical current flow in an alternating-current system.

inhibitory postsynaptic potential—Decrease in membrane potential in a postsynaptic cell as a result of an increase in permeability to potassium, induced by a neurotransmitter released from a presynaptic cell.

initial segment—The section of the motor nerve adjacent to the axon hillock where inhibitory and excitatory inputs to the motor nerve are summed.

insulin—A hormone, secreted by the pancreas, that promotes the uptake of glucose from the blood into the muscle and liver. Artificial insulin is available to inject into persons with diabetes to help control blood glucose. This injected insulin is available in several short-acting and long-acting formulations.

insulin-like growth factor I (IGF-I)—Released from skeletal muscle cell as part of muscle hypertrophy. Increases transport of amino acids into cell.

integrative zone (also known as trigger zone)—An area of the nerve axon where receptor potentials are summed and, if of high enough magnitude, initiate a train of action potentials.

integrin—A large protein embedded in the cell membrane of skeletal muscle fibers that provides a connection between the Z lines of sarcomeres inside the fiber and the endomysium outside the cell.

intermittent claudication—Intervals of vascular spasm causing occlusion of blood flow and pain in the affected area.

interneuron—Neuron lying in the spinal cord between sensory afferent and motor efferent neurons whose actions are confined within a restricted area.

intimal—The innermost part of the vascular wall.

intrafusal—Within the muscle spindle.

ischemia—Little or no blood flow.

isokinetic contraction—A muscle contraction in which the angular velocity is regulated and the force varies.

isometric contraction—The muscle force and the resistance applied to the muscle are equal, so no change in muscle length occurs.

isotonic contraction—A dynamic contraction in which muscle force remains constant.

isovolumic stage—The phase of the cardiac cycle when the ventricles are contracting but all valves are closed. The volume inside the ventricles remains constant.

Karvonen formula—Used to calculate target heart rates for exercise programs where target heart rate = ([maximal heart rate–resting heart rate] × desired % maximal heart rate) + resting heart rate.

Kernig's sign—A test for meningitis. Pain is induced when a supine patient attempts to raise the head and legs with the hands cupped behind the head. This causes a stretching of the meninges, which, if inflamed (meningitis), will cause pain.

ketoacidosis—Increased metabolic production of ketone bodies resulting in a metabolic acidosis.

ketone bodies—Quantities of ketones (e.g., acetoacetic acid, acetone or beta-hydroxybutyric acid) found in the blood.

kinesthesia—Sense of limb movement.

Krebs cycle—A mitochondrial metabolic cycle that produces energy (ATP) and high energy compunds (NADH and $FADH_2$) that can be processed through the electron transport chain to produce ATP. (Also citric acid cycle or tricarboxylic acid cycle).

lactic acid—An acid produced as a byproduct of glycolytic metabolism under conditions when the pyruvic acid produced by this cycle does not enter the oxidative cycle.

lean body mass—The mass of the body that consists of tissues other than fat. We generally think of lean body mass as representing muscle mass, but it also includes bone.

limbic system—A combination of several structures that encircle the brain stem that control emotion and motivation.

lipase—An enzyme involved in the mobilization of fat from adipose tissue. Lipase splits fatty acids from triglycerides.

lipid—Fat. A cellular substance that serves as a metabolic fuel and an insulator. Lipids are part of cell membranes and form a sheath around some nervous tissue.

lipoproteins (high, low and very low density)—A complex of fat and protein. The protein enables fat, which is insoluble in water, to be transported in the blood. A high amount of fat in the complex makes the lipoprotein a lower density while a low amount makes it a higher density.

longitudinal study—Observation of a group of people over an extended period, usually years.

low-frequency fatigue—Decrement in muscle force related to low-frequency stimulation. Fatigue is specific to slow fibers, and recovery of force-generating capacity is prolonged.

low-density lipoprotein and high-density lipoprotein—Carry lipids in the bloodstream.

M line—Structural proteins in skeletal muscle that link together myosin filaments in the central region of the filament.

macrophages—A class of mononuclear phagocytic cell involved in the inflammatory process.

mechanoreceptors—Receptors, such as those in the joints, that sense force. These joint receptors can stimulate increases in ventilation with movement.

medullary center—A part of the medulla oblongata, the most caudal part of the brainstem, where a major part of respiratory control takes place.

meningitis—A disease resulting in inflammation of the meninges of the brain.

mesoderm—The middle layer of the three primary germ layers of the embryo

MET—One resting metabolic equivalent. 1 MET is equal to 3.5 ml O_2/kg/min of oxygen consumption. Exercise oxygen consumption can be expressed as multiples of 1 MET.

minute ventilation—The total amount of ventilation breathed in one minute, expressed as L or ml/min.

mitochondria—A cell organelle that contains the constituents to perform oxidative metabolism.

mobilization—The process of releasing stored lipids so that the fatty acids become available for metabolic fuel production.

mono, di and polysaccharides—Carbohydrates that include single (mono), double (di), or many (poly) sugar molecules.

monocyte—A large leukocyte. *See* phagocyte.

mononuclear cells—Cells with a single nucleus.

monosynaptic—Making a single connection.

motivational nervous system—Centered in the limbic system and has direct and indirect effects on the quality of movement.

motor end plate—Specialized region of a skeletal muscle fiber where the motor nerve makes contact.

motor neuron pool—Motor neurons that innervate the same muscle are grouped together in the spinal cord.

motor unit heterogeneity—Human skeletal muscles contain a wide variety of motor units, all with different characteristics, which form a functional continuum within the muscle.

motor unit—Consists of a motor nerve and all the muscle fibers it innervates.

motorsensory—Involving both the motor and sensory systems.

multiple sclerosis—An autoimmune disease resulting in the demyelination of the central nerves.

muscle pump—The process whereby blood is propelled through the veins by alternate contraction and relaxation of skeletal muscle.

muscle spindles—Proprioceptors in skeletal muscle that provide feedback about the rate and magnitude of stretch.

muscle twitch—The mechanical response of skeletal muscle fiber to a single electrical stimulus.

muscular dystrophies (MD)—Muscle diseases (i.e., Duchenne's, Becker's) associated with muscle loss and weakness due to a lack of production of the dystrophin protein.

mutation—A change in gene structure that results in altered cell constituents.

myasthenia gravis—An autoimmune disease affecting the neuromuscular junction resulting in weakness and loss of muscle function.

myelin—Lipid-rich layer that surrounds nerve axons.

myoblasts—Single nucleated, primitive muscle cells.

myocardial infarction—Necrosis of heart muscle cells caused by an occlusion of a coronary blood vessel.

myogenesis—Formation of new muscle fibers.

myosin—One of the contractile proteins in skeletal muscle whose globular head attaches to actin, forming the actomyosin complex necessary for muscle contraction.

myotactic reflex—Contraction of a muscle in response to stretch, due to activation of muscle spindles.

myotubes—Immature muscle fibers formed by the fusion of myoblasts.

NAD$^+$/NADH—An electron acceptor involved in the production of ATP through the mitochondrial electron transport chain.

neural control of ventilation—That part of ventilation that is stimulated by mechanoreceptors.

neurofibrillary tangles—Tangled masses of nervous system tissue, including the cell body, dendrites, and axons.

neuromuscular junction—Where the motor axon terminal joins with the motor end plate.

neurotransmitter—Chemical substance released by a presynaptic cell that diffuses across the synapse to activate the post-synaptic cell.

nodes of Ranvier—Unmyelinated regions of an axon that provide areas of low resistance on the cell membrane.

nonaerobic capacity—The same as the anaerobic capacity. A measure of the ability to work at exercise intensities that rely primarily on anaerobic metabolism.

obstructive pulmonary diseases—Diseases of the lung that are associated with airway blockage, or obstruction, including chronic bronchitis, emphysema, and cystic fibrosis.

optic neuritis—Inflammation of the optic nerve.

osteoporosis—Loss of bone commonly associated with aging.

output signal—The amount of transmitter that is released from the pre-synaptic cell.

overfat—An increase in body fat, measured through several indirect techniques, above normal.

overload—The practice of taxing physiological systems beyond some critical level in order to achieve an increase in function.

overweight—An increase in body weight above normal as defined by standard height and weight tables.

oxidation—The process of energy production where hydrogen and its associated electrons, released during chemical reactions, are combined with oxygen.

oxygen consumption ($\dot{V}O_2$)—The volume of oxygen consumed to produce energy. The oxygen is consumed in the energy-producing processes of metabolism.

oxygen free radicals (OFR)—Altered forms of oxygen that react with and break down organic compounds.

pallidotomy—Surgical destruction of the globus pallidus to reduce involuntary movements.

palsies—Paralysis or paresis (incomplete paralysis) of the muscles.

parallel processing—Similar information is delivered to, or from, brain centers by different pathways.

parasympathetic nervous system (PNS)—Division of the autonomic nervous system that prepares the body for rest by lowering heart rate, reducing blood pressure, and restoring homeostasis.

paresis—Incomplete paralysis of the muscles.

paresthesia—Abnormal sensation such as burning or tingling.

Parkinson's disease—A disease resulting in rhythmical tremor from a progressive loss of the neurotransmitter dopamine.

peak work capacity—The peak ability to perform exercise. Usually measured as the point at which a person stops an exercise test.

pennation—When skeletal muscle fibers are arranged at an angle relative to the direction of force generation.

perimysium—Layer of connective tissue encircling muscle fascicles.

peripheral fatigue—Decrement in performance related to a dysfunction at the neuromuscular junction and/or in the muscle fiber.

phagocytes—Cells capable of ingesting foreign particles. They include leukocytes and macrophages.

phagocytize—To ingest bacteria, other cells, and foreign particles.

phase I-IV rehabilitation—The standard phases of cardiac rehabilitation from in-patient to out-patient unmonitored activity.

phosphofructokinase—An enzyme of the glycolytic pathway of metabolism. PFK catalyzes the transfer of a phosphate.

phospholipid—A lipid with phosphorus (e.g., lecithins).

phosphorylase—An enzyme involved in the breakdown of long chain glucose molecules (glycogen) so that the individual glucose molecules can be used for the production of energy in the glycolytic metabolic pathway.

phosphorylation—The chemical process of adding a phosphate to a compound.

plaque—An area of abnormal tissue on the surface of the skin or vessel. Plaque is involved in the process of atherosclerosis.

plasmapheresis—Removal of whole blood, which is then separated so the plasma can be reinfused. Reinfusion prevents loss of blood volume.

plasticity—Ability to change characteristics.

plyometric—A movement in which an eccentric muscle contraction is followed immediately by a concentric contraction.

postsynaptic—Distal to the synapse.

power—An expression of the work rate, or the amount of work done per unit time (Watts).

preload—Ventricular filling pressure or amount of blood filling the heart.

pressure overload—An increase in resistance to blood flow from the heart that causes the heart to work harder to eject blood.

presynaptic—Proximal to the synapse.

primary (essential) hypertension—An increase in blood pressure that is unexplained. Primary, or unexplained, pulmonary hypertension is associated with an increase in blood pressure in the pulmonary artery.

primary aging—Deteriorization of cellular function as a result of alterations in the synthesis of the cell's various constituents, accumulated over time.

progressive resistance exercise—The principle used in weight training to achieve continual increases in strength. Muscles are overloaded by increasing the weight lifted over a period of weeks or months.

proprioception—The sense of limb and body position and of limb movement.

proprioreceptors—Receptors in the muscles, joints, and skin that sense movement or the force of movement.

pseudohypertrophy—A hypertrophy, or enlargement, of the muscle that is not due to normal increase in the contractile elements of the muscle (i.e., actin and myosin). Pseudohypertrophy is seen in the early stages of some muscular dystrophies.

pulmonary function—Measures made to characterize how well the lung is functioning. Common measures are made with a spirometer. Others are made with more sophisticated equipment, including the body plethysmograph.

pulse pressure—Difference between systolic and diastolic blood pressures.

pyruvic acid—The end product of glycolytic metabolism. Pyruvic acid can be further processed to produce energy in the oxidative cycle or, if this cycle can not accept the pyruvic acid, to form lactic acid.

radiation—A method of heat exchange in the form of electromagnetic waves. Heat loss or gain is related to the temperature difference between objects in the surrounding environment.

rating of perceived exertion (RPE, Borg scale)—A scale, originally from 6 to 20, designed to measure exercise intensity as perceived by an individual.

receptor potential—Depolarization of sensory nerve endings in a graded fashion. The magnitude of the depolarization is directly proportional to the magnitude of the input signal.

red blood cells—Transport oxygen and carbon dioxide in the blood.

refractory—The cell is resistant to generation of an action potential.

rehydration—Replacement of lost body water.

repetition maximum (RM)—The maximum weight that can be lifted a single time.

repolarize—Restoration of charge separation across the cell membrane after depolarization.

reserve—The difference between a maximum available quantity and that amount being used at any point in time.

residual volume—The volume left in the lungs at the end of a maximal exhalation.

resorption—Reabsorption of osseous tissue.

respiratory quotient (RQ); respiratory exchange ratio (RER)—The ratio of carbon dioxide produced to oxygen consumed ($\dot{V}CO_2 : \dot{V}O_2$) that is a reflection of the metabolic substrate being consumed to produce energy. A RER of 1 indicates that only carbohydrates are being consumed, 0.7 indicates that only fats are being consumed. Ratios that lie between these values indicate the proportion of each substrate that is being consumed.

restrictive pulmonary diseases—Lung diseases associated with a stiffening, or restriction, of the lung or chest wall.

risk factors—Things that are associated with an increased risk of disease. For example, smoking is a risk factor for both lung and heart disease.

SA node—Pacemaker cells in the right atrium where electrical signal for excitation of myocardium is initiated.

saccadic—A quick jump of the eye from one fixation point to another, such as occurs in reading.

saltatory conduction—The electrical current in a myelinated nerve axon jumps from node to node, thus increasing the speed of the transmission.

sarcolemma—The cell membrane of a skeletal muscle fiber.

sarcolemmal "wounding"—Microscopic tears in the cell membrane of skeletal muscle fibers thought to initiate hypertrophy.

sarcomere—Basic functional unit of skeletal muscle containing contractile proteins actin and myosin, bordered by rigid Z lines.

sarcoplasmic reticulum—A network of tubules within the muscle fiber that store and release calcium.

satellite cells—Myoblasts that become trapped between the basal lamina and cell membrane of skeletal muscle fibers during formation of myotubes. These cells lie dormant until stimulated by fiber injury or mechanical stress, then divide and proliferate and fuse together to repair fibers or add to the mass already present (hypertrophy).

Schwann cells—Cells that provide the myelin sheath surrounding axons.

second wind—A little understood phenomenon whereby one experiences a sense that the exercise has suddenly become easier. It is usually associated with a sense of breathing easier.

secondary aging—Decline in functional capacity as a result of disease, injury, or environmental influences.

secondary hypertension—An increase in blood pressure that can be attributed to known cause(s).

senescence—Aging.

septa—Thin segments of the alveoli that create a large alveolar surface.

serotonin—Involved in inflammation associated with skeletal muscle injury.

shear forces—A deformation caused by movement of two adjacent structures in opposite directions.

shuttle tests—Field tests designed to measure peak work capacity by having the individual move between two points at incremental rates until they can no longer continue.

somites—Paired cell masses which will become vertebrae, formed in the fourth week of development.

spasticity—Increased muscular tone.

spatial summation—Adding together of two or more electrical signals that arrive from different sources simultaneously at a postsynaptic cell.

specific airway conductance (sGaw)—A measure of the conductance of the airways, corrected for the lung volume at which it was measured. The conductance is the reciprocal of the airway resistance.

specificity—The training principle that states in order to achieve maximum training results, muscles should be conditioned in a manner similar to which they are to perform, including the exact pattern of movement.

spirometry—Measurement of pulmonary function using a device that measures air volume and flow.

static contraction—A muscle contraction in which no change in length occurs.

stenosis—A narrowing. An aortic stenosis impedes blood flow from the left ventricle.

sterols—A steroid with one OH group attached (e.g., cholesterol).

stitch in the side—A pain in the side associated with exercise. The mechanism is not understood.

storage fat—Fats that can be stored in adipose tissue and therefore do not have to be ingested as regularly as essential fats.

strain—Change in length relative to resting length.

strength—The maximum ability of a muscle or muscle group to generate force.

stress—Force per cross-sectional area.

stroke—A partial or complete blockage of a cerebral vessel resulting in specific neurological dysfunction, depending on the site.

stroke volume—Amount of blood ejected from a ventricle of the heart (in ml) in a single beat.

substrate—The carbohydrate or fat being used to produce energy (ATP).

summation—When action potentials are delivered to a muscle fiber in quick succession, the force of the second contraction is added to the first to produce a higher level of force.

supplementary motor cortex—An area of the brain that projects onto the motor cortex and helps with planning of complex movements.

supraventricular dysrhythmias—An arrhythmia of the heart that is observed in the atria.

symptom-limited work capacity—The peak work capacity measured at the point where a patient's symptoms prevent continuing the test.

synaptic cleft—Gap between the motor nerve and motor end plate.

synaptic potential—Graded receptor potential in motor nerves.

synthetase—An enzyme that enhances the storage of glucose in the cell. The glucose is stored in long chains called glycogen.

systole—Heart is contracting, blood is ejected.

systolic blood pressure—Heart is contracting, pressure is at its highest point.

temporal summation—Adding together of synaptic potentials that overlap in time.

tetanic force—Maximal contraction of skeletal muscle produced by fusion of individual muscle twitches.

tetanus—*See* tetanic force.

thalamus—Processes and relays most sensory input from the lower regions of the brain on the way to the cerebral cortex.

thermogenesis—Heat production.

thermoneutral—An ambient temperature that does not result in heat or cold stress.

thermoregulation—Control of body temperature.

thrombosis—The formation or presence of a blood clot within a blood vessel.

tidal volume—The volume of each breath, expressed in ml or L at body temperature and pressure saturated with water vapor (BTPS).

time-to-peak tension—Upon activation, the time required by a muscle to reach peak force.

titin—A protein that provides stability to the myosin filament in skeletal muscle by linking the end of the filament with the Z line.

torque—The rotational force generated on a moment arm.

total peripheral resistance (TPR)—Resistance to blood flow through the entire systemic circulation.

trains—Groups of action potentials.

transformed—Changing from one skeletal fiber type to another.

transient ischemic attacks (TIAs)—An acute episode of cerebral ischemia.

transit time—The time that a red blood cell spends in the pulmonary or tissue capillary.

tricarboxylic acid cycle—A mitochondrial metabolic cycle that produces energy (ATP) and high energy compounds (NADH and $FADH_2$) that can be processed through the electron transport chain to produce ATP. (Also citric acid cycle or Kreb's cycle).

trigger zone—*See* integrative zone.

triglyceride—A glycerol with three fatty acids attached to each of its hydroxyl groups.

tropomyosin—In skeletal muscle, a rod-shaped protein that covers myosin binding sites on actin.

troponin—In skeletal muscle, a globular protein that sits on top of tropomyosin which, when activated by calcium, rotates and pulls tropomyosin off the myosin binding site.

twitch—*See* muscle twitch.

unacclimatized—Not adapted to adverse environmental conditions such as heat and altitude.

unipolar—Having only one pole.

unmyelinated—Axons that do not have a myelin sheath.

utilization—A term used to describe the consumption of a metabolic substrate (e.g., fat or carbohydrate).

utilization—The process of breaking down lipids to produce energy (ATP) through β-oxidation and oxidative metabolism.

$\dot{V}O_2$ max—The maximum oxygen consumption or maximum aerobic capacity.

Valsalva—A forced exhalation performed with the glottis closed.

varicose veins—An abnormal, permanent dilation of the veins.

vasodilation—An increase in the diameter of blood vessels.

venoconstriction—Reduction in diameter of veins.

ventilation/perfusion matching—The ratio of the amount of ventilation to the amount of blood flow in a particular alveolus, lung region, or whole lung.

ventral horn—Gray matter of the spinal cord lying toward the front.

volume overload—An increase in venous return, which fills the heart more and leads to an increase in the volume of the ventricular chambers.

waist-to-hip ratio—The ratio of waist to hip circumference that is used as an indicator of fatness or obesity.

windchill index—The cooling effect of wind on exposed skin. Related to wind velocity.

work—A measure of amount of force produced and the distance over which that force is exerted (kilogram meters, kgm, or kpm).

Z line—Rigid network of interconnecting proteins in sarcomeres of skeletal muscle in which actin filaments are anchored.

Z-line streaming—Distortion of the Z line in skeletal muscle characteristic of eccentric contraction–induced injury.

Index

Note: The italicized *f* and *t* following page numbers refer to figures and tables, respectively.

About the Authors

Frank Cerny, PhD, has been involved in the clinical/applied aspects of physiology since the beginning of his 30-year career as an exercise physiologist. In 1985, he joined the faculty of the department of physical therapy and exercise science at the State University of New York at Buffalo. As chair of the department since 1995, he has been involved in many clinical aspects of physical therapy. An award-winning teacher, Dr. Cerny has an established scientific reputation in the areas of understanding the exercise response in patients with cystic fibrosis and in developing programs for these patients as part of their treatment. He earned his PhD from the University of Wisconsin at Madison and conducted postdoctoral research at the Sports Medicine Clinic in Freiburg, Germany. Dr. Cerny is a fellow in the American College of Sports Medicine (ACSM).

Harold Burton, PhD, has been teaching exercise physiology for 14 years. Currently, he is with the State University of New York at Buffalo where he is an associate professor and has been director of the program in exercise science for the past six years. He is a consultant to sports medicine clinics, professional sports teams, and many physical therapists. His research has been published in leading journals in the field. Dr. Burton is a fellow in the ACSM.